Multilayers: Synthesis, Properties and Non-Electronic Applications

MATERIALS RESEARCH SOCIETY SYMPOSIUM PROCEEDINGS   VOLUME 103

Multilayers: Synthesis, Properties and Non-Electronic Applications

Symposium held December 2-4, 1987,
Boston, Massachusetts, U.S.A.

EDITORS:

T. W. Barbee, Jr.
Lawrence Livermore National Laboratory, Livermore, California, U.S.A.

F. Spaepen
Harvard University, Cambridge, Massachusetts, U.S.A.

L. Greer
Cambridge University, Cambridge, Unite Kingdom

M|R|S MATERIALS RESEARCH SOCIETY
Pittsburgh, Pennsylvania

CAMBRIDGE UNIVERSITY PRESS
Cambridge, New York, Melbourne, Madrid, Cape Town,
Singapore, São Paulo, Delhi, Mexico City

Cambridge University Press
32 Avenue of the Americas, New York NY 10013-2473, USA

Published in the United States of America by Cambridge University Press, New York

www.cambridge.org
Information on this title: www.cambridge.org/9781107411029

Materials Research Society
506 Keystone Drive, Warrendale, PA 15086
http://www.mrs.org

First published 1988
First paperback edition 2012

Single article reprints from this publication are available through
University Microfilms Inc., 300 North Zeeb Road, Ann Arbor, MI 48106

CODEN: MRSPDH

ISBN 978-1-107-41102-9 Paperback

This work was supported in part by the U.S. Army Research Office under Grant Number
DAAL03-88-G-0004. The views, opinions, and/or findings contained in this report are those
of the authors and should not be construed as an official Department of the Army position,
policy, or decision unless so designated by other documentation.

This work was supported by the Office of Naval Research under Grant Number
N00014-87-G-0233. The United States Government has a royalty-free license throughout
the world in all copyrightable material contained herein.

Contents

*Invited Paper

*Invited Paper

Preface

The symposium on Multilayers: Synthesis, Properties and Non-
Electronic Applications at which the papers included in this
volume were presented, was organized because this is one of
the most rapidly expanding areas of solid state science. This
growth draws on significant advances in a number of other scientific
and technological arenas. In particular, thin-film deposition
techniques have in the past two decades progressed to the point
that it is now possible reproducibly to prepare multilayer
structures of high perfection relative to superlattice characteristics, individual layer structures, and interfacial structure.
It is also important to recognize that the broad range of sophisticated and sensitive characterization tools and approaches that
are now available have been central to the rapid advance of
multilayer research. It is now clear that these new materials
can be synthesized with structures allowing the scientific investigation of microstructure sensitive physical phenomena, and the
optimization of the properties of technologically useful materials.
In summary, multilayer structure research and technology represents
a significant part of the forefront of what may be termed
Microstructure Engineering.

We are indebted to the symposium authors, in part for
presenting excellent reviews and summaries of the many areas
involved in multilayer research, for presenting the results of
recent investigations, and also for relating these results to
fundamental properties that are important in this area of materials
science. The editors also thank all participants, session chairs,
reviewers, and MRS staff and program officials for their help and
guidance in organizing the symposium and publishing this volume.

The symposium was sponsored by the Office of Naval Research,
the Army Research Office--Durham, Lawrence Livermore National
Laboratory, Harvard University, and Ovonic Synthetic Materials
Corporation.

Troy W. Barbee, Jr.
Lawrence Livermore National
    Laboratory

Frans Spaepen
Harvard University

Lindsay Greer
Cambridge University

MATERIALS RESEARCH SOCIETY SYMPOSIUM PROCEEDINGS

ISSN 0272 - 9172

MATERIALS RESEARCH SOCIETY SYMPOSIUM PROCEEDINGS

MATERIALS RESEARCH SOCIETY SYMPOSIUM PROCEEDINGS

MATERIALS RESEARCH SOCIETY CONFERENCE PROCEEDINGS

Tungsten and Other Refractory Metals for VLSI Applications, R. S. Blewer, 1986; ISSN: 0886-7860; ISBN: 0-931837-32-4

Tungsten and Other Refractory Metals for VLSI Applications II, E.K. Broadbent, 1987; ISSN: 0886-7860; ISBN: 0-931837-66-9

Ternary and Multinary Compounds, S. Deb, A. Zunger, 1987; ISBN:0-931837-57-x

Tungsten and Other Refractory Metals for VLSI Applications III, Victor A. Wells, 1988, ISSN 0886-7860; ISBN 0-931837-84-7

Atomic and Molecular Processing of Electronic and Ceramic Materials: Preparation, Characterization and Properties, Ilhan A. Aksay, Gary L. McVay, Thomas G. Stoebe, 1988, ISBN 0-931837-85-5

PART I

Synthesis

THIN FILM GROWTH

R. W. VOOK
  Syracuse University, Physics Department, 201 Physics Bldg., Syracuse, NY
13244-1130

ABSTRACT

   A review of the experimental and theoretical results describing thin
film growth modes is presented. Thermodynamic criteria for determining
which growth mode might be expected to occur in a particular case along
with some kinetic considerations are given. The characteristics of each of
the three principal growth modes, namely Frank and van der Merwe (layer),
Stranski-Krastanov (layer plus island), and Volmer-Weber (island), are
discussed. Lastly, the requirements favoring the growth of epitaxial
multilayers are briefly considered.

INTRODUCTION

   The morphology, structure and microstructure of thin films can take
on an almost limitless variety of forms [1]. Nevertheless the modes of
growth can be categorized into four basic types: island, layer, layer plus
island, and columnar. The latter case arises generally at relatively low
temperatures when surface diffusion is completely inhibited [2]. Computer
modeling has shown that it arises when atoms or molecules strike a
substrate and do not leave the point of impingement [3]. These films tend
to be porous, highly defective, and generally of poor quality. They will
not be discussed further in this review.

   The more common and generally more interesting growth modes are the
first three listed. In island growth, also designated as Volmer-Weber
growth (VW), nucleation occurs followed by three dimensional growth of the
nuclei, which eventually coalesce leading to a continuous film. The
topography of such films is therefore initially quite rough on an atomic
scale. In thick continuous films that originally grew in this mode,
however, the surfaces may be quite smooth and flat. On such surfaces layer
growth may take place if the films are epitaxially oriented.

   When layer or Frank and van der Merwe (FM) growth occurs, atoms or
molecules impinging on a substrate surface diffuse, usually to step, kink,
or defect sites in such a way as to extend the stepped layer. Thus ideally
the crystal grows one layer at a time. A variant of this growth mode
occurs when growth on subsequent layers is initiated before the prior
layer has been completed. Such growth could occur, for example, when some
"entity" on a surface promotes layer nucleation at a rate that exceeds the
rate at which impinging atoms can diffuse to a growing step. A very nice
possible example of such a case has been given in the literature for Ag
growing on MoS_2 [4].

   Stranski-Krastanov (SK) growth typically means that growth is
initiated in the FM mode; but then after one or more (whole or fractional)
layers is formed, island growth nucleates on top of the initial whole or
fractional layers. The morphology of the islands can vary widely from more
or less hemispherical shape [5] to crystallographically faceted islands
[6] to extended, more or less flat-topped "layers" [1,7].

THIN FILM GROWTH MODES

A. Thermodynamic Basis

Bauer [8] showed that the three major thin film growth modes can be predicted from a consideration of the net surface and interfacial energies that are involved, namely $\Delta\gamma_n$:

$$\Delta\gamma_n = \gamma_{on} + \gamma_{in} - \gamma_s \qquad (1)$$

where γ_{on} is the surface energy of the overgrowth material, γ_{in} is the interfacial energy of overgrowth - substrate interface, and γ_s is the substrate surface energy. The subscript n refers to the situation after n equivalent layers have been deposited. VW, FM, and SK growth are predicted when $\Delta\gamma_n > 0$, ≤ 0, and ≤ 0 respectively. The difference between FM and SK growth lies mainly in the elastic strain energy associated with the overgrowth and incorporated in the γ_{in} term [9]. In the case of homoepitaxy (growth of material A on A), the strain energy is zero. However, for heteroepitaxy there is always some strain energy because of the inevitable lattice misfit f_o and in some cases structure differences between the film of material B growing on the substrate A. The misfit is defined as $f_o = (a_s^* - a_o^*)/a_o^*$ where a_s^* and a_o^* are the unstrained lattice parameters of the substrate and overgrowth respectively. If this strain energy is small relative to γ_{on}, then FM growth is likely. Conversely, large strain energies favor SK growth. Qualitatively, these thermodynamic results can be seen in terms of what happens during nucleation and growth: namely surface and interfacial energies ($\gamma_{on} + \gamma_{in}$) replace the surface energy γ_s of the substrate. If the difference between these two quantities $\Delta\gamma_n$ is positive, an increase in energy occurs. In order to reduce this increase as much as possible, the film will tend to "ball up", so as to decrease the surface atom to volume atom ratio. On the other hand if $\Delta\gamma_n$ is negative, the net energy can be reduced by increasing the surface atom to volume atom ratio. In this case the overgrowth would tend to grow by layers.

When large misfit strain energies are involved in layer growth, the strain energy contribution to γ_{in} increases until a critical n = n* value is attained, at which point the sign of $\Delta\gamma_n$ is reversed. Then $\Delta\gamma_n > 0$ and island growth is favored. In this way the initial FM growth is converted to VW growth as thickening proceeds [9].

If interdiffusion occurs where $\gamma_A < \gamma_B$ so that A atoms have diffused into the overgrowth B layer then $\gamma_A < \gamma_{AB} < \gamma_B$ and FM growth is more favored for the alloyed AB layer than if it were pure B. If the next layer is pure B, then an increase in surface energy occurs, thereby favoring island growth [1].

An alternative explanation of SK growth was given by Matthews et al. [10] and applied to the case of (111)Cu_2O growing on (111)Cu [1,7]. This explanation is based on the idea that if misfit dislocations can nucleate (for some reason) in particular regions on the surface of the bilayer, then the overgrowth will be less strained there than where there are no misfit dislocations. Thus more rapid film growth would be favored in the less strained regions, a condition that would lead to SK growth.

B. Supersaturation Effects and Kinetic Considerations

Various experimental results have indicated that higher deposition rates and/or lower substrate temperatures during deposition cause the formation of films with flatter surfaces [11,12]. Some of this evidence suggested that by increasing the supersaturation the initial mode for film growth may be altered such that a normal island growth mode may be converted to layer growth. To interpret these results theoretically, Markov and Kaischew [13] re-derived Bauer's thermodynamic relations to include a supersaturation (S) term. S is given by P/P_e, where P and P_e refer to the vapor pressure of the incident supersaturated molecular beam and the equilibrium vapor pressure of the incident material at the substrate temperature. The modified thermodynamic equations are given as follows:

$$\text{Island growth: } \gamma_s < \gamma_{hkl} + \gamma_i - \frac{\Delta\mu}{2k_{hkl}b^2} \tag{2}$$

$$\text{Layer growth: } \gamma_s \geq \gamma_{hkl} + \gamma_i - \frac{\Delta\mu}{2k_{hkl}b^2} \tag{3}$$

where $\Delta\mu = kT_s \ln S$, k = Boltzmann's constant, T_s = substrate temperature, b = nearest neighbor distance in the substrate, and $k_{hkl}b^2$ is the area of an atom in the contact plane ($k_{100} = 1$, $k_{111} = \sqrt{3}/2$). γ_{hkl} is the surface energy of the (hkl) face of the overgrowth. From eqn. (2) we see that increasing S and therefore $\Delta\mu$ will eventually reverse the inequality at some critical $\Delta\mu_c$, thereby favoring layer growth.

Bauer and van der Merwe [9] have recently criticized this derivation and suggested that there really is no definitive experimental evidence for the VW to FM transformation at high supersaturations, even though much experimental evidence suggests that it occurs. The problem arises from a consideration of kinetic factors. Thus Bauer and van der Merwe suggest that only with extremely high nucleation rates and negligible mobility (which would avoid subsequent clustering of the monolayer high nuclei into their equilibrium, three dimensional shapes) could monolayer growth occur. Thus, they agree, one should not include what are really kinetic factors in the original thermodynamic equations, as was done in (2) and (3). Nevertheless, these kinetic factors are extremely important and do allow for layer growth to occur under thermodynamically unfavorable circumstances as the formation of strained layer superlattices attests.

GROWTH MODE CHARACTERISTICS

A. Volmer-Weber Growth

The essential characteristics of the various growth modes will be illustrated by examples rather than through a detailed theoretical analysis. For recent theoretical and experimental reviews see references [14-16]. When islands are nucleated, the binding energy of the overgrowth B atoms to each other is generally greater than that for the B atoms to the A atoms of the substrate. Atoms or molecules incident on a substrate surface diffuse until they either desorb or find site to which they become bound. Nucleation is followed by growth and coalescence until a continuous, hole-free film is formed. The early stages of this type of growth is illustrated for (111)Cu islands growing on a flat epitaxial (111)Ag substrate at 210°C in Figs. 1a and b. In Fig. 1a the average Cu layer thickness is 5 $\overset{\bullet}{A}$, and the atoms have nucleated as tiny islands. Fig. 1b shows a later stage of this growth when an average 33 $\overset{\bullet}{A}$ of Cu has been deposited. Clearly the islands have become larger and are beginning to contain a visible microstructure. As more Cu is deposited, the islands will eventually coalesce and form a continuous, hole-free film.

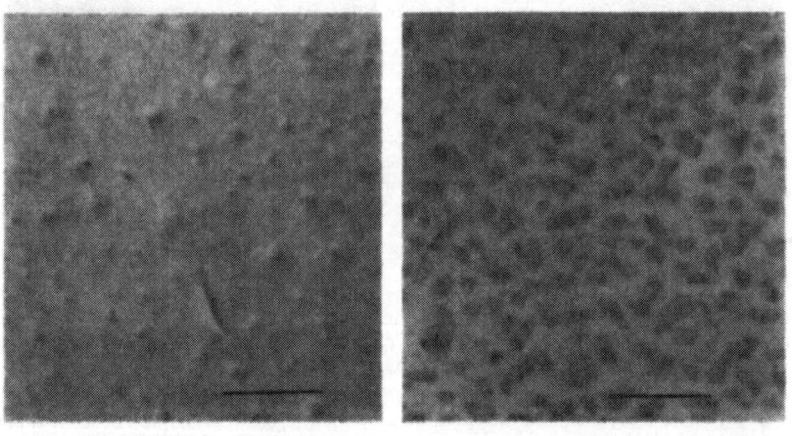

a   b

Figure 1: Epitaxial island growth of (111)Cu on (111)Ag at 210°C: a) 5 $\overset{\bullet}{A}$ Cu, b) 33 $\overset{\bullet}{A}$ Cu. (Micrographs taken by C.T. Horng). Marker: 0.2μm.

B. Frank and van der Merwe Growth

The essential features of this theory are illustrated in Fig. 2. For a lattice misfit f_o, the overgrowth is pseudomorphic up to a critical thickness h_c. That is, the overgrowth film is strained to the lattice parameter and structure of the substrate up to a thickness h_c. Beyond that, misfit dislocations are introduced into the interfacial region and the elastic strain ϵ decreases approximately as $1/h$ where h is the film thickness. For a greater misfit f'_o, the critical thickness h'_c is smaller. Thus if the misfit is large enough, misfit dislocations can be expected to be introduced during the formation of the first monolayer [17].

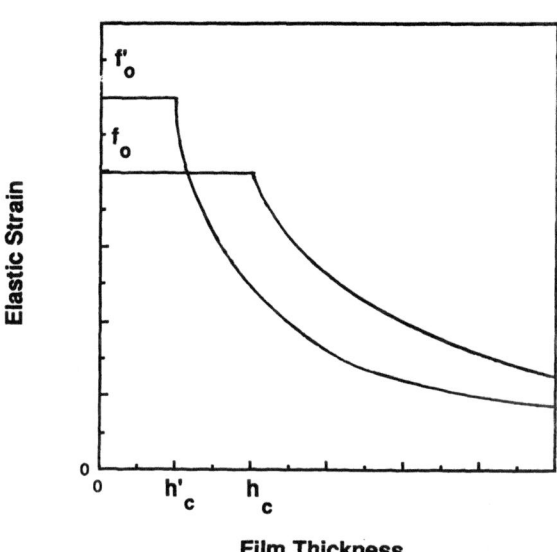

Figure 2: Elastic strain in an epitaxial bilayer as a function of overlayer thickness h for two misfits f_o and f'_o.

Misfit dislocations may be introduced by slip from the edges of monolayer high islands, by atomic reorganization at the interface, by nucleation from the surface of the film or possibly at the interface, or by the glide of threading dislocations which were present in the original substrate and grew up into the film as it was formed. Figure 3 illustrates these last two processes [18].

Figure 4 is an example of the case shown in Fig. 3b. In the sample shown 27 Å of (111)Pt was grown epitaxially on an epitaxial (111)Au film. Three sets of straight line misfit dislocations can be seen. They lie in the three {11Ī} planes which intersect the (111) interface and along <1Ī0> directions in that plane. Slip most likely occurred in the Pt overlayer, thereby removing the corresponding threading dislocations from the region of the Pt film shown in the figure.

NUCLEATION

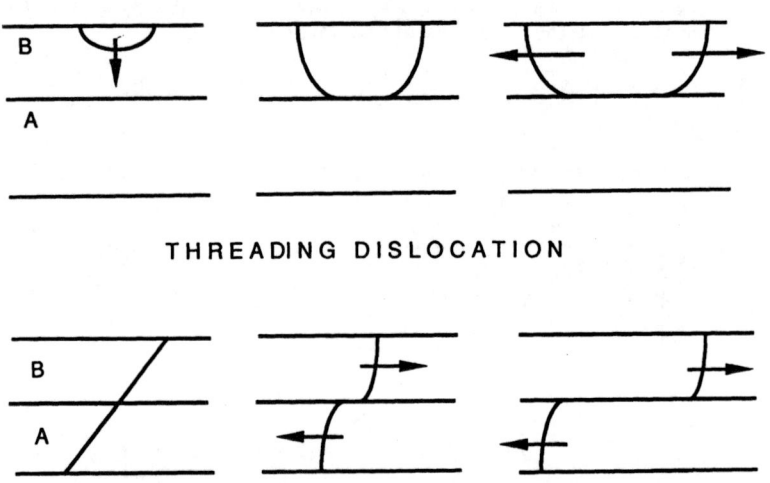

THREADING DISLOCATION

Figure 3: Processes whereby misfit dislocations may be introduced into a bilayer: a) surface nucleation and glide and b) by the glide of threading dislocations.

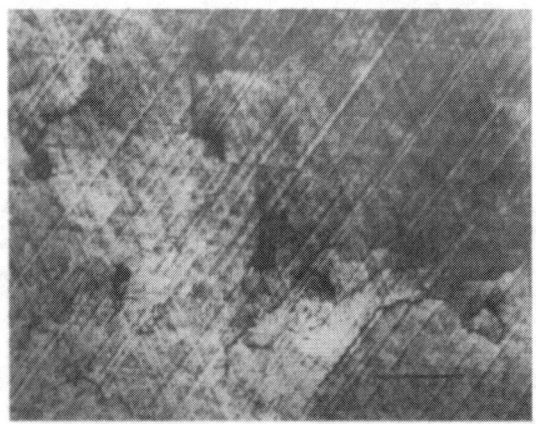

Figure 4: Misfit dislocations in a 27 Å (111)Pt/(111)Au bilayer (work by S.S. Chao and E.-A. Knabbe). Marker: 0.5μm.

C. Stranski-Krastanov Growth

In this growth mode, the formation of one complete monoatomic layer may be followed by one or more fractional monolayers before island growth is initiated on top of these layers. The first clear observation of such a morphology in vapor deposited thin films was made by Macur [19]. Subsequently many other S-K systems have been observed [1]. An example from the epitaxial $Cu_2O/(111)Cu$ system is shown in Fig. 5. For this system flat topped Cu_2O islands which have been partially stress relieved by the misfit dislocations at the Cu_2O/Cu interface, grew up out of the surface [7]. There is also a tendency for the islands to be truncated tetrahedra. In some cases, the islands became extended parallel to the plane of the film, as is the case in Fig. 5. The three sets of straight line misfit dislocations probably arose from the glide of threading dislocations as illustrated in Fig. 3b.

Figure 5: SK growth of Cu_2O on (111)Cu (work by J.H. Ho). Marker: $1\mu m$

D. Epitaxial Multilayers

Layered films are often called superlattices. A layered superlattice is one in which the (hkl) plane of the substrate A is parallel to the (HKL) plane of the overgrowth film B. However, there may be azimuthal misorientation across the interface plane. An example would be (111)Ag growing epitaxially on (001)Cu in which the silver overgrowth is in the form of two sets of (111) islands oriented at 90° to each other and elongated along <110> in the (001) Cu plane [20]. Thus an attempt to grow a superlattice from this system might result in a "layered superlattice".

On the other hand a crystalline superlattice would be one in which all of the interface region has the same epitaxial orientation. In this case, if one wants to avoid the introduction of misfit dislocations, then each layer must have a thickness less than the critical thickness for that layer [18]. However, it is possible for threading dislocations to be propagated into the superlattice. Matthews and Blakeslee [18] have shown that there are mechanisms which may reduce the threading dislocation density as the layers grow.

In growing a strained layer superlattice (SLS) there are a number of important considerations that need to be addressed [9]. Figure 6 illustrates a typical SLS. One starts with a substrate crystal and grows a buffer layer on it, the top of which has a lattice parameter equal to the average value for the A-B superlattice [18]. Clearly the growth of such a SLS requires that layer growth occur for B on A as well as for A on B. This requirement is not possible on the basis of the thermodynamic principles discussed earlier, namely that $\Delta\gamma \leq 0$ for all n. See eqn. (1) and the following discussion. One therefore needs to rely on kinetic

factors to make such growth possible. One first makes sure that $\gamma_A \approx \gamma_B$ and that γ_{in} is small. Then by going to higher supersaturations S, namely larger deposition rates R and lower substrate temperatures T_s, one may be able to achieve metastable conditions that will allow the growth of SLS's. Unfortunately increasing S tends to introduce more defects into the lattices during growth so that it is not assured that high quality SLS's can always be grown in this manner.

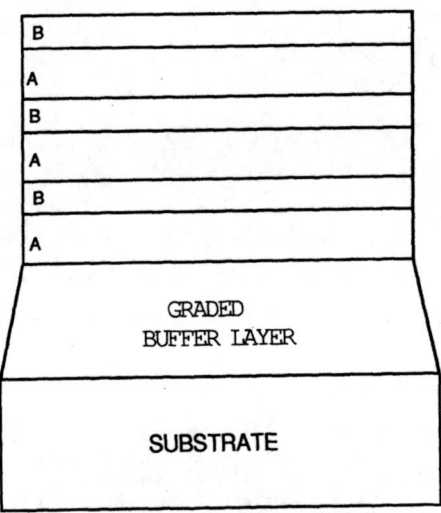

Figure 6: Strained layer superlattice

ACKNOWLEDGMENTS

This work was supported by the U.S. Department of Energy on grant number DE-FG02-84ER45139.

REFERENCES

1. R.W. Vook, International Metals Reviews 27, 209 (1982); Optical Engineering 23, 343 (1984).
2. R.F. Bunshah, in Deposition Technologies for Films and Coatings, edited by R.F. Bunshah (Noyes Publications, Park Ridge, NJ, 1982), p. 83.
3. K.H. Guenther, Proc. SPIE 346, 9 (1982).
4. K. Takayanagi, Proc. 10th International Congress on Electron Microscopy, Hamburg (Deutsche Gesellschaft fur Electron Mikroskopie, Frankfurt, FRG, 1982) p. 287.

5. J.E. Macur and R.W. Vook, Thin solid Films 66 ,371 (1980).
6. K. Hartig, A.P. Janssen, and J.A. Venables, Surface Sci. 74, 69 (1978).
7. J. H. Ho and R. W. Vook, Phil. Mag. 36, 1051 (1977); J. Crystal Growth 44, 561 (1978).
8. E. Bauer, Z. Kristallographie 110, 372 (1958).
9. E. Bauer and J. H. van der Merwe, Phys. Rev. B33, 3657 (1986).
10. J. W. Matthews, D. C. Jackson, and A. Chambers, Thin Solid Films 26, 129 (1975).
11. C. T. Horng and R. W. Vook, J. Vacuum Sci. and Technol. 10, 160 (1973); 11, 140 (1974).
12. U. Gradmann, W. Kummerle, and P. Tillmanns, Thin Solid Films 34, 249 (1976); U. Gradmann and P. Tillmanns, Phys. Stat. Solid. 44a, 539 (1977).
13. I. Markov and R. Kaischew, Thin Solid Films 32, 163 (1976); Kristall Techn. 11, 685 (1976).
14. E. Bauer, in The Chemical Physics of Solid Surfaces and Heterogeneous Catalysis, edited by D. A. King and D. P. Woodruff (Elsevier, Amsterdam 1984), Vol. III B, P.1.
15. G. E. Rhead, Contemp. Phys. 24, 535 (1983).
16. I. Markov and S. Stoyanov, Contemp. Phys. 28, 267 (1987).
17. F.C. Frank and J.H. Van der Merwe, Proc. Roy. Soc. London, A198, 205 (1949); 198, 216 (1949); 200, 125 (1949); J. Appl. Phys. 34, 117, 123 (1963).
18. J.W. Matthews and A.E. Blakeslee, J. Cryst. Growth 32, 265 (1976).
19. J.E. Macur, 33rd Ann. Proc. Electron Microscopy Soc. Amer., Las Vegas, Nev. (G.W. Bailey, ed.) 98 (1975).
20. R.W. Vook and S.S. Chao, Thin Solid Films 58, 203 (1979).

ROLE OF FILM-SUBSTRATE INTERACTION
IN DETERMINING
THE MORPHOLOGY OF THIN FILMS

MARCIA H. GRABOW AND GEORGE H. GILMER
AT&T Bell Laboratories, Murray Hill, New Jersey 07974.

ABSTRACT

The equilibrium structure and metastable states of thin films have been investigated using molecular dynamics computer simulations. The energy as a function of coverage for a variety of film/substrate systems has been determined, and this information has been used to determine the growth mode and the conditions under which the film is dislocation free. In particular, the influence of the strength of the film-substrate binding will be discussed.

I. INTRODUCTION

In this paper we report results of computer simulations that bring out the importance of the strength of the film-substrate binding. As presented for the monolayer case, increasing the strength of the film-substrate binding allows arbitrarily large misfit film/substrate systems to remain dislocation free and uniform. In this paper we discuss the more general case of multi-layer films.

In the next section we briefly summarize the simulation technique that we have used and some of the terminology. In section III the stability of the uniform film relative to clusters is discussed. Section IV will present the results for the density of dislocations that enter at equilibrium. A summary and some conclusions follow in section V.

II. MODEL AND TERMINOLOGY

The motion of particles of the film and substrate were calculated by standard molecular dynamics techniques that have been described elsewhere [1]. In the simulations discussed here, our purpose is to calculate equilibrium or metastable configurations of the system, at zero Kelvin unless otherwise stated. For this purpose, we have applied random and dissipative forces to the particles. Finite random forces provide the thermal motion that allows the system to explore different configurations, and the dissipation serves to stabilize the system at a fixed temperature. The potential energy minima are obtained by reducing the random forces to zero, thus permitting the dissipation to absorb the kinetic energy.

For most of the simulations discussed here, the configuration will be based on a face-centered cubic (FCC) lattice, and the film and substrate atoms interact via a Lennard-Jones potential. For comparison, results for a diamond cubic (DC) lattice, with the film and substrate atoms interacting via a Stillinger-Weber (SW) potential for silicon will also be included.

The Lennard-Jones (LJ) potential, for particles i and j with separation r_{ij} is

$$\phi_{ab}(r_{ij}) = 4\,\epsilon_{ab}\,[(\sigma_{ab}/r_{ij})^{12} - (\sigma_{ab}/r_{ij})^6], \tag{1}$$

which is smoothly truncated at $r_{ij} = 2.5\sigma_{ab}$, and the subscripts a and b indicate the types for particles i and j. The energy of the system will be expressed in units of ϵ_{ss}, the well depth of the potential between two substrate particles. In addition, we assume that the depth of the film-film potential, ϵ_{ff}, is equal to ϵ_{ss}, and we will vary the film-substrate value ϵ_{fs}. The relative strength of the film-substrate potential is

$$W \equiv \epsilon_{fs}/\epsilon_{ff}. \tag{2}$$

Large film-substrate interactions are represented by $W \gg 1$. The dimensionless unit of length is taken to be σ_{ss}, a measure of the atomic diameter of the substrate particles. The degree of misfit, η, is the difference between the film and substrate lattice constants, normalized by the substrate lattice constant. In terms of the LJ parameters,

$$\eta = \sigma_{ff}/\sigma_{ss} - 1, \tag{3}$$

and the film particles are always larger than those of the substrate, or $\eta > 0$. The film-substrate interaction diameter is assigned the average value $\sigma_{fs} = (\sigma_{ff} + \sigma_{ss})/2$.

The SW potential for silicon is composed of a pair potential, similar to the LJ expression described above, and a three-body term that favors the tetrahedral angle between bonds. The pair component is given by

$$f_2(r_{ij}) = \epsilon A [B(\sigma/r_{ij})^4 - 1] \exp[(r_{ij}/\sigma - 1.8)^{-1}], \tag{4}$$

for $r_{ij} < 1.8\sigma$; otherwise $f_2(r_{ij}) = 0$. With values $A = 7.04955$ and $B = 0.602224$, the depth of the potential well is ϵ, as happens for the LJ potential. We have omitted the subscripts a and b on f_2 and the parameters ϵ and σ, although the potential is, of course, dependent on the identities of the interacting species just as shown in eq. (1) for the LJ potential. The three-body component is given by

$$f_3(\vec{r}_j, \vec{r}_i, \vec{r}_k) = 21\epsilon(\cos\theta_{jik} + 1/3)^2 \exp[1.2(r_{ij}/\sigma - 1.8)^{-1} + 1.2(r_{ik}/\sigma - 1.8)^{-1}], \tag{5}$$

for r_{ij} and r_{jk} less than 1.8σ, otherwise it is zero. This expression corresponds to the triplet of particles j, i and k with θ_{jik} measured with particle i at the vertex. Similarly, components for the same triplet with j and k at the vertex must be included in the sum for the total energy. We have chosen to assign ϵ for eq. (5) in terms of the two-body parameters as $\epsilon_{bac} = \sqrt{\epsilon_{ab}\epsilon_{ac}}$, where a, b and c indicate the types for particles i, j and k, respectively. If the substrate is taken to be silicon, $\sigma = 2.0951\text{A}$.

The substrate atoms were arranged to expose a (001) face of the FCC crystal for the LJ potential, and a (111) or (001) face of the DC crystal for SW. Periodic boundary conditions for the (001) systems were applied to the x, $<110>$, and y, $<\bar{1}10>$ directions. Typical dimensions of an elongated simulation cell were $120d_{(110)}$ along the y axis and $2d_{(110)}$ along the x, where $d_{(110)}$ is the spacing between (110) layers. The z dimension of the cell was selected to allow expansion caused by strain. Motion was constrained along the x direction. We vary the density of atoms in a (110) layer by introducing an array of misfit dislocations uniformly spaced along the y axis, with the dislocation lines parallel to the x axis. A similar procedure is followed for the DC (111) face, with the x axis along $<11\bar{2}>$ and y along $<1\bar{1}0>$. Initial configurations with misfit dislocations, for example, are created by simply removing a plane of atoms from the coherent film for each dislocation. These are then annealed by the molecular dynamics procedure to obtain a local energy minimum. The equilibrium structure is obtained by a comparison of the film chemical potentials for various local energy minima as described below. All data reported here were obtained with a rigid substrate (substrate atoms fixed at the bulk equilibrium spacing). The relaxation of the substrate can change the energies of misfit dislocations and other defects a significant amount, but the qualitative features are unchanged.

III. UNIFORM FILMS AND CLUSTERS

Uniform thin films are obtained for only a limited number of film/substrate material combinations [2]. The more common experience is that the deposited material aggregates into three-dimensional clusters. The clusters may form directly on the bare substrate, in the Volmer-Weber growth mode, or on top of a very thin but uniform film of the deposit, the Stranski-Krastanov growth mode illustrate in Fig. 1. Our molecular dynamics results have been presented in detail elsewhere [3-5], so only a summary will be given here. First we describe the method for determining the equilibrium configuration of the deposited material, and then we consider the nucleation of clusters on top of a uniform but metastable film.

A) Equilibrium

The equilibrium configuration of a deposit can be ascertained from a plot of the system energy as function of the coverage. In addition to the bulk energy, there are two contributions to the energy of the system. First, the misfit between the film and the substrate leads to a strain energy that will be proportional to η^2 and the thickness of the film. Second, there is a film-substrate interaction that typically acts over a very short range ($\approx 10A$).

A large three dimensional cluster has approximately the bulk film energy. The lattice constant approaches that of the bulk film, and has a negligible contribution from the surface energy because of a low surface to volume ratio. Therefore, the strain energy serves as a driving force for cluster generation for any case of non-zero misfit.

The only driving force for uniform films is a strong interaction between the film and substrate, which requires that $W > 1$. Even if there is a strong interaction, it is only when the film atoms interact directly with the substrate that there will be a driving force for layer-by-layer growth, so it is effective only up to a limited film thickness for non-zero misfit. If the film-substrate interaction is weak ($W \leq 1$), there will be no driving force for layer-by-layer growth.

The boundary between Stranski-Krastanov and Volmer-Weber growth has been determined from our molecular dynamics results, and is shown in Fig. 2 for

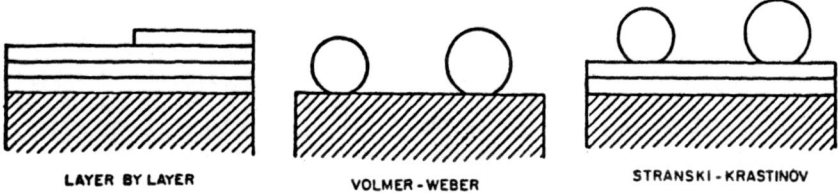

LAYER BY LAYER   VOLMER-WEBER   STRANSKI-KRASTINOV

Fig. 1. Schematic illustration of several structures observed when material is deposited onto a crystalline substrate.

the LJ and SW potentials. Above the curve (larger W) there is a sufficiently large film-substrate interaction for the first monolayer to be stable, and Stranski-Krastanov growth is predicted; below the curve clusters are predicted to grow directly on the bare substrate in the Volmer-Weber mode. Layer-by-layer growth is confined to the W-axis at $W = 1$ and above.

Therefore, we conclude that for a system with non-zero misfit, a uniform film with a thickness greater than several monolayers is not the true equilibrium state. The system will have a lower chemical potential if clusters form. Clusters will form on either the bare substrate (Volmer-Weber mode; any finite misfit with $W \leq 1$ and large misfits if $W > 1$) or on a few layers of uniform film (Stranski-Krastanov mode; up to moderate misfits with $W > 1$). This will be true for any system without long-range (e.g. electrostatic) forces.

B) Metastability

The stability of a uniform film against shape perturbations of small amplitude will be reviewed here. If perturbations can grow monotonically with time, a metastable film is not possible. Bruinsma and Zangwill [6] have shown that thermally roughened film surfaces will permit infinitesimal perturbations to grow under certain conditions.

The situation is very different if the temperature is below the roughening temperature, which is the case with the growth of many metallic and most semiconductor films. Then the deviations from planarity of the film surface will create well defined steps on the surface. Below the roughening temperature, steps have a positive edge free energy and contribute a positive term to the total surface free energy. For perturbations over the entire range of length scales this term will cause the perturbation to decay. Under these conditions, we do expect an activation barrier to forming clusters, because of the larger surface area associated with the initial stages of cluster growth.

We have used nucleation theory to determine the temperature dependence of the critical cluster size and the misfit at which nucleation of the clusters becomes possible. We have treated the case $W > 1$, where we expect the cluster to form on a few layers of uniform film. Cluster formation will then be independent of the strength of the film-substrate interaction, because the atoms in the cluster will be outside the range of that interaction, but cluster nucleation will depend on the misfit and the temperature. We have determined the metastable limit of the uniform film elsewhere [4], and simply give the results here. For both the LJ and SW films, we have plotted in Fig. 3 the misfit η for which there is sufficient driving force to nucleate a critical-sized cluster on an experimental time scale. In particular, we find that uniform films remain in a metastable state for misfits up to ≈0.10, and this result is not very sensitive to temperature. In practice, clustering is generally observed for $\eta \leq 0.05$, and the nucleation theory provides an indication of the intrinsic limits on planar growth on perfect substrates without active impurities. We have not considered the stability against clustering of planar growth in the Volmer-Weber regime of Fig. 2, where the film-substrate interaction parameter W plays an important role. The theory of this process has been discussed by Venables (7).

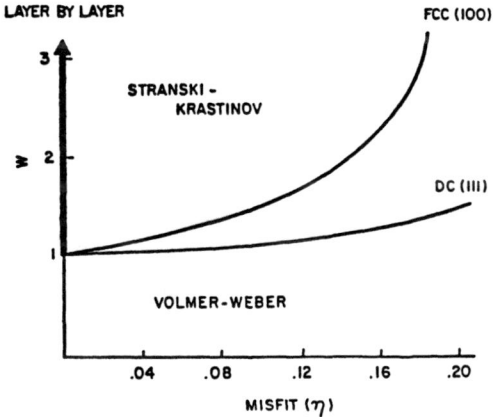

Fig. 2. Equilibrium growth modes, as a function of misfit (η) and strength of film-substrate interaction (W) for LJ and SW potentials. The solid lines separate the regions of the Volmer-Weber mode (clusters on bare substrate) and Stranski-Krastanov mode (clusters on a few layers of uniform film); layer-by-layer growth is confined to the region $\eta=0.0$ with $W>1$.

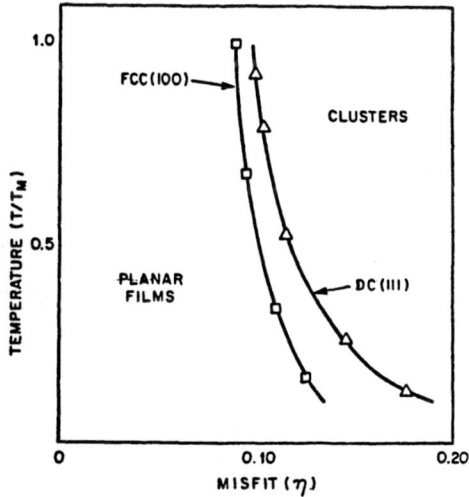

Fig. 3. Plot of the misfit at which clusters will spontaneously form on a uniform film as a function of the temperature, relative to the melting point, for the LJ and SW potentials ($W>1$).

IV. COHERENT INTERFACES AND MISFIT DISLOCATIONS

Theories predicting the coherency of film-substrate interfaces were first developed for monolayer films [8-10], for which exact analytic expressions could be obtained, and later extended to multilayer systems [11,12]. To study the effect of more realistic potentials or higher dimensionality, computer simulations are necessary [3,13-17].

For monolayer films, Frank and van der Merwe (FM) determined the critical misfit, η_c, the maximum amount of misfit between film and substrate that can be accommodated in a coherent monolayer, as a function of W [8,9]. On a phase diagram of η versus W, FM theory predicts two phases: a coherent phase and an incoherent phase that consists of a regular array of misfit dislocations. The boundary between the two phases, the critical misfit, scales as $W^{\frac{1}{2}}$. Molecular dynamics computer simulations of monolayer films interacting with LJ forces [16] allowed us to refine the phase diagram. The films are still predicted to have coherent and incoherent phases, but the misfit dislocations in the incoherent film are of two distinct types, which we called continuous and localized. Another defect that is consistent with the FCC(001) surface is a Shockley partial dislocation, and this was found to be lower in energy over a large portion of the phase diagram, even down to zero misfit for weak substrate interactions, $W \leq 0.51$. Although the details of the phase diagram of η vs W are modified from FM theory, we still find that the critical misfit can be arbitrarily large if W is made large enough.

For multilayer films, we're interested in the critical thickness, h_c, at which it becomes energetically favorable to introduce misfit dislocations for a film/substrate system of given η and W. Matthews developed a theory that can be expressed in terms of bowing forces on dislocations [11], or in terms of the strain that minimizes the energy [12]. This theory only holds for edge dislocations and $W=1$. To study $W>1$ or other types of defects (analogous to the Shockley partials in the monolayer case), we turn to molecular dynamics computer simulations.

The method for determining the critical misfit has been discussed elsewhere [3], and will simply be stated here. Because the incoherent state (with misfit dislocations) has a different number of atoms than the coherent state (no misfit dislocations), the thickness at which it will become energetically favorable to introduce misfit dislocations will be based on the relative chemical potential of the two states. The chemical potential at a given coverage is simply the slope of the energy versus coverage curve that was mentioned in Section III.A. The critical thickness, h_c, is the largest value of h at which the coherent state has the lower chemical potential; for larger h, the incoherent state will have a lower chemical potential than the coherent state.

Fig. 4 shows are results for large film-substrate binding, $W=4$, and for $W=1$ with two types of defects. The open squares in 4a and 4b are the critical thickness at which it becomes energetically favorable to introduce a pure edge dislocation at the film-substrate interface. Because of the large film-substrate interaction at $W=4$, the critical thickness in 4b far exceeds that of 4a. However, in Fig. 4b we now consider introducing an edge dislocation between the

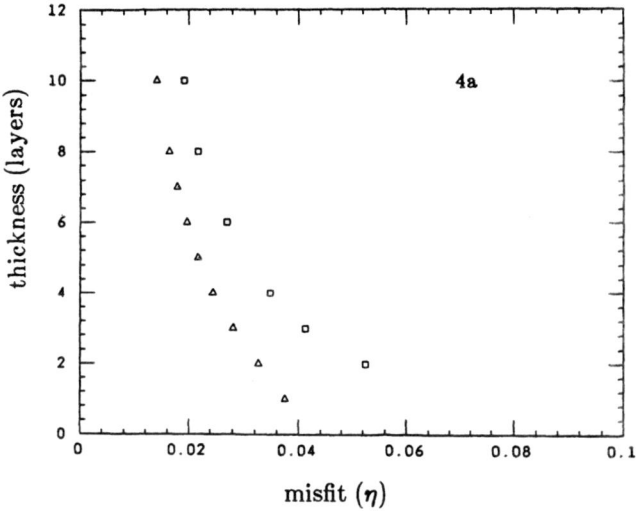

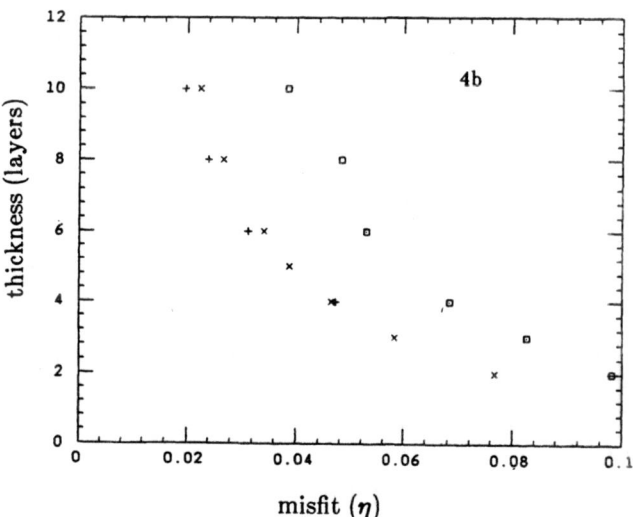

Fig. 4   Critical thickness as a function of misfit for face-centered cubic films
with LJ potentials. a) $W=1$. Open square: edge dislocation; open tri-
angle: capped Shockley partial dislocation. b) $W=4$. Open square:
edge dislocation at the film-substrate interface; "x": edge dislocation
between 1st and 2nd film layers; "+": edge dislocation between 2nd and
3rd film layers.

2nd and 3rd layers of the film. Because of the short-range nature of the film-substrate interaction, this is almost equivalent to the case of and edge dislocation at the film-substrate interface for $W=1$. Therefore, the critical thickness for a system with large W can exceed that of a system with $W=1$ by only a few atomic layers (≈ 10A). The actual critical thickness for $W=4$ with the edge dislocations between the 1st and 2nd layers of the film ("x") and between the 2nd and 3rd layers of the film ("+") is shown in Fig. 4b.

As in the monolayer case, we also need to consider defects other than pure edge dislocations. For the FCC(100) films, a dislocation directly related to the Shockley partial was found to have a lower energy than the pure edge dislocation, so the actual critical thickness predicted for $W=1$ is shown as the triangles in Fig. 4a.

V. SUMMARY

The structure and stability of thin epitaxial film/substrate have been investigated using molecular dynamics computer simulations. Both the misfit, η, and the strength of the film-substrate binding, W, have been varied.

The simulation results show that the uniform film is never the equilibrium state for a film (thickness $> \approx 10$ A) with non-zero misfit. We have predicted the values of η and W that lead to the Stranski-Krastanov mode (cluster forming on a few layers of uniform film) rather than the Volmer-Weber mode (cluster forming on a bare substrate). The strength of the film-substrate binding is critical in determining which of these modes is the lowest energy state. Although uniform films are never the equilibrium state, if $W>1$ they can persist in a metastable state at moderate misfits ($<10\%$ at half the melting point) because of the large nucleation barrier to the formation of clusters.

We have also calculated the critical thickness, h_c, at which it becomes energetically favorable to introduce misfit dislocations. If the only possible defects are edge dislocations at the film-substrate interface, there will be a strong dependence of the critical thickness on the strength of the film-substrate interaction. However, because of the short-range nature of the film-substrate interaction, the edge dislocation will be energetically favorable at a point 1-3 layers above the film-substrate interface, so the critical thickness for $W>1$ can only vary by 1-3 layers from the critical thickness predicted for $W=1$. In addition, if other types of misfit dislocations are considered, the critical thickness may be lower than that predicted by considering edge dislocation only.

REFERENCES

(1) J. C. Tully, G. H. Gilmer, M. Shugard, J. Chem. Phys. **71**, 1630 (1979).

(2) M. Bienfait, J. L. Seguin, J. Suzanne, E. Lerner, J. Krim, and J. G. Dash, Phys. Rev. **B29**, 983 (1984).

(3) M. H. Grabow and G. H. Gilmer, R. Hull, eds., Symposium of the 1987 Spring Meeting of the Materials Research Society, North-Holland, (1987).

(4) M. H. Grabow and G. H. Gilmer, to be published in Surf. Sci.

(5) G. H. Gilmer and M. H. Grabow, J. Metals **39**, 19 (1987).

(6) R. Bruinsma and A. Zangwill, preprint,and J. Physique **47**, 2055 (1986).

(7) J. A. Venables and G. L. Price, *Epitaxial Growth, Part B*, ed. J. W. Matthews, (Academic Press, New York, 1975), chapter 4.

(8). F. C. Frank and J. H. van der Merwe, Proc. Roy. Soc. (London) **A198**, 205 (1949); **A198**, 216 (1949).

(9) F. C. Frank and J. H. van der Merwe, Proc. Roy. Soc. (London) **A200**, 125 (1949).

(10) J. H. van der Merwe, J. Appl. Phys. **41**, 4725 (1970).

(11) J. W. Matthews and A. E. Blakeslee, J. Crystal Growth **27**, 118 (1974).

(12) J. W. Matthews, *Epitaxial Growth, part B*, (Academic Press, New York, 1975), chapter 8.

(13) T. Halicioglu, J. Cryst. Growth **29**, 40 (1975).

(14) B. W. Dodson, Phys. Rev. **B30**, 3545 (1984).

(15) G. H. Dienes, K. Sieradzki, A. Paskin, and B. Massoumzadeh, Surf. Sci. **144**, 273 (1984).

(16) M. H. Grabow and G. H. Gilmer, J.M. Gibson, G.C. Osbourn and R.M.Tromp, eds., Symposium of the 1985 Fall Meeting of the Materials Research Society, North-Holland, (1986).

(17) B. W. Dodson, Apl. Phys. Lett. **49**, 642 (1986).

PHOTON-CONTROLLED GROWTH OF MULTILAYERED STRUCTURES

DOUGLAS H. LOWNDES, D. B. GEOHEGAN, D. ERES, D. N. MASHBURN and S. J.
PENNYCOOK; Solid State Division, Oak Ridge National Laboratory, Oak Ridge,
TN 37831-6056

ABSTRACT

Pulsed ArF excimer laser (193 nm) photolysis has been used to deposit
entirely amorphous and mixed amorphous/polycrystalline superlattice structures containing Si, Ge and Si_3N_4. High resolution in situ optical reflectivity measurements were used to monitor and/or control deposition. Transmission electron microscope cross-section views demonstrate that amorphous
superlattice structures having highly reproducible layer thicknesses (from
about 50 to several hundred Å), and sharp interlayer boundaries, can be
deposited at low substrate temperatures under laser photolytic control.

INTRODUCTION

The growing number of applications for structures composed of alternating thin layers of crystalline or amorphous materials has given new impe-
tus to the search for alternative low temperature thin film deposition
techniques. For semiconductors, low temperatures are needed when multilayered structures contain highly doped adjacent layers, in order to avoid
dopant interdiffusion during growth. For semiconductors and other
materials, low deposition temperatures minimize in-diffusion of unwanted
impurities from the surroundings, prevent interdiffusion with the substrate
or other adjacent materials, and provide unique access to any low temperature phases that may exist. However, conventional thermally driven
(pyrolytic) chemical vapor deposition (CVD) film growth reactions usually
are limited to very low, or even negligible, film growth rates at low temperatures.

In this paper we describe an alternative to pyrolytic growth, namely,
direct photolysis of precursor gas molecules, using pulsed ArF (193 nm)
excimer laser radiation. The principal advantage of pulsed-laser photolysis
for fabrication of amorphous multilayered structures is that high resolution
can be obtained in layer thickness, because deposition is inherently
"digital": Each laser pulse produces much less than a monolayer of film, on
average. Nevertheless, the high pulse repetition rate of excimer lasers
permits high deposition rates to be achieved at low temperatures. We also
describe an optical reflectivity technique that provides sub-monolayer resolution film thickness monitoring. The combination of high resolution optical monitoring with pulsed laser photolysis can, in principle, provide
monolayer accuracy in controlling the average thickness of deposited layers.
We report results of experiments in which we have explored deposition of
superlattice structures containing amorphous semiconductor and dielectric
thin film layers, using these techniques.

EXPERIMENTAL CONDITIONS

Multilayer growth experiments were carried out in a deposition chamber
[1] based on a six-way stainless steel cross and equipped with 2-in.-diam
Suprasil windows. All depositions were made onto 2.5-cm square (100)
crystalline (c) Si substrates; the substrate temperature was measured with
an infrared radiation thermometer (IRCON Type W). The ArF laser beam (15 ns

FWHM pulse duration) was unfocused and parallel to the substrate; it was passed through a rectangular slit before entering the chamber, to more precisely define its cross section (6 x 20 mm^2), and the bottom edge of the beam was set 1 mm above the substrate's surface.

Amorphous (a) Si was deposited by photolysis of disilane (Si$_2$H$_6$), Si$_3$N$_4$ by photolysis of a mixture of disilane and ammonia (NH$_3$), and a-Ge by photolysis of germane (GeH$_4$). In addition to these gases, He was used to flush the inside of the two windows through which the excimer laser beam entered and exited the deposition chamber. He flow conditions were found [1] that would nearly completely eliminate the rapid deposition of Si or Si$_3$N$_4$ films, which will otherwise occur on these windows, resulting in a rapid decrease of laser power and preventing accurate monitoring of the laser power within the chamber. The He flush did not prevent a very gradual decay of laser power due to Ge film deposition on the windows. However, we found that the Ge film buildup is a strong function of the incident laser intensity; by operating at low fluence during the a-Ge depositions it was possible to forestall Ge window-film buildup.

Because the absorption cross-section of germane is very small [2] at 193 nm, our a-Si/a-Ge multilayers were fabricated in two different ways. Structures containing relatively thin (<100 Å) a-Ge layers were grown entirely by photolysis, but thicker a-Ge layers were grown by pyrolysis, in order to obtain reasonable growth times and to minimize the gradual loss of laser power due to Ge film deposition on the windows. However, pyrolytic growth of a-Ge required working at a higher substrate temperature of 390°C. The completely photolytic a-Ge/a-Si structure was deposited at 250°C. Its a-Si layers used 40 sccm of a 10% disilane/90% He mix at p=5 Torr, 40 Hz laser rep rate and 200 mW of transmitted laser power. The a-Ge layers were deposited using 140 sccm of a 10% germane/90% He mix at p=50 Torr, 60 Hz rep rate and 420 mW transmitted power. A He window purge of 450 sccm (total) was used continuously. The only major differences for the structure with pyrolytically grown a-Ge layers were that the flow rate of germane mix was 20 sccm with p=5 Torr and no window flush, and the carrier gas for both silane and germane was H$_2$ rather than He. In this case, we observed a long "incubation time" (5–10 min) before pyrolytic growth of each a-Ge layer could begin. We speculate that this may have been due to H$_2$ covering the previous a-Si layer and preventing nucleation of a-Ge. We also found that photolytic deposition of a-Ge from GeH$_4$/H$_2$ mixtures was much slower than from GeH$_4$/He.

Several a-Si/Si$_3$N$_4$ multilayers were grown at 350°C and p=5 Torr. For the a-Si layers, a 10% disilane/90% H$_2$ mixture was used at a flow of 20 sccm; for the SiN$_3$N$_4$ layers 60 sccm of NH$_3$ and 20 sccm of the disilane mix were used (~30:1 NH$_3$:Si$_2$H$_6$ ratio). At 350°C, measurements that are reported elsewhere [1] show that the ratio of our photolytic and "background" pyrolytic deposition rates for a-Si is about 1000:1, so that film deposition is completely photonically controlled.

Monitoring and Control of Multilayer Growth

We have developed an optical reflectivity technique for rapidly and precisely measuring film deposition rates under varying conditions [1]. A low-power HeNe (632.8 nm) laser beam is reflected from the gas-film and film-substrate interfaces at near-normal incidence and the reflected beam is detected using a large-area Si photodiode. Using this techique, it is possible to monitor film deposition in real time with a resolution of ~0.002Å (where the film thickness is averaged over the HeNe laser's ~0.5 mm diam unfocused spot size). Thus, submonolayer resolution easily is obtained and deposition rates <0.001 Å/min can be measured [1].

Figure 1 shows the reflectivity signal obtained during deposition of an a-Si film by photolysis of disilane. This figure illustrates that essentially complete "photonic" control of deposition can be obtained:

Deposition ceases when the laser is turned off, and film thickness steps of only a few Angstroms are introduced easily. In fact, the mean deposition rate of ~1 Å/sec in this experiment was obtained at a laser pulse repetition rate of 40 Hz, so that each laser pulse corresponds to deposition of only a very small fraction (~0.02 Å) of a monolayer (averaged over the HeNe beam spot). The substrate temperature is sufficiently low that the "background" thermal growth rate is negligible [1]. Thus, direct "photonic" control appears to be a feasible method for fabricating amorphous superlattice structures.

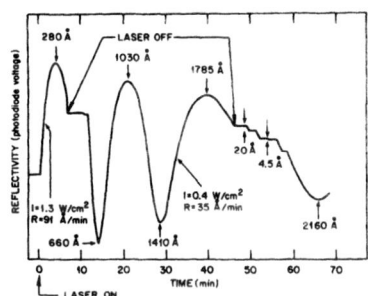

Fig. 1. Photo-controlled deposition of a-Si on an oxidized c-Si substrate. (ArF laser: 40 Hz and 1 W/cm^2; 10% disilane/90% He at 20 sccm; 5 Torr, 350°C.)

The results shown in Fig. 1 suggest two methods for controlling the growth of multilayered and superlattice structures: (1) "open loop" control, in which the average laser power transmitted into the chamber and the number of laser pulses are monitored, with corresponding layers deposited such that the product of (power)×(no. of pulses) is kept constant; (2) "closed loop" control, in which the actual film thickness is monitored and used to determine the end point for each layer. Open loop control assumes that the deposition rate for each material can be calibrated in advance. It then depends upon accurate measurement of the laser power and upon non-laser-related parameters, such as gas flows and pressure, remaining constant during deposition. The practical limit for this case is set by the stability and reproducibility of power meters, mass flow controllers, pressure controllers, etc., and by the accuracy with which the UV laser power in the chamber can be measured. Closed loop ("feedback") control places less stringent requirements on controllers and meters, but requires continuous, accurate in situ monitoring of the actual film thickness. For this to be possible, it is necessary to calculate the reflectivity signal that will be obtained during deposition of a multilayered structure, i.e., the optical constants of the various layers must be known under the actual deposition conditions. In addition, it must be assumed that interlayer interactions (diffusion, chemical reactions, phase transformations) either do not occur during growth or occur in ways that can be modeled. Considering the latter constraint, deposition at low substrate temperatures has obvious advantages.

Laser Photochemical Vapor Deposition (LPVD) Reactions

Although deposition from disilane, germane and ammonia mixtures has been studied under a variety of conditions, no definitive studies of the reaction chemistry leading to deposition have been presented in the literature. Deposition of Ge by photolysis of GeH$_4$ relies upon a very small (< 4 × 10^{-19} cm^2) absorption cross section at 193 nm [2]. Nevertheless, at temperatures < 275°C and high pressures (~200 Torr), photolytic deposition at modest laser fluences (15 mJ/cm^2) dominates the pyrolytic deposition rate [3]. While germylene (GeH$_2$) is probably produced at the surface by pyrolysis of germane, growth of the hydrogenated germanium film may involve a series of insertion reactions involving GeH$_2$ followed by desorption of H$_2$ [4]. In addition to GeH$_2$, the photolytic reaction may produce other intermediates such as GeH, GeH$_3$, Ge$_2$H$_6$ and atomic Ge [3].

Amorphous Si deposition from disilane by pyrolysis and by 193 nm photolysis is discussed in detail in another paper [1]. The absorption cross section at 193 nm ($\sim 3 \times 10^{-18}$ cm^2) is significantly larger than for GeH$_4$ [2]. The primary products of photodecomposition of Si$_2$H$_6$ are thought to be SiH$_3$SiH and H$_2$ while secondary gas-phase reactions produce such intermediates as SiH, SiH$_3$ and Si$_2$H$_6$, which have for their final product (SiH$_2$)$_n$ deposits on the surface. Above 200°C, H$_2$ desorbs from the surface, leaving silicon. Thermal growth of silicon involves the decomposition of Si$_2$H$_6$ into SiH$_4$ and SiH$_2$, followed by a complicated series of silylene insertion reactions yielding (SiH$_2$)$_n$ deposits, as in the photolytic case.

The reaction chemistry leading to growth of Si$_3$N$_4$ by photolysis of ammonia and silane (or disilane) mixtures is even more complex and has not yet been determined. Under irradiation at 193 nm, NH$_3$ is strongly absorbing ($\sigma_{193} = 1.7 \times 10^{-17}$ cm^2) [6] and photodissociates predominately to NH$_2$ (X ^{2}B$_1$) + H with only a small percentage (2.5%) going to NH$_2$ (A ^{2}A$_1$). The excited NH$_2$ (A ^{2}A$_1$) state can absorb another photon, dissociating to NH (A $^3\Pi$) + H but this is energetically impossible for the majority of dissociated NH$_2$ in the (X ^{2}B$_1$) state [5]. Considering the variety of intermediates formed during photolysis of Si$_2$H$_6$, the chemistry becomes quite complicated. In work with mercury-photosensitized reactions of SiH$_4$/NH$_3$ mixtures, however, Wu identified several Si-N-containing intermediates including silylamine (SiH$_3$NH$_2$), disilazane ((SiH$_3$)$_2$NH) and disilaneamine (Si$_2$H$_5$NH$_2$) [7]. Polymeric solids are suspected to result from such intermediates, with successive SiH$_2$ insertions resulting in films of higher Si and N content. These solids are known to transform to silicon nitride under prolonged heating.

RESULTS AND DISCUSSION

Figure 2 is a TEM cross section view of a photolytically grown nine-layer a-Si/a-Ge structure. The average thickness of the first, third and fourth a-Ge layers is 54 (±2) Å, while for the four a-Si layers the average thickness is 107 (±4) Å. The thickness of successive layers was controlled by keeping the product of laser power and deposition time a constant for corresponding layers. The second a-Ge layer was deposited with the power-time product increased by ~24%, resulting in a layer 30% thicker (70 Å) than the average. These results illustrate the precise control over layer thickness, and the very sharp boundaries between layers, that can be achieved by laser photolysis. The HeNe reflectivity signal was monitored during deposition and was

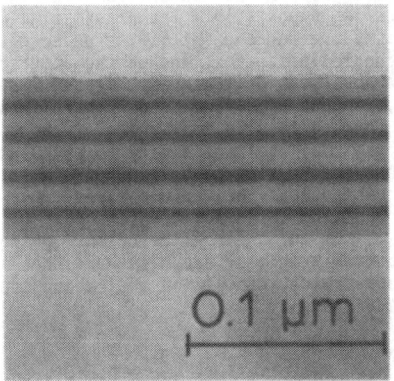

Fig. 2. Nine-layer a-Si/a-Ge structure deposited on (100) c-Si by ArF laser photolysis at 250°C.

compared with model calculations. The experimental and calculated reflectivities were in semi-quantitative agreement (i.e., relative heights of successive reflectivity peaks and valleys were usually in the correct relationship), as a function of layer thickness. However, we must emphasize again that quantitative differences did occur between the calculated and actual reflectivities, presumably because n and k values for a-Si and a-Ge were not sufficiently well known under our deposition conditions. Nevertheless, these results show that very high precision can be maintained during photolytic deposition of amorphous superlattice structures, and that

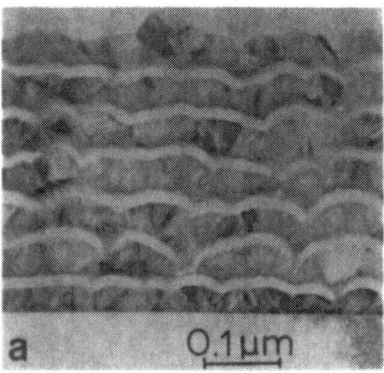

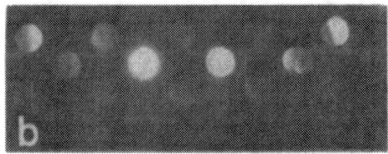

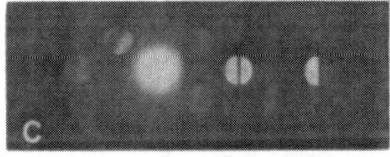

Fig. 3. (a) Fourteen-layer poly-Ge/a-Si structure deposited at 390°C, together with diffraction patterns illustrating epitaxial orientation of the bottom Ge layer (b) with the (100) c-Si substrate (c).

Fig. 4. TEM cross section view of 32-layer Si_3N_4/a-Si amorphous superlattice structure deposited by ArF laser photolysis at 350°C. Strong Fresnel fringe contrast is seen at the Si/Si_3N_4 interfaces.

model calculations provide at least semi-quantitative on-line verification that deposition is proceeding correctly.

Figure 3 shows the seven-period Ge/Si structure in which the Ge layers were deposited pyrolytically. In this case, the 390°C deposition temperature resulted in crystallization of all of the Ge layers, and in some resultant buckling of the multilayered structure when crystallization occurred. However, the bottom Ge layer is epitaxially oriented with the (100) c-Si substrate, though it still contains large numbers of twin boundaries. The Ge layers are 330 (±70) Å thick, while the photolytic a-Si layers are about 130 (±12) Å thick, again with very sharp boundaries and uniform thicknesses.

Figure 4 illustrates the result of photolytic deposition of a 32-layer (16-period) superlattice in which amorphous semiconductor (a-Si) and ceramic (Si_3N_4) layers are alternated. The overall thickness of the complete structure is about 6,260 Å. After the first layer, the a-Si layer thicknesses are 133 (±4) A, while the Si_3N_4 layer thicknesses are 266 (±20) A, as determined from TEM cross-section views. The optical reflectivity signal that was calculated before deposition of this structure gave an accurate picture of the relative heights and depths of successive reflectivity minima and maxima, respectively. However, the depth of modulation of the experimental reflectivity signal was still somewhat larger than in the calculations, which were based on optical parameters of single-layer a-Si and Si_3N_4 films that had been deposited in earlier experiments at temperatures that were only slightly different than the actual superlattice deposition temperature of 350°C. Thus, at this time we cannot rule out the possibility that at least part of the modulation of the reflectivity signal was due to interactions (interdiffusion, chemical reactions) occurring at or near the interface between alternate layers.

SUMMARY AND CONCLUSIONS

Because film deposition is controlled photochemically and not thermally, laser photolysis has certain advantages for the fabrication of artificially structured (multilayered) materials, as follows. (1) Film deposition can be carried out at low temperatures, minimizing dopant and impurity diffusion and resulting in atomically sharp interfaces. (2) The laser photon fluence provides excellent "on-off" control over film deposition. The use of high repetition rate pulsed excimer lasers results in precise "digital" control over film thickness, at the sub-monolayer level on average (per laser pulse), while maintaining attractive overall deposition rates. Since the film deposition rate also can be monitored optically with comparable resolution, then actual control over film growth can be achieved at the monolayer level. (3) The photolytic film growth concept appears to be broadly applicable to the growth of semiconductor, ceramic (dielectric) and metal thin films, using a variety of precursor gases, and thus to fabrication of a wide variety of multilayered structures.

ACKNOWLEDGMENTS

We would like to thank P. H. Fleming, J. T. Luck and C. W. Boggs for their assistance with sample preparation. This work was sponsored by the Division of Materials Science, U.S. Department of Energy under Contract DE-AC05-840R21400 with the Martin Marietta Energy Systems, Inc.

REFERENCES

1. See D. Eres, D. H. Lowndes, D. B. Geohegan, D. N. Mashburn and S. J. Pennycook, "Laser Photochemical Growth of Amorphous Silicon at Low Temperatures and Comparison with Thermal CVD," Symposium B of this meeting.
2. U. Itoh, Y. Toyoshima, H. Onuki, N. Washida and T. Ibuki, J. Chem. Phys. **85**, 4867 (1986).
3. K. K. King, V. Tavitian, D. B. Geohegan, E.A.P. Cheng, S. A. Piette, F. J. Scheltens and J. G. Eden, Mat. Res. Soc. Symp. Proc. **75**, 189 (1987).
4. T. Motooka and J. E. Greene, J. Appl. Phys. **59**, 2015 (1986).
5. V. M. Donnelly, A. P. Baronavski and J. R. McDonald, Chem. Phys. **43**, 271 (1979).
6. K. Watanabe, J. Chem. Phys. **22**, 1564 (1964).
7. C.-H. Wu, J. Phys. Chem. **91**, 5054 (1987).

THE SPUTTER DEPOSITION OF METAL MULTILAYERS

R.E. SOMEKH, R.J. HIGHMORE, K. PAGE, R.J. HOME and Z.H. BARBER
Department of Materials Science and Metallurgy, University of Cambridge, Pembroke Street, Cambridge, CB2 3QZ, UK.

ABSTRACT

We describe the strategy that we are using to make precision metal multilayers. Differential scanning calorimetry has been used with the Ni/Zr system to study the abruptness of the interface as a function of the sputtering pressure. For 10nm period multilayers there is a monotonic increase in the width of the interface with increasing sputtering pressure. W/Si multilayers have been studied as a function of both the sputtering pressure and the relative thicknesses of tungsten and silicon. At reasonably low sputtering pressures a well textured (110) tungsten X-ray peak is seen which is compatible with the expected thickness of the tungsten layers.

Finally, we report some preliminary work on sputtering from tungsten and silicon targets which are at different distances from the substrate so that the degrees of bombardment on the growing layers of the film can be independently varied.

INTRODUCTION

In the last few years the possibility of depositing a metal multilayer in which there is exact atomic registry from top to bottom of a thin film has come closer to reality. In the pursuit of such an objective we have been developing means of preparing precisely stacked multilayers. Various levels of ideality can be envisaged, the ultimate objective being to prepare a single crystal multilayer with atomically flat interfaces. This would require a deposition rate stability and accuracy of 0.01% through the thickness of, for example, a 300 nm thick film in order to guarantee near-perfect atomic register. Polycrystalline or amorphous multilayers may be produced which would necessarily have less well defined interfaces but which could have equally good registry.

Whether one can make atomically sharp interfaces with a single crystal system will probably ultimately depend upon the purity of the deposition system. Very high surface mobilities at modest temperatures minimise the surface/sub-surface diffusion process, as described by Schneider *et al.* [1]. Single crystal multilayers rely for their perfection on being prepared under near equilibrium thermodynamic conditions and minimal energy must be injected to ensure the best quality [2]. This is not true for polycrystalline and amorphous systems in which energetic bombardment during deposition aids mobility and promotes smoothness.

In the case of single crystal multilayers smoothness and flatness arise from thermally induced surface mobility, and the distance between ledges in the layer by layer growth process is determined by thermal kinetics. This process is disrupted by any very energetic atoms which may create defects at the surface of the growing film. However, in order to deposit polycrystalline or amorphous multilayers mobility must be promoted to keep the film flat as it grows. This can be achieved by bombarding the surface with energetic particles. In the case of sputter deposition this occurs via reflected neutrals and sputtered atoms [9]; in the case of evaporation ion beams may be used. Atomic mixing on some scale is expected during sputter deposition and there may be real difficulties in preventing surface/sub-surface diffusion occurring in the single crystal case. One may pose the question: which type of multilayer can ultimately be made with the sharpest interfaces?

In this paper we describe our work on the UHV sputter deposition of Ni/Zr and W/Si multilayers in which we try to address some of the problems of precise deposition of non-single crystal systems. This work has the view of learning about the deposition techniques and some of the processes which will eventually lead to the deposition of ideal single crystal multilayers.

DEPOSITION SYSTEM

The vacuum system and magnetrons employed have been described elsewhere [3]. We use a UHV compatible, laboratory-produced rectangular dc magnetron with a 32x11 mm race-track. The only effective means of control which we have evolved is that of precise, suitably corrected, computer control of the power to these magnetrons. Our thesis is that, if accumulated thickness errors and layer thickness variance are to be kept to an accuracy of 0.1% or better, one must rely on 0.1% accuracy in the deposition control on the scale of the deposition of 1 nm. Since no thickness monitor can measure such small thicknesses with this accuracy we have abandoned direct measurement of the deposition rate and, instead, have opted for a 'software' solution to the problems of accurate control throughout a deposition run. We believe that if sufficient short term power control (better than 0.1%) can be achieved, with a long term power control of better than 0.01%, then all one needs to correct for are target erosion and pressure fluctuations.

The magnetron power supply consists of a direct mains driven pre-regulator driving a switched mode, transformer isolated, high voltage dc supply. This supply is capable of delivering 1 kV at up to 1 A. The output current and voltage are monitored and feedback is introduced to give an overall stability of $< 50 ppm/°C$. The supply can operate in either constant current or constant voltage mode with automatic selection dependent on the lowest setting. In addition to the manual controls, remote operation by a computer is possible. Under remote operation the current and voltage are read to a resolution of 14 bits and controlled to a resolution of 16 bit full scale. This leads to a resolution of 0.06 V and 0.06 mA. The systems typically operate at 300 V and 100 mA and thus, reading to reading (every second), a noise fluctuation of 0.1% is seen. However over a 10-100 s time scale the power fluctuations are only of the order of 0.01% under good conditions, *i.e.* operating at 30 W or more.

The usefulness of such an interface rests in the potential software control it allows. In order to control the deposition rate through a complete multilayer run, the power can be adjusted as a function of time to allow for the erosion of the targets. The erosion rate must be measured, but once established the race track erosion rate is a simple function of the magnetron design, as shown by Gurvitch [4]. (In our case the depositon rate increases with time, in contrast to that seen by Gurvitch.) Our sputter deposition system has a pressure control which, though adequate in controlling the pressure to within 1%, is not good enough to achieve 0.1% second-by-second control. To overcome this we use the continually read sputtering voltage as a measure of the pressure and correct the deposition rate accordingly, knowing how the deposition rate varies with pressure to first order [3]. These two software corrections can be used to reduce the errors by an order of magnitude. It is only by more extensive work that we will be able to control the deposition rate to better than 0.1%.

EXPERIMENTAL WORK

Ni/Zr Multilayers

Ni-Zr multilayers have been prepared for research into solid state amorphisation [5,6]. The films, of up to 5 μm total thickness, were deposited on glass and rocksalt substrates by using a crystal-controlled stepper motor to rotate a copper substrate holder situated 40 mm beneath two magnetrons of a design which has been described previously [3].

Interdiffusion caused by heating effects was observed as the period of the Ni/Zr multilayers was reduced to 10-20 nm; consequently the magnetron powers were each reduced to about 20 W. This power level results in heating which is counteracted by the cooling effect of the nitrogen-cooled sputtering chamber. For the two lowest periods studied, nominally 10 and 5 nm, we have examined the amorphous content of the layers as a function of sputtering pressure, although for purposes of comparison one should consider not the sputtering pressure alone but the product of this pressure, P, with the target-substrate distance, D, as a measure of the total number of collisions which the sputtered atoms and reflected neutrals have as they travel from the target to the substrate [7,8]. We note that the films are only just in the compressive regime described by Hoffman [9], even at the lowest pressures used.

We have observed a pronounced increase in the level of texturing of the layers, (111) for Ni and (0002) for Zr, as the pressure is reduced. This observation is to be expected; as the pressure is lowered bombardment of the film increases and the structure has a greater chance to reach its preferred, close-packed growth plane. A more interesting result is the apparent degree of mixing and amorphisation as a function of the product PD; contrary to our expectations we have observed a decreasing amount of mixing at interfaces with decreasing PD. This is illustrated in Figure 1, in which the amount of material which has been amorphised during deposition (calculated from the magnitude of the amorphisation peak in the DSC trace) is plotted. Results are shown which are predominantly for the 10 nm period structures with just two points from the 5 nm period study. These latter samples encounter the problem that their layer thickness is comparable with the thickness that mixes during deposition. However, their agreement with the results of the 10 nm period is very good. There are two possible explanations for our results, which may both be

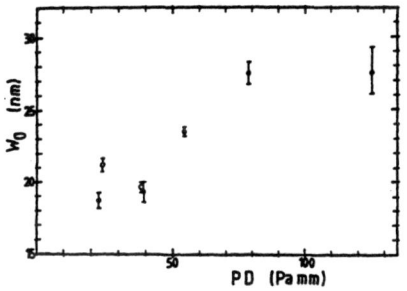

Figure 1.
W_0, the width of the intermixed Ni/Zr layer, as a function of the product of the sputtering gas pressure and the target-substrate distance, PD, during deposition.
●5/5 nm layers; ○2.2/2.2 nm layers

operating. Firstly, the films become rougher as the pressure is increased (this is partly born out by the diminishing number of superlattice lines which are seen) and thus there are more atoms in contact at the interfaces. Secondly, there is a decrease in density and accompanying increase in the concentration of defects (e.g. voids and vacancies) as the pressure is increased, leading to greater diffusion during film growth. Although the diffusion of the near bulk layers is low (the activation energy is about 1.1 eV [5]), if the activation energy is reduced by a factor of the order of 40% then diffusion during deposition can explain the observations. We note that Clemens [10] deduced the formation of amorphous layers approximately 2.1 nm thick at interfaces in Ni/Zr multilayers sputtered using a pressure of 0.4 Pa (which is less than the lowest pressure which we used; 0.65 Pa) and an unreported target-substrate distance.

We expected to see atomic mixing at low pressures, due to bombardment of the deposit with particles which have energies of up to 100 eV [11]. However, one would expect this to give an interface mixing of only 0.5 nm, which is less than the thinnest mixed interface observed. We are therefore unable to illustrate this effect, if it is present.

W/Si Multilayers

This may be an ideal X-ray mirror system; it is also a practical system for studying materials science, since very good X-ray diffraction data is produced. Furthermore, the two elements are easy to sputter, although the high electron contrast may lead to problems with transmission electron microscopy. We have performed a series of experiments on the deposition of these multilayers. A multilayer with a period of 2.61 nm has been tested at the synchrotron source at Daresbury, UK, and has been shown to have a 1% efficiency for the second order line [12] for use with an RKAP crystal in the 900-2000 eV range.

Most of our studies have been based upon production of a set of samples using the rotating substrate holder described above, with shields in front of the magnetrons to give variations of layer thickness and composition. The substrates used are 5 pieces of high quality (111) Si (1 cm square) placed along a radius. These can then be looked at as complete structures or, if the variation across each is large (as, deliberately, it often is), they can be cut tangentially into quarters 2.5 mm wide or into eighths approximately 1.25 mm wide. As the signal is invariably very strong even small samples give useful superlattice lines. Simple comparisons can be made from sample to sample as they are of the same size and are placed in the same orientation with respect to the X-ray beam. Our major problem is in making a satisfactory assessment of the multilayers which we have produced, since we only have access to a Philips vertical diffractometer; in practice this can only be used with 1/4 degree slits with Cu K_α radiation. This means that at 5 degrees 2θ

17 mm is exposed to the X-ray beam, which is much greater than the lengths of our samples. Rudimentary corrections are made for this. We have used the concept of the Debye-Waller factor [13] to obtain a rapid estimate of the roughness of the layers. In essence this says that the higher the angle for which superlattice lines exist, the smaller the roughness, and that the lines should decay as

$$\frac{1}{(\sin\theta)^4}\exp\left(-2\left[\frac{2\pi}{\lambda}\sigma\sin\theta\right]^2\right)$$

Typical roughnesses observed are in the range .25 to .5 nm.

X-ray traces of two good samples, with periods of 1.4 and 2.7 nm, are shown in Figure 2. The background level is obtained by measuring the sample upside-down. The shorter period

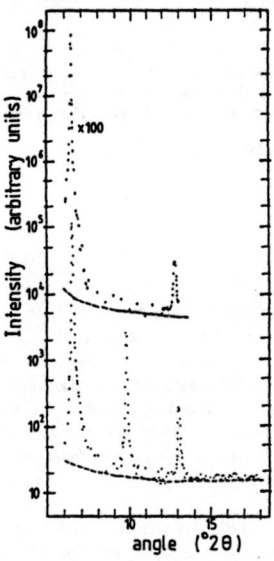

Figure 2.
X-ray step-scans of the 1st and 2nd order peaks of a nominally 1.4 nm period W/Si multilayer (top); and 2nd, 3rd and 4th order peaks of a 2.7 nm period multilayer. The background level in each case is shown as a dotted line.

multilayers have very respectable first order peaks, which for the very best samples are only 2 to 3 times less good than the best second order peaks of multilayers of twice the period. They do, however, show more roughness in that their second order peaks tend to be much lower than the fourth order peaks of their double period counterparts. We have found that multilayers with periods in the 1.5 to 2.0 nm range are good for this sort of study, as their first order peaks are easier to measure than for periods which have 1st order peaks close to grazing incidence. They are also particularly sensitive to the amount of interface roughness, which is one of our areas of interest. The strength of this type of experiment, comparing samples from one deposition run, is that the thickness errors will be the same for all samples, irrespective of the origins of the errors (accumulative or random), and all the observed differences will be due to variations in the relative thicknesses of the W and Si layers and to the relative positions of the samples with respect to the magnetrons. This latter factor becomes more important for experiments in which the samples are very close to the magnetrons.

Several simple preliminary experiments have been performed, both in the system described above and in a larger system which can hold seven samples across the substrate holder radius. In this latter system we are able to change the relative target-substrate distances, D, for the two targets. Films have been deposited with varying periods and at varying sputtering pressures. Experiments have also been performed in which the thickness of one of the components has been varied whilst compensating with the thickness of the other in order to keep the overall period constant.

In an experiment in which the Si thickness was varied along the radius of the substrate holder, the major variation was related to the distance that the individual substrates were from being directly below the magnetrons; D was routinely 32 mm for these runs. The roughness was seen to increase from 0.38 to 0.53 nm as D was increased by 50%. In a comparison between two runs in which the pressure, P, was increased from .55 to 1.0 Pa, the roughness increased from 0.24 to 0.38 nm. This work suggested that the best strategy was to try to reduce P and D as much as possible. However, since we believe that the bombardment of the growing film is directly related to the atomic mass of the target material, it would seem reasonable to reduce D for Si to a greater extent than for W [9]. In this way the silicon would be under less tension. Two series of runs have been performed: one in which the Si was brought closer to the substrates (22 mm) and the W further away (92 mm); another with the W at 22 mm and the Si at 62 mm. To date, the tentative conclusion to this work is that there are improvements to the quality of the layers as the targets are brought closer to the substrates. It is doubtful whether any special advantage comes from reducing the target-substrate distance of the less massive element.

The most revealing experiment has been one in which the W layer thickness was varied from 1.0 to 0.25 nm across the radius of the substrate holder whilst at the same time holding the overall period roughly constant at between 2.3 and 2.7 nm. We observe a strong and pronounced (110) W line, the width of which correlates well with the measured thickness of the W layers (see Figure 3). This shows that the W is being deposited as crystallites on the scale of the W layer thickness, rather than as an amorphous layer, as observed by other workers [14].

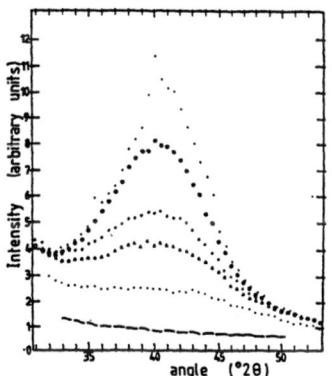

Figure 3(a).
The W (110) diffraction peaks from W/Si multilayers. Decreasing line width corresponds to increasing W layer thickness.

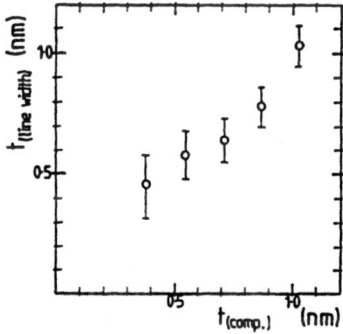

Figure 3(b).
W layer thickness calculated from the line width of the (110) diffraction peak versus layer thickness calculated from the composition of the multilayered film.

Another observation is that there is a correlation between the quality of the superlattice lines and the proximity of the W layer thickness to an integral number of (110) atomic planes, which have a spacing of about 0.224 nm. Our principal evidence for the benefit conferred by an integral number of atomic planes comes from the observation that, in a sample which has a range of periods (of say 5%), one finds just one period appearing strongly at the high angles due to a smaller Debye-Waller factor reduction. The need for integral numbers of atom planes has been observed by Ziegler et al. [14] and by Evans and Kent [15], although they claim that a layer thickness equal to four or five atomic planes is required. We suggest that, although the presence of an integral number of atom layers will often improve the multilayer's quality, the improvement is not as marked if the quality is already very high due to other factors, such as energetic bombardment during deposition.

CONCLUSIONS

This preliminary work shows that the smoothest interfaces are obtained when we deposit multilayers using the lowest values of PD, the product of gas pressure and target-substrate distance, which can be attained in our system (*i.e.* less than 20 Pamm). In a different sputtering geometry one may enter a regime where an excessively low sputtering pressure causes the interfaces to be disrupted, but as yet we have not encountered this phenomenon.

In our clean deposition system tungsten is microcrystalline; this is obviously not ideal if smooth interfaces are desired. We are initiating a programme in which we try to amorphise the W layers by making additions such as hafnium, and possibly platinum.

ACKNOWLEDGEMENTS

We thank J.E. Evetts, A.L. Greer and J.A. Leake for useful discussions, K. Butler for EDS analysis of our multilayers, P.V. Evans for help with this manuscript and E. Spiller for encouragement and tuition at the NATO Summer School on multilayered structures. R.J. Highmore is grateful for a SERC CASE studentship with GEC Hirst Research Centre.

REFERENCES

[1] M. Schneider, I.K. Schuller and A. Rahman, in *Interfaces, Superlattices and Thin Films*, Ed. J.D. Dow and I.K. Schuller, (Mater. Res. Soc. Proc. 77, Pittsburgh, PA 1987) pp. 91-98.

[2] I.K. Schuller and C.M. Falco, *Surf. Sci.* 113, 443 (1982).

[3] R.E. Somekh and Z.H. Barber, *UHV Sputter Deposition With A Research Scale DC Magnetron* submitted to *J. Phys. E.*

[4] M. Gurvitch, *J. Vac. Sci. Technol. A* 2, 1550 (1984).

[5] R.J. Highmore, J.E. Evetts, A.L. Greer and R.E. Somekh, *Appl. Phys. Lett.* 50, 556 (1987).

[6] R.J. Highmore, R.E. Somekh, A.L. Greer and J.E. Evetts, *Mat. Sci. and Eng.* in press.

[7] R.E. Somekh, *J. Vac. Sci. Technol. A* 2, 1285 (1984).

[8] T. Motohori, *J. Vac. Sci. Technol. A* 4, 189 (1986).

[9] D.W. Hoffman, in *Proc. 7th Int. Conf. on Vacuum Metallurgy* (The Iron and Steel Inst. of Japan, Tokyo, 1982) p.145.

[10] B.M. Clemens, *Phys. Rev. B* 33, 1715 (1986).

[11] R.E. Somekh, *Vacuum* 34, 987 (1984).

[12] J. West (private communication).

[13] E. Spiller and A.E.Rosenbluth, *Proc. SPIE* 563, 221 (1985).

[14] E. Ziegler, P. Houdy and L. Nevot, *Proc. SPIE* 563, 306 (1985).

[15] B.L.Evans and B.J.Kent, *Proc. SPIE* 733, 361 (1986).

MBE GROWTH OF COMPOSITIONALLY MODULATED CERAMICS*

R.A. MCKEE, F.A. LIST, AND F.J. WALKER
Oak Ridge National Laboratory, Oak Ridge, Tennessee 37831

ABSTRACT

An ultra high vacuum system fitted with thermal effusion cells that heat 20cc of material to temperatures as high as 2000°C is being used to synthesize thin film and multilayer ceramics from their constituent elements. A molecular source for oxygen used in conjunction with these effusion cells allows oxides of Al, Ti, Cu, and Y to be grown with compositional modulation controlling growth rate and stoichiometry.

INTRODUCTION

Multilayer, structured materials research has been cited[1-3] as an area with significant potential for major advances in physics and materials science. We are attempting to synthesize a new class of ceramic materials using molecular beam epitaxy and composition modulation techniques and to determine whether the synergistic modulation effects that have been found in metal/metal layered systems[4-6] can be exploited in ceramic materials; recent evidence suggests that this may indeed occur[7]. The most significant problem to overcome in this effort was the development of a high temperature effusion cell that would allow materials like Ti, Ni, Cu, Si, and Y to be used as constituents in combination with oxygen or nitrogen to grow the ceramic materials. We have solved this problem, and report data for effusion cell operation heating 20cc crucibles to 2000°C in UHV environments. We have grown both metal and ceramic thin film and multilayer structures using molecular sources for Ti, Al, Cu, Y, and O.

EXPERIMENTAL

An ultra high vacuum system consisting of a growth chamber, analysis

*Research sponsored jointly by the U.S. Air Force Office of Scientific Research under interagency agreement DOE NO. 40-1514-84, USAF No. AFOSR-ISSA86-033 and both the Office of Energy Utilization Research, Energy Conversion and Utilization Technologies (ECUT) Program and the Division of Materials Sciences, U.S. Department of Energy, under contract DE-AC05-840R21400 with the Martin Marietta Energy Systems, Inc.

chambers, and interconnecting sample introduction and transfer chambers has been constructed primarily for MBE synthesis of ceramic materials, i.e. metal oxides and metal nitrides. This system provides for 9cm samples and uses thermal effusion cells as the sources for the metals and a molecular source for the oxygen or nitrogen. The unique aspect of the system is the operational characteristics of the effusion cells. Figure 1 shows a power vs temperature curve for a 20 cc effusion cell. These data illustrate cell operational characteristics that far exceed those for any commercially available effusion cell[8]. This effusion cell will heat 20 cc quantities of material to 2000°C in UHV environments and can be used for thermal evaporation of materials like Si, Ti, Ni, Cu, and Y at flux rates that can be sensitively controlled in the .001 to 10 nm/s range.

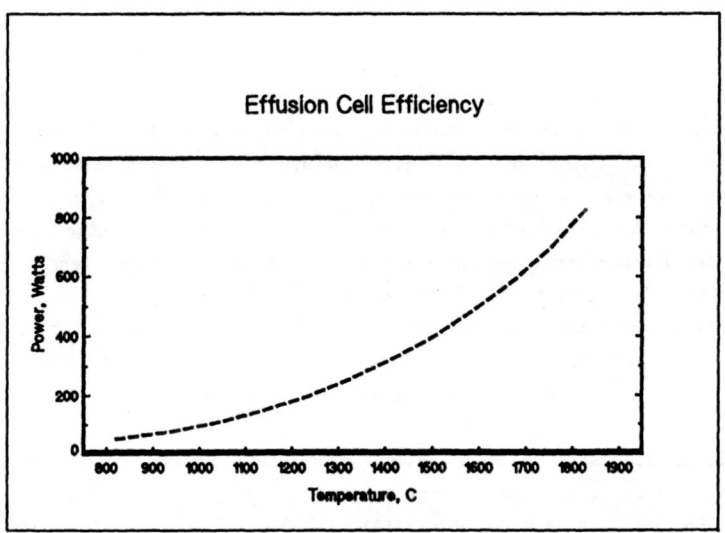

Figure 1: Power vs Temperature for High Temperature Effusion Cell

RESULTS AND DISCUSSION

As illustration of the effusion cell/system performance, Fig 2 shows data for Si evaporation rate vs temperature. These data were collected for a source-to-substrate distance of approximately 40 cm and give a uniformity of ± 2% over 6 cm for a nonrotating substrate; the flux scales with the square of source-to-substrate distance and would increase by a factor of 4 to the 1 nm/s rate at a distance of 20 cm. Similar data have been collected for Ti, Al, Cu, Ni, and Y. Several single film and multilayer structures of these metals and/or their oxides have been grown in our system.

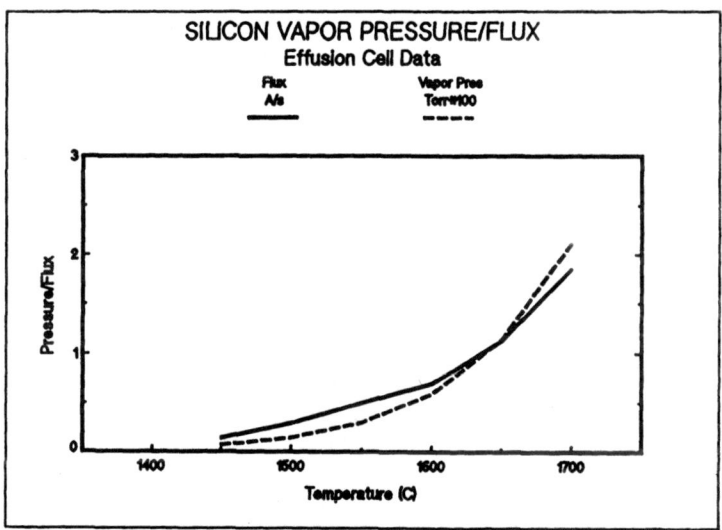

Figure 2: Flux data for Silicon evaporation

Nickel/Titanium Multilayer

We have been using our MBE facility to study the Nickel/Titanium alloy system and its tendency to form intermetallic alloys and amorphous structures. Composite NiTi structures are grown in the MBE chamber by codeposition of Ni and Ti on silicon to a thickness of 50 nm and then alternately growing 10 nm layers of Ti and Ni for a combined thickness of 50 nm. The codeposited material forms an amorphous structure, but the Ti and Ni layers are crystalline. We have synthesized such samples to allow careful microstructural characterization of the coexisting crystalline and amorphous phases in this system. The unique capability of the effusion cells provided the Ni flux at .05 nm/s with cell temperature at 1750°C and the Ti flux at .05 nm/s with the cell temperature at 1850°C; appropriate shutter sequencing either produces the NiTi amorphous alloy or the pure constituent layer. There are two interesting results that have come from this study. We have synthesized the amorphous NiTi codeposited material, and this is consistent with expectations based on previous work[5], but we also find a thin (.7-1.4 nm) amorphous region between the Ti and Ni pure layers that forms at the interface without the process of codeposition being responsible. Associated with this is the crystalline growth of either Ni or Ti off this amorphous surface. This will ultimately give us specific information about the structural dynamics of the amorphous NiTi material. The other very interesting feature of this study is the cystalline form of Ti that we find in this multilayer. Titanium metal in pure bulk form is either hcp at low temperatures or bcc at high temperatures. The Ti in this multilayer material is face centered cubic. This is a result that has not been previously known about this material, and its detailed study is of interest to us as it relates to the overall question of metastability of structures in the Ni/Ti system.

Titanium-Copper-Oxygen and Yttrium-Copper-Oxygen

Attempts have been made to synthesize a multilayer structure of TiO_2/CuO and a plane-by-plane reaction of $YCuO_2$. These experiments were unsuccessful because the Cu did not oxidize at the low substrate temperature of 100°C that was used in combination with the flux of Cu. In both cases metallic Cu was found dispersed in a matrix of either TiO_2 or Y_2O_3. A detailed study of the growth of Cu oxides during the codeposition of Cu and O is being undertaken to quantify the optimum conditions for the more complex oxide formation.

Al_2O_3/TiO_2 Multilayer

A wedge shaped sample has been grown on Si with the wavelength of the multilayer (Al_2O_3 thickness + TiO_2 thickness) varying from 1 to 10 nm and the wedge thickness varing from approximately 100nm to 900nm. The substrate surface was maintained at 750°C; the Al and Ti fluxes were set at .1 and .05 nm/s respectively; and the oxygen cell supplied oxygen for the reaction at approximately 1 monolayer/s coverage. These conditions are appropriate for the plane-by-plane reaction of Al and Ti to their respective oxides. The wedge was grown by coordinating the substrate shutter movement with the timing sequence for the shutters on the Ti and Al effusion cells. This sample has been grown to allow us to probe the multilayer microstructure for its optimum in mechanical properties; these studies are in progress but are not complete at this time.

CONCLUSION

We have developed an ultra high vacuum system for the synthesis of ceramics from their constituent elements. This sytem is based on a high temperature effusion cell design that operates in UHV to 2000°C with a 20 cc crucible. Fluxes can be sensitively controled in the .001 to 10 nm/s range and several ceramic materials have been grown using this concept.

ACKNOWLEMENTS
Research sponsored jointly by the U.S. Air Force Office of Scientific Research under interagency agreement DOE No. 40-1514-84, USAF No. AFOSR-ISSA86-0033, and both the Office of Energy Utilization Technologies (ECUT) Program and the Division of Materials Sciences, U.S. Department of Energy, under contract DE-AC05-840R 21400 with the Martin Marietta Energy Systems, Inc.

REFERENCES

1. V. Narayanamunti, Phys. Today, 24 (October, 1984).

2. "Report of Research Briefing Panel on Selected Opportunities in Physics," Hans Frauenfelder and Mildred S. Dresselhaus, Co-Chairman, p. 107 in Research Briefings, 1984.

3. Physics Today, 21, (January, 1985).

4. D. Baral, J.B. Ketterson, J.E. Hilliard, p. 465 in Proceedings of NATO Advanced Study Institute on Modulated Structure Materials. Elsevier, Oxford, (1984).

5. W.M.C. Yang, T. Tsakalakos, and J.E. Hilliard, J. Appl. Phys., 48, 876 (1977).

6. R.W. Sprugir and D.S. Catlett, Thin Solid Films, 54, 197 (1978).

7. U. Helmersson, S. Todorova, S.A. Barnett, J.E. Sundgren, L.C. Markert, and J.E. Green, J. Appl. Phys., 62, 481 (1987).

8. UltraTherm Inc. owns the cell design and is in the process of patenting it for commercial development.

9. B.M. Clemens, Phys. Review B, 33, 7615 (1986).

GROWTH AND STRUCTURE OF TUNGSTEN CARBIDE-TRANSITION METAL SUPERLATTICES

T.D. Moustakas*, J.Y. Koo and A. Ozekcin
*Boston University College of Engineering,
Boston, MA 02215
Exxon Research and Engineering Company
Annandale, N.J. 08801

ABSTRACT

Superlattices between ceramic materials, such as tungsten carbide, and transition metals have been synthesized for the first time. The growth and structure of these superlattices were investigated by low angle X-ray diffraction and TEM lattice imaging and microdiffraction. The data show that the low temperature process of forming these two dimensional composites leads to unique crystal structures and morphology in the nanoscale range.

INTRODUCTION

Artificially layered strucrues made of pure metals, such as Au-Ni(1), Cu-Pd(1), Cu-Ni(2), Ag-Pd(3), Cu-Au(3), Nb-Cu(4), Mo-Ni(5), Ni-V(6), etc. have been studied extensively the past few years. Some of these systems, for example, Cu-Ni, form solid solutions in their binary phase diagram and have matching lattice constants, while others, for example Nb-Cu, have a eutectic binary phase diagram (8) and do not have matching lattice constants. Growth and epitaxial phenomena in these systems were primarily investigated by X-ray diffraction studies (9,10) and by Transmission Electron Microscopy (11). The work has primarily been motivated by early reports that some of these systems exhibit enhanced elastic modulii (1-3,12) and other novel physical properties (9,13,14). In addition, such composition-modulated films are suitable systems for studying stability and critical phenomena in solid solutions (15).

Superlattices between ceramic materials, such as refractory metal carbides, and metallic materials, such as transition metals, have been prepared only recently (16). The work has been motivated by the need to form two dimensional composites, which in analogy with the family of cemented carbides (17) should combine the high hardness and wear resistance of the carbide layers with the mechanical and thermal shock resistance of the metallic layers. In the present paper we report on phenomena related to crystal growth and structure of tungsten carbide and cobalt superlattices. More specifically, we examined how epitaxy and strain affect the structure and sharpness of the individual layer. These studies were carried out by using both low angle X-ray diffraction and modern analytical electron microscopy, including high resolution latice imaging and microdiffraction.

EXPERIMENTAL METHODS

Tungsten carbide-cobalt superlattices were formed by RF sputtering from cobalt and hexagonal tungsten carbide (WC) targets, utilizing a diode system with a rotating substrate table. The system was pumped by a combination of a turbomolecular pump, roots pump and a mechanical pump to a total leak rate of $3 \times 10^{-6} \, Torr\text{-}l/sec$, and sputtering was performed at $9 \, mTorr$ of argon. The power in each target was adjusted for approximately the same deposition rate, namely $55\text{Å}/min$ for the cobalt and $60\text{Å}/min$ for the tungsten carbide. The induced DC voltage on the WC target was 1900 Volts while that on the Co target was 1300 Volts. The superlattices were deposited on a Si (111) substrate.

Before the run, the substrate was mounted under the tungsten carbide target and was maintained at ambient temperature. During the run, the substrate table was rotated sequentially under each target and the residence time under each target was regulated to grow layers of desired thickness.

The structure of the superlattices was investigated by X-ray diffraction at low angles and TEM microscopy. The X-ray diffraction studies were performed in an ordinary powder diffractometer with Cu-Ka($\lambda = 1.54$Å) radiation. The high resolution electron microscopy study was conducted in a Philips EM430 microscope, equipped with a LaB_6 source, and the microdiffraction investigation was performed using a Philips EM400T microscope, fitted with a field emission gun. The TEM study was done on the cross-section of the superlattice. The TEM thin foil preparation involved sandwiching the superlattice with an appropriate epoxy, mechanical thinning, ultrasonic cutting of a 3mm disc, dimpling and ion beam milling.

EXPERIMENTAL RESULTS AND DISCUSSION

Structure of Thick Tungsten Carbide and Cobalt Films

We reported previously (16) that sputtering from a hexagonal tungsten carbide target in pure argon leads to films having the structure of beta tungsten carbide ($\beta - WC_{0.51}$). This material has the FCC structure of NaCl with carbon vacancies and a lattice constant 4.248Å. Although, this structure can only be produced under equilibrium conditions at temperatures higher than 2500°c (18) we were able to kinetically stabilize it at low temperatures. We also reported (16) that films produced at ambient substrate temperatures and argon pressures $20\,mTorr$ or higher are polycrystalline with a random orientation of crystallites. However, increasing the substrate temperature to 300° C or decreasing the argon pressure to less than $10\,mTorr$, results in films with strong (002) preferred orientation.

We also reported (16) that the structure of the cobalt films depends on the substrate temperature. Films produced at ambient substrate temperature have an HCP structure with $ao = 2.51$Å and $co = 4.07$Å, while those produced at substrate temperature 300°c or higher have a FCC strucutre with a lattice constant $ao = 3.545$Å.

Tungsten Carbide-Cobalt Superlattices Formed at Ambient Substrate Temperature

Figure 1 is a TEM micrograph showing the cross-section of a typical tungsten carbide-cobalt superlattice formed at ambient substrate temperature. This superlattice is made of 137 periods consisting of tungsten carbide layers 37Åthick and of cobalt layers 32Åthick, with the tungsten carbide layer grown first onto the substrate. A higher magnification of the same TEM cross-section, showing five periods next to the silicon substrate, is given in Fig.2. Note that the superlattice was grown onto the native SiO_2 layer, approximately 15Åthick, and thus the growth was not influenced by the crystal structure of the substrate. The data of Figures 1 and 2 indicate that the superlattice is uniform throughout and that interfaces between the individual layers are atomically abrupt. It is also apparent from Figure 2 that the top interface of each tungsten carbide layer is smoother than the bottom interface. We propose that this asymmetry in the two interfaces results from more intermixing during the sputtering of the tungsten carbide layer, due to higher kinetic energy of the W and C atoms sputtered from the WC target.

Figure 1. TEM bright field image showing a cross-section view of the WC-Co superlattice. Dark bands are WC layers (37Å thick) and light bands are Co layers (32Å thick).

Figure 2. Higher magnification view of Figure 1. Note the presence of SiO₂ phase between WC phase and Si substrate.

The X-ray diffraction data at low angles for the same superlattice, shown in Fig.3, corroborate the findings in Figures 1 and 2. The period of the superlattice L is determined from the Bragg relation

$$2L\sin\theta = m\lambda \qquad (1)$$

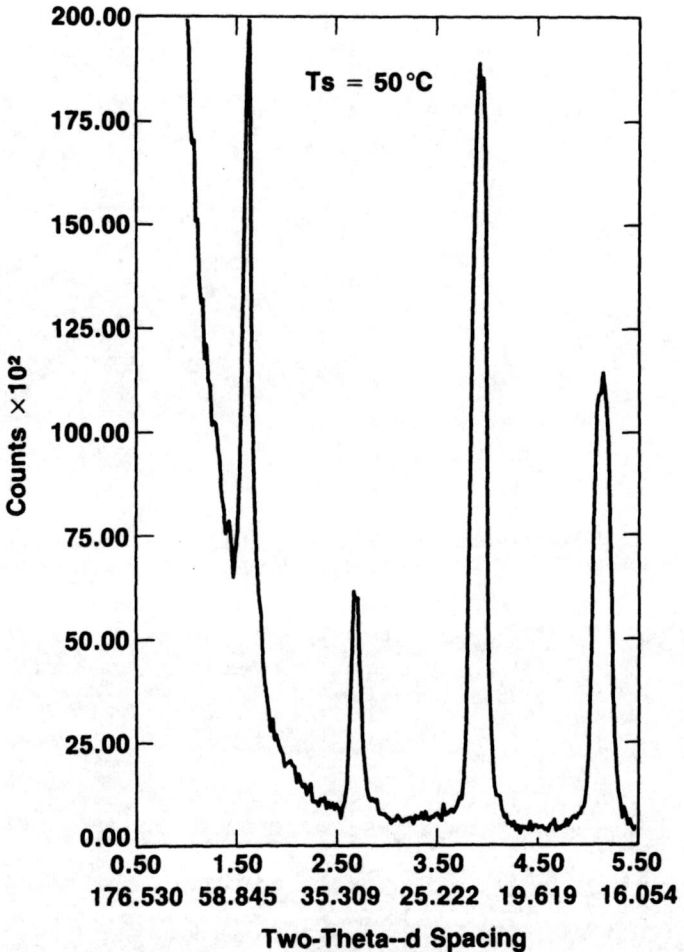

Figure 3. X-ray diffraction of the Tungsten Carbide-Cobalt superlattice discussed in Figures 1 and 2.

From the position of the first four harmonics we calculated that the values of L is 54Å, 64.6Å, 67.2Å, 69Å. The values calculated from the first harmonic and to a lesser degree from the second and third harmonics need to be corrected for refraction effects (19), since these peaks ride on the total external reflection curve. The period of 69Å, determined from the 4th harmonic, is in good agreement with the period determined from deposition rate data and TEM cross-section studies. For ideally abrupt interfaces (square wave modulation), and for equal layer thickness only the odd orders $m = 1, 3, 5 \ldots$ are permitted reflections (19). The persistence of even order reflection can qualitatively be accounted for by the small difference in the thickness of the two layers. A quantitative analysis of these results will be given elsewhere.

The structure of the individual layers was studied by combining TEM lattice imaging and microdiffraction studies. Details of these studies will be published elsewhere (20). However, some preliminary results are briefly discussed in this paper. Figure 4 shows a TEM lattice imaging of the previously discussed superlattice. The image was taken by axial illumination mode with a large objective aperture encompassing a transmitted beam and many diffraction spots. The exsistence of interference fringes suggests that both the tungsten carbide and the cobalt layers are highly crystalline. The d-spacings from the fringes were measured using the Si(111) fringes as internal reference. The lattice constant of the tungsten carbide layer agrees with that of the beta tungsten carbide. The structure of the cobalt layer was found by the microdiffraction studies to vary across its thickness. The first 10-15Å on the top of the tungsten carbide have the FCC structure and the rest of the layer converts into the HCP structure, which is the thermodynamically stable phase at low temperatures. The FCC to HCP transition occurs in the plane $(111)FCC\|(0001)HCP$. This plane has the same atomic configuration in both phases, resulting in atomic coherency. Presumably, the first few monolayers were stabilized in the FCC structure due to epitaxial effects induced by the underlying FCC beta tungsten carbide layer.

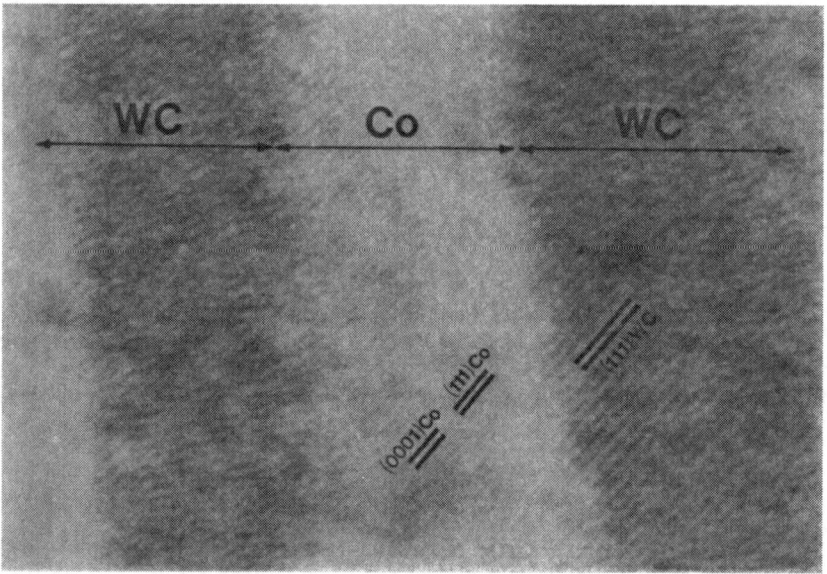

Figure 4. Lattice fringe image of the WC-Co superlattice showing the constituent phases.

CONCLUSIONS

The results reported in this paper indicate that superlattices between tungsten carbide and cobalt can be formed on room-temperature substrates. The sharp interface indicated in the TEM data as well as the low angle X-ray scattering suggest little intermixing at the interfaces and square wave compositional modulation. This finding is in agreement with the eutectic nature of the phase diagram of the constituents of the superlattice.

When the thickness of the individual layers are more than 30Å the layers are polycrystalline. The tungsten carbide crystallizes in the $\beta-WC_{1-x}$ form, which has the FCC structure of NaCl with carbon vacancies. The cobalt grows on the top of the tungsten carbide in its FCC structure initially and then converts in its HCP structure, which is the thermodynamically stable phase at low temperatures.

When the thickness of the individual layers are less than 20Å, we reported previously (16) that the x-ray diffraction peaks at large angles are broad, suggesting that the individual layers are made of very small crystallites. However, part of the peak braodening could be due to interphase strain (approximately 20% lattice mismatch).

Thus, these two-dimensional composites, contrary to the family of cemented carbides whose mocrostructure is coarsened due to grain growth and coalescence during high temperature sintering, have microstructures in the nanoscale range.

References

1. W.M.C. Yang, T. Tsakalakos and J.E. Hilliar, J. Appl. Phys. 48, 876 (1977).
2. T. Tsakalakos and J.E. Hilliard, J. Appl. Phys. 54, 734 (1983)
3. G.E. Henein and J.E. Hilliard, J. Apply. Phys. 54, 728 (1983)
4. I.K. Schuller, Phys. Rev. Lett. 44, 1597 (1980)
5. M.R. Khan, C.S.L. Chun, G.P Felher, M. Grimsditch, A. Kueny, C.M. Falco and I.K. Shuller, Phys. Rev. B 27, 7186 (1983)
6. R. Danner, R.P. Huebener, C.S.L. Chun, M. Grimsditch and I.K. Schuller, Phys. Rev. B 33, 3696 (1986)
7. T. Tsakalakos, Ph.D. Thesis, Northwestern University, 1977
8. R.P. Elliot, Constitution of Binary Alloys (McGrawn-Hill, New York, 1965)
9. Charles M. Falco and Ivan K. Schuller, Novel Materials and Techniques in Condensed Matter, edited by G.W. Crabtree and P.D. Vashishta (North-Holland, New York, 1982) p.21
10. H. Homma, Y. Lepetre, J.M. Murdock, I.K. Schuller and C.F. Majkrzak, Applications of Thin Film Multilayered Structures to Figured X-ray Optics, SPIE Vol 563, p. 150 (1985)
11. Y. Lepetre, I.K. Schuller, G. Rasigni, R. Rivoira, R. Philip and P. Dhez, Applications of Thin-Film Multilayered Structure to Figured X-Ray Optics SPIE Vol. 563, p. 258 (1985)
12. T. Tsakalakos, Thin Solid Films, 75, 293 (1981)
13. T.R. Werner, I.Banerjee, Q.S. Young, C.M. Falco and I.K. Schuller, Phys. Rev. B 26, 2224 (1982)
14. N.K. Flevaris, J.B. Ketterson and J.E. Hilliard, J. Apply. Phys. 53, 8046 (1982)
15. T. Tsakalakos, Thin Solid Films, 86, 79 (1981)
16. T.D. Moustakas, H.W. Deckman, J.Scanlon, R. Friedman and J.A. McHenry, in High Temperature Structural Composites: Synthesis, Characterization and Properties Ed. by M.J. Luton (TMS, 1988)
17. Fritz V. Lenel, Powder Metallurgy Principles and Applications (Metal Powder Industries Federation, Princeton, New Jersey) Ch. 16
18. E. Rudy, S.T. Windisch and J.R. Hoffman, Fundamentals of Refractory Compounds, edited by Henry H. Hausner and Melvin G. Bowman (Plenum, New York, 1968) p37
19. J.H. Underwood and T.W. Barbe, in Low Energy X-ray diagnosis-1981, edited by D.T. Atwood and B.L. Henke, AIP Conference Proceedings No. 75, (American Institute of Physics, New York, 1981) p. 170
20. J. Koo, T.D. Moustakas and A. Ozekcin (to be published)

PHENOMENOLOGICAL APPROACH TO MULTILAYER GROWTH AND STABILITY

ANDREW ZANGWILL
School of Physics, Georgia Inst. of Tech., Atlanta, GA, 30332

ABSTRACT

I review recent theoretical studies designed to examine the effect of strain, facetting and growth conditions on multilayer stability. A phenomenological approach is used to investigate both morphological and crystal structural phase stability.

INTRODUCTION

The fascinating and technologically useful properties of artificial multilayer structures fabricated from metals, semiconductors, and insulators have driven experimental studies of these systems for a decade or more. Accordingly, a significant body of lore has accrued that specifies under what sort of experimental conditions one should operate in order to guarantee "good", i.e, layer-by-layer growth of a heterostructure with a particular desirable crystal structure. Deviations from such wel-defined "recipes" can lead to unusual and poorly understood growth structures. In this article, I focus on three archetypical issues that arise in connection with this situation: solid-on-solid wetting, morphological instability and pseudomorphy. In particular, I review recent attempts to address these phenomena by means of macroscopic and/or phenomenological models that do not require detailed material-specific information. The goal is to provide a framework to guide further experiments and more microscopic theoretical investigations.

SOLID-ON-SOLID WETTING

It is well known that thin films grown via standard vapor phase deposition techniques such as molecular beam epitaxy (MBE) or chemical vapor deposition (CVD) typically adopt one of three familiar "growth modes". So-called Frank-van der Merwe (FM) growth proceeds in a regular layer-by-layer manner. Volmer-Weber (VW) growth corresponds to nuleation of bulk-phase clusters onto the substrate directly from the vapor whereas the Stranski-Krastanov (SK) mode is said to occur when bulk phase clusters nucleate only after a few monolayers have adsorbed in layer-by-layer fashion. A clear review of the general issues recently has been supplied by Gilmer and Grabow [1] who, in addition, focus attention on the theoretical method of molecular dynamics as a means to address the thermodynamic stability of the FM growth morphology. In what follows, I discuss an alternative approach to the same issue following the phenomenological analysis of Bruinsma and Zangwill [2].

It is convenient to regard the occurence of a particular growth mode in an epitaxial system as a problem in morphological selection. To that end, we characterize the growing overlayer by a function h(x,y,t) that describes the local height of the deposited material above the substrate as a function of time. To begin, let us ignore the dynamic (growth) aspects of the problem and simply ask: what shape function h(x,y) represents the thermodynamic ground state of a system of N particles epitaxially deposited onto a fixed substrate? To answer this question we

must specify the free energy of the overlayer as a functional of
h(x,y) and then minimize. If, for example, the solution is
h(x,y)=constant, we should expect the FM mode under conditions
of slow growth.

The dominant contributions to the free energy we require
are (1) a term $S = \gamma_s - \gamma_i - \gamma$ that measures the relative surface
tensions of the substrate-vacuum (γ_s), substrate-overlayer (γ_i)
and overlayer vacuum (γ) interfaces; ; (2) a term $\gamma(\nabla h)^2$ that
counts the extra energy cost if a fixed substrate area A is
covered by an overlayer of surface area > A, i.e., if h(x,y) is
anything but a constant (see Figure 1); (3) a term P(h) that
counts the local elastic strain in the overlayer due to misfit
between the substrate and overlayer lattice constants; and (4)
a term that recognizes that curved surface profiles expose
crystal facets whose energy depends on crystallographic orientation.

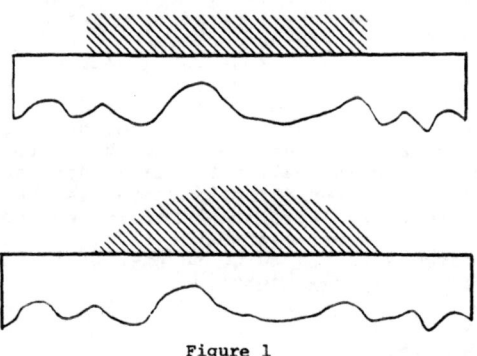

Figure 1

Specializing to a model two-dimensional system where h=h(x)
only, an appropriate free energy that includes all these features is:

$$F = \int dx \left[-S + \gamma(dh/dx)^2 + P(h) - y_0\cos(2\pi h) + \lambda h \right] \quad (1)$$

The first four terms are precisely as described above. P{h(x)}
is a concave, monotonically increasing function of its argument
that includes the effect of both interfacial misfit dislocations
and residual elastic strain. It is described thoroughly in conventional treatments of the effect of strain during epitaxial
growth [3]. Sophisticated theories of the so-called "roughening
transition" (which model the destruction of crystal facets by
thermal fluctuations) show that the sinusoidal term in Equation
(1) is sufficient to account for the increased energy of vicinal
surfaces inclined at a small angle with respect to the substrate
plane [4]. The combination $\sqrt{\gamma y_0}$ is proportional to the energy of
a monatomic step. Finally, the quantity λ is a Lagrange multi-

plier that guarantees that only surface profiles h(x) which
preserve total particle number may be considered in the varia-
tional minimization of the free energy.

The details of this minimization are non-trivial and may be
found elsewhere [2,5]. On the other hand, the final results can
be understood quite simply. Consider first the limit where y_0
vanishes. This case resembles that of fluid wetting except that
the overlayer can support a finite strain. In agreement with the
simplest theory [6], Volmer-Weber "droplet" formation is found
if S<0. The system minimizes the effect of a relatively costly
substrate-overlayer surface tension by minimizing the area of
contact. There is no wetting.

Adhesion between substrate and overlayer is favored if S>0
and we expect at least one complete monolayer to wet the substrate. However, the calculations show that further layer-by-
layer growth is never favorable. That is, the Stranski-Krastanov
morphology is __always__ the ground state for non-zero misfit and
S>0. This surprising result arises from the concavity of P(h)
and the fact that the "footprint" of a three-dimensional droplet
of atoms is very much smaller ($\sim 10^{-6}$) than the area covered by
the same number of atoms when spread out into a monolayer. The
integrated strain energy within the droplet is considerably
smaller than the corresponding quantity for the layer.

This conclusion persists for moderate, non-zero values of
the step energy. True FM growth is at best __metastable__ in this
regime. If one prepares such a morphology by a non-equilibrium
process (as is routinely done in the laboratory!), we predict
that the overlayer will transform to the SK morphology if left
undisturbed indefinitely. This result agrees with that of the
aforementioned microscopic calculation [1]. Nonetheless, to
determine if the transformation is observable, one must specify
the quantitative height of the thermal activation barrier and
the mode of morphological evolution. Within the framework of
Equation (1), a dynamical calculation [2,5] implicates a
sinusoidal modulation of the crystal surface that is formally
similar to the phenomenon of spinodal decomposition. The barrier
energy is related to both the step energy and the Peierls energy
for misfit dislocation motion.

Morphological evolution from FM to SK effectively is suppressed when the step energy becomes very large, or, more correctly, when the surface energy of a finite contact angle facet
(see Figure 1) becomes very large. In fact, it is possible (in
that limit) that the layer-by-layer morphology actually sup-
plants SK growth as the thermodynamically stable configuration
for some range of particle number. The details of the crossover
have yet to be worked out.

MORPHOLOGICAL INSTABILITY DURING GROWTH

I now suppose that droplet nucleation from the vapor has
been suppressed as indicated above and a multilayer is to be
grown by deposition of some specified number of layers. At this
point, strict adherence to an empirical growth "recipe" is important since it is well known that small variations in external
control parameters (deposition rate, temperature, etc.) can have
large morphological consequences. Recent theoretical studies of
the CVD process illustrate this fact dramatically.

Chemical vapor deposition involves two distinct physical
steps. First, flowing gas must diffuse down through a non-

hydrodynamic "stagnant" layer to bring reactant molecules into the immediate vicinity of the growing surface. Second, some activated process (such as dissociative chemisorption) must occur that actually incorporates the material into the solid. Let us characterize the former by a diffusion constant D and the latter by a rate constant k. Quasi-static growth of the interface at velocity v(x,y)≡v(s) then can be modelled [7] by the following macroscopic equations:

$$\nabla^2 c(r) = 0$$
$$c(\delta) = c_o$$
$$k\{c(s)-c_{eq}(s)\}^o = D\nabla c \cdot n|_s \qquad (2)$$
$$v(s) \propto \nabla c \cdot n|_s$$

That is, the concentration of diffusing reactant c(r) is determined (in steady state) by Laplace's equation subject to two boundary conditions: a fixed "inlet" concentration at a distance δ above the surface and a mixed boundary condition at the growing interface that depends on the relative importance of the two transport processes noted above. The driving force for growth depends on the difference between the reactant concentration at the surface c(s) and its equilibrium value $c_{eq}(s)$. The latter is determined by the surface tension γ and the local radius of curvature of the interface.

Equation (2) constitutes a moving boundary value problem. One solves the equations for a given shape of the surface and then advances each surface point according to its local growth velocity. Iteration of this two-step procedure traces out the growth profile as a function of time. The problem is highly nonlinear and not amenable to simple analytic analysis. Consequently, we [8] have applied an essentially exact numerical technique to extract solutions for various values of the characteristic dimensionless parameter ν=kδ/D.

To focus on the question of instability, consider deposition onto a patterned substrate as might be encountered for semiconductor applications. Figure 2 illustrates the predicted morphology (after identical growth periods) for three different growth conditions. The results are striking. When incorporation into the surface is the rate limiting step (ν<<1) we observe smooth layer-by-layer "template" growth; the interface follows the morphology set by the substrate. By contrast, when deposition is limited by reactant diffusion (ν>>1) an instability of the Mullins-Sekerka variety [9] develops that evolves nonlinearly into a pattern reminiscent of dendritic growth. Details will be presented elsewhere [8].

The reader should not suppose that the limiting cases shown below reflect unphysical values of the parameters. In fact, the top and bottom panels of Figure 2 very nearly reproduce cross-sectional micrographs obtained by van den Brekel [7] for silicon growth at T=1010 °C and T=1180 °C under otherwise identical conditions.

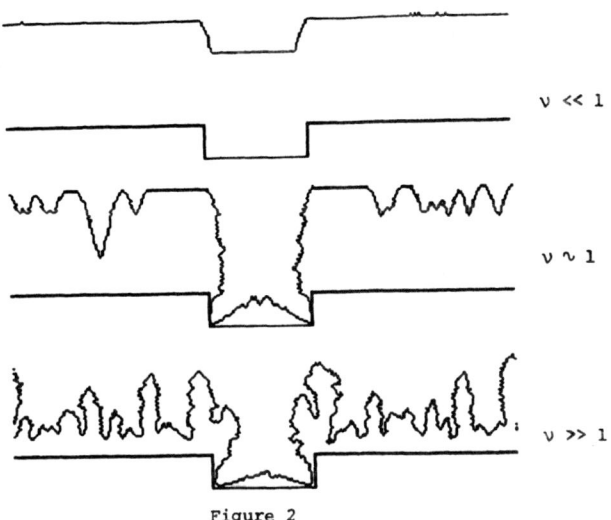

$\nu \ll 1$

$\nu \sim 1$

$\nu \gg 1$

Figure 2

PSEUDOMORPHY

I now suppose that FM growth in the "template" mode has been achieved and turn to my final example of metastability relevant to multilayer growth: pseudomorphy. The general phenomenon is not new; there are many documented cases [10] of the ability of a substrate to induce an overlayer to adopt a crystal structure that differs from its bulk-stable structure. Renewed interest has been stimulated by Prinz' successful growth of surpisingly thick films of BCC cobalt on GaAs(110) [11].

The essential physics of this phenomenon has been analyzed in some detail by Bruinsma and Zangwill [12] using a phenomenological free energy model of epitaxial growth. The key idea is to extend the classic analysis of van der Merwe [13] to the case where the overlayer can respond non-linearly to a shear stress applied by the substrate. In brief, an overlayer may adopt a new crystal structure (and lattice constant) if too great a misfit energy would be incurred if epitaxy were required with the bulk-stable crystal structure and lattice constant. Evidently, an energetic trade-off is involved: energy gained at the interface must exceed energy lost by each overlayer unit cell which finds itself in an "unatural" crystal structure. A simple surface-to-volume argument then demonstrates that a phase transition back to the bulk-stable phase must occur for sufficiently thick film growth. On the other hand, multilayers provide an attractive means to "trap" a sandwiched layer in the pseudomorphic phase.

To be more quantitative, suppose that the free energy density (free energy/unit cell) of the overlayer f(ε) resembles Figure 3. The independent variable represents strain along some "easy" axis. The absolute minimum corresponds to the bulk-stable crystal structure, FCC for example. But, since the FCC structure is crystallographically equivalent to a body-centered tetragonal structure, a strain of sufficiently large magnitude along the tetragonal axis converts the original FCC structure to BCC. The existence of a metastable minimum at this value of strain (and the barrier at intermediate values of strain) can be deduced from general principles [12].

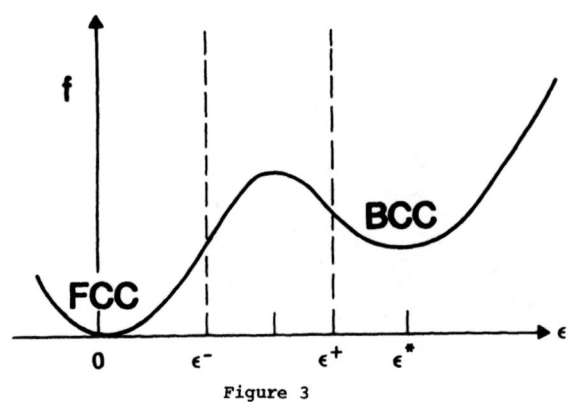

Figure 3

Now let $f=(b-a)/a$ denote the misfit between the substrate lattice constant (b) and the overlayer bulk-stable phase lattice constant (a). Then, within the Peierls model of epitaxy for (again) a model two-dimensional system, the total free energy per unit length of a film of height h can be written:

$$F = \int dx \left[hf(\partial u/\partial x) - \kappa\cos[2\pi(u - fx)/b] \right] \qquad (3)$$

The strain ϵ is related to the local displacement of the overlayer $u(x)$ by $\epsilon = \partial u/\partial x$ and κ is a constant that is proportional to an interfacial shear coefficient. Equation (3) presumes a thin-film aproximation wherein each layer adds an identical bulk energy per length up to the film height h. Observe also that $u(x)=fx$ if the overlayer bulk-stable phase is uniformly strained into commensurability with the substrate.

The conventional theory of epitaxy [13] concerns itself with small excursions around $\epsilon=0$ in Figure 3. If registry can be achieved at the monolayer level, some critical height h_c exists above which the strain in the film becomes excessive and must be relieved by the introduction of misfit dislocations. Here, we are concerned with the <u>large</u> misfit regime where the oscillatory Peierls term in Equation (3) induces an average strain $\epsilon \approx \epsilon^*$. In that case, epitaxy which retains the overlayer bulk-stable crystal structure is prohibitively costly in energy and does not occur. Instead, a structural transition occurs to the pseudomorphic (BCC) phase where the new lattice constant a_* is well matched to the substrate and the residual misfit f_* is small. The linear theory may then be applied to determine whether the overlayer is merely strained (coherent) or contains misfit dislocations (incoherent).

A complete (mean field) analysis of Equation (3) yields the structural phase diagram illustrated in Figure 4 for the case of strong substrate-overlayer adhesion ($\kappa \gg 1$). Near $f=0$ and $f=f^*$ we observe a coherent-incoherent transition with respect to the bulk stable and pseudomorphic phases, respectively. These transitions are continuous. More interestingly, a massive <u>first-order</u> transition occurs between the pseudomorphic phase and the <u>bulk</u>-stable phase as a function of height. This is the triumph of bulk thermodynamics over epitaxial adhesive forces alluded to earlier. However, because the transition is first order, there is a nucleation barrier to transformation. A crude estimate of the barrier height in a typical case yields [12] $E_b \approx 0.1$ eV/mono-layer – a substantial number compared to room temperature. Thus, one may expect pseudomorphic layers to exceed equilibrium critical heights due to kinetic constraints.

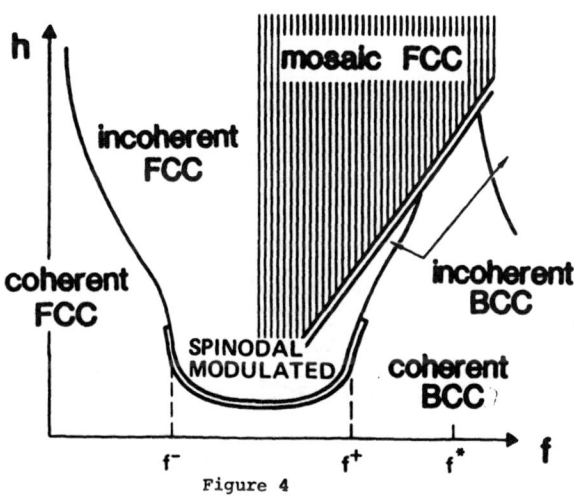

Figure 4

Two features of our phase diagram deserve special mention. First, a characteristic domain or mosaic structure appears in the overlayer once it has transformed back to its bulk-stable structure. The reason is simple: we now have <u>forced</u> two structures with large misfit to share a common planar boundary. Crystallographic twins (or some related defect structure) must develop in order to relieve macroscopic strain along the epitaxial interface. This prediction follows from a direct analogy to the phenomenology of martensitic phase transformations [14].

The second point of interest concerns the region of Figure 4 between the points labelled f^- and f^+. This corresponds to a situation where the Peierls stress of the overlayer maintains an average strain in the overlayer between ε^- and ε^+ in Figure 3. There is no problem at the one or two monolayer growth level where interfacial adhesive forces dominate the total energy. However, for larger values of h, the overlayer begins to realize that it possesses negative elastic constants (due to the concavity of $f(\varepsilon))$! The system is mechanically unstable and must evolve towards some stable strucure. Bruinsma and Zangwill [12] suggest one possiblity: a spinodal type instability where some parts of the overlayer distort towards pure BCC while others distort towards pure FCC. To date, this prediction remains untested.

CONCLUSION

This report has reviewed some recent phenomenological studies of the structural and morphological stability of thin crystalline films with emphasis on those aspects germane to multilayer growth. I have focused entirely on macroscopic concepts such as surface tension, diffusion, and elastic strain in order to illustrate that much can learned without recourse to specific microscopic models. So far, these methods have been applied more or less independently to equilibrium and near-equilibrium situations of wetting, facetting, pseudomorphy and growth. Future work aims toward combining all of these features into a single comprehensive model of direct relevance to experiment.

REFERENCES

1. G.H. Gilmer and M.H. Grabow, J. Metals <u>39</u>, 19 (1987).

2. R. Bruinsma and A. Zangwill, Europhys. Lett. <u>4</u>, 729 (1987).

3. R. Kern, in <u>Interfacial Aspects of Phase Transformations</u>, edited by B. Mutaftschiev (North-Holland, Amsterdam, 1982), p. 287.

4. J.D. Weeks, in <u>Ordering in Strongly Fluctuating Condensed Matter Systems</u>, edited by T. Riste (Plenum, New York, 1980), p. 293.

5. A. Zangwill and R. Bruinsma, to be published.

6. E. Bauer, Z. Krist. <u>110</u>, 372 (1958).

7. C.H.J. van den Brekel, Philips Res. Rpts. <u>32</u>, 118 (1977).

8. G.S. Bales and A. Zangwill, to be published.

9. R.F. Sekerka, in <u>Crystal Growth - An Introduction</u>, edited by P. Hartman (North-Holland, Amsterdam, 1973), p. 403.

10. R.F.C. Farrow, J. Vac. Sci. Tech. <u>B1</u>, 222 (1983).

11. G.A. Prinz, Phys. Rev. Lett. <u>54</u>, 1051 (1985).

12. R. Bruinsma and A. Zangwill, J. Phys. (Paris) <u>47</u>, 2055 (1986).

13. J.H. van der Merwe, J. Appl Phys. <u>34</u>, 117 (1963).

14. Z. Nishiyama, <u>Martensitic Transformation</u> (Academic, New York, 1978).

FORCES ON DISLOCATIONS IN MULTILAYER STRUCTURES

SAMIR V. KAMAT[*], J.P. HIRTH[*] AND B. CARNAHAN[**]
[*]Ohio State University, Dept. of Metallurgical Engineering, Columbus, Ohio, 43210
[**]University of Michigan, Dept. of Chemical Engineering, Ann Arbor, Michigan, 48109

ABSTRACT

In multilayer structures, dislocation forces that arise from coherency strains are well understood in both the isotropic and anisotropic elastic cases. Image forces produced by elastic inhomogeneity have also been developed for single layers surrounded by extended phases. Image forces for the multilayers are shown to reduce to the simpler single embedded layer case to a good approximation. Image forces are shown to lead to a " stand-off " position of interface dislocations from the interface, with several implications for the properties of multilayer structures.

INTRODUCTION

Multilayer structures are of current interest as an example of nanometer-scale, artificially constructed microstructures (1). For strained multilayer structures in particular, the mechanical properties are important in connection with the stability of the structures. In the strained multilayers, forces arising from coherency strains are well understood in both the isotropic and anisotropic elastic cases (2-5) and have been included in analyses of the vulnerability of the multilayers to damage by dislocation or crack injection (5).

In all of these treatments, effects arising from the elastic inhomogeneity of the component layers have been neglected. In a recent analysis (6), the image forces associated with elastic inhomogeneity have been derived for screw dislocations in multilayer structures. The results showed that multiple image terms were important within a given layer, but that, to a good approximation (within five percent) the multilayer situation could be represented by a single layer bounded by two semi-infinite layers.

In this present work, we use this three-layer approximation with its simple analytical form, to determine the equilibrium " stand-off " position of a dislocation from the interface in a strained multilayer structure. We treat the case of an infinite straight screw dislocation with its axis parallel to the interfaces of a given layer. The stand-off distance is determined by a balance of the forces caused by image forces and coherency forces.

FORMULATION

The geometry of the multilayer system is as shown in Fig. 1. The layer thickness is h, the dislocation is at position λ in the material with the smaller shear modulus, and the reduced distance is defined by the parameter $a = (\lambda / h)$. The image force in the three-layer approximation for the isotropic case (6) is

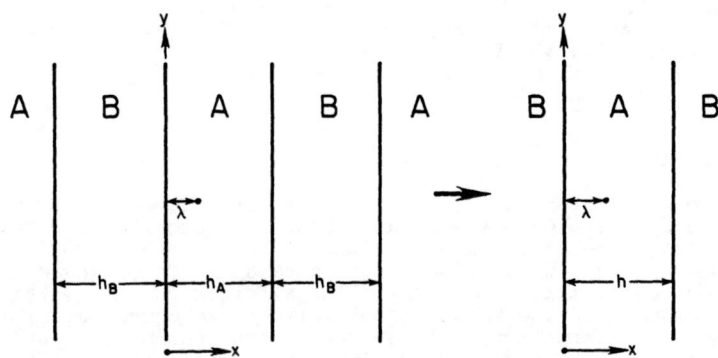

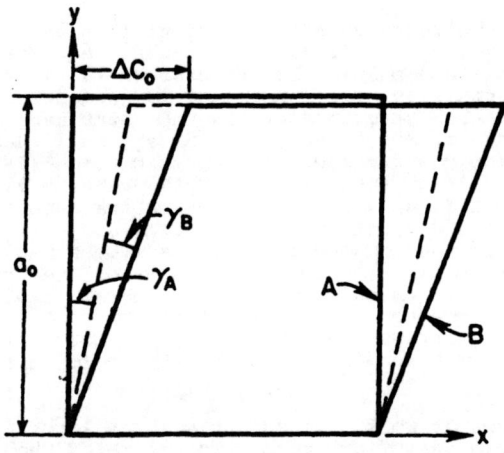

Fig. 1 a. Multilayer structure and its three layer approximation
   b. Unstrained A and B layers and strained configuration
      (dashed lines).

$$F_I/L = \sigma_{yz} b_z = \sigma b \qquad [1]$$

where
$$\sigma(x = \lambda) = \frac{\mu_A b}{4 \pi h} \qquad [2]$$

and
$$\phi = (1 - 2a) \sum_{n=1}^{\infty} \frac{\beta^{2n-1}}{(n + a - 1)(n - a)} \qquad [3]$$

Here $\beta = (\mu_B - \mu_A) / (\mu_B + \mu_A)$ is a factor related to the respective shear modulii μ_B and μ_A of materials B and A. The dislocation is repelled from the interface by the force F_I, which acts in the x direction.

In order to create the appropriate coherency strains for the screw dislocation case, we imagine a monoclinic structure aligned to a cubic structure as shown in Fig. 1b. In the strained state the engineering coherency strains are γ_A and γ_B as shown. The coherency strains would be relieved by a set of right-handed screw dislocations with line directions parallel to the x axis. Mechanical equilibrium requires that

$$\mu_A \gamma_A h_A = \mu_B \gamma_B h_B \qquad [4]$$

while the geometry of Fig. 1b requires that

$$\gamma_o = \frac{\Delta c_o}{a_o} = \gamma_A + \gamma_B \qquad [5]$$

Together, equations [4] and [5] yield the result

$$\gamma_A = \gamma_o h_B \mu_B / (h_A \mu_A + h_B \mu_B) \qquad [6]$$

The coherently strained state is equivalent to that of a set of continuous infinitesimal left-handed screw dislocations lying in the interface (7). The interaction of this set with the dislocation at position λ produces the coherency force, acting in the x direction in such a manner as to attract the dislocation to the interface,

$$F_C/L = -\mu_A \gamma_A b \qquad [7]$$

At local equilibrium, the total force on the dislocation must vanish. Thus, the condition $F_I + F_C = 0$ together with equations [1] and [7] gives the following implicit relation for the stand-off position of the dislocation

$$\phi = 4\pi \gamma_A (h/b) \qquad [8]$$

RESULTS AND DISCUSSION

Fig. 2 is a plot of ϕ versus a for several values of β. On this plot, the quantity $4\pi \gamma_A (h/b)$, which is independent of a, is also shown for several values of (h/b). The point of intersection the two lines for a given value of (h/b) gives the reduced equilibrium stand-off position a_e for the dislocation in the multilayer structure. Fig. 3 is a plot of a_e versus β for several values of (h/b). Other cases can be determined from Figs. 2 and 3 by interpolation. Fig. 4 is a plot of the absolute equilibrium stand-off position λ_e versus β for a case where the Burgers vector b = 0.5 nm. The trends in this figure are in qualitative agreement with physical expectation. For small values of h, multiple image terms weaken the net image force and the equilibrium position of the dislocation is near the interface. For large values of h, the simple image force, which is larger, becomes dominant near the interface and the equilibrium position increases. One can see this in Fig. 4 by looking at the trend in λ_e as a function of (h/b) for a particular value of β.

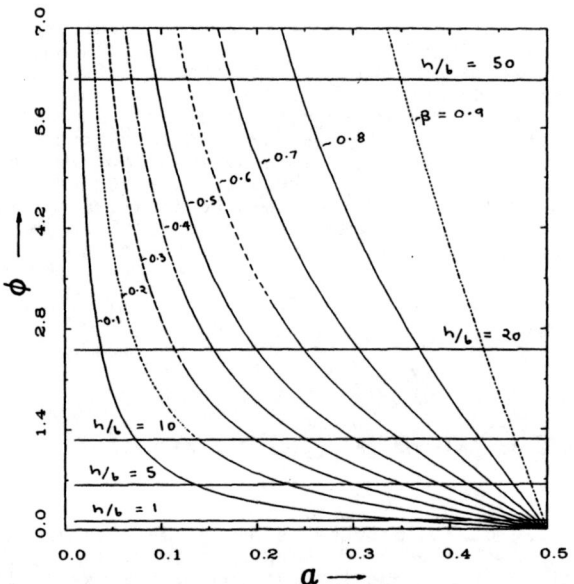

Fig. 2 Image force factor ϕ as a function of a .

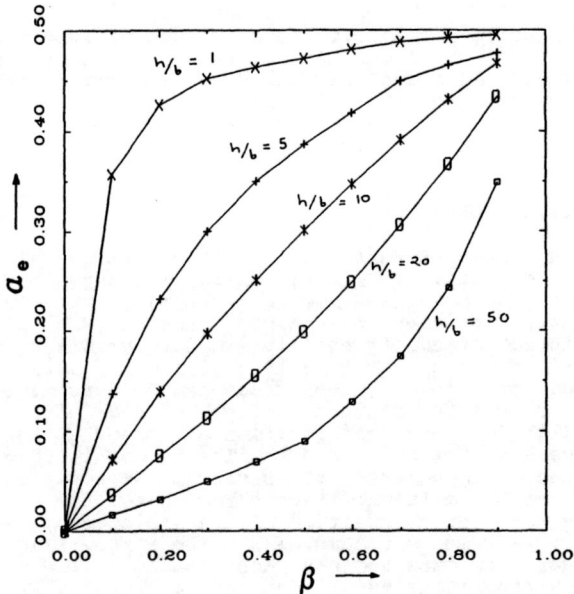

Fig. 3 Equilibrium reduced stand-off distance a_e as a function
of β .

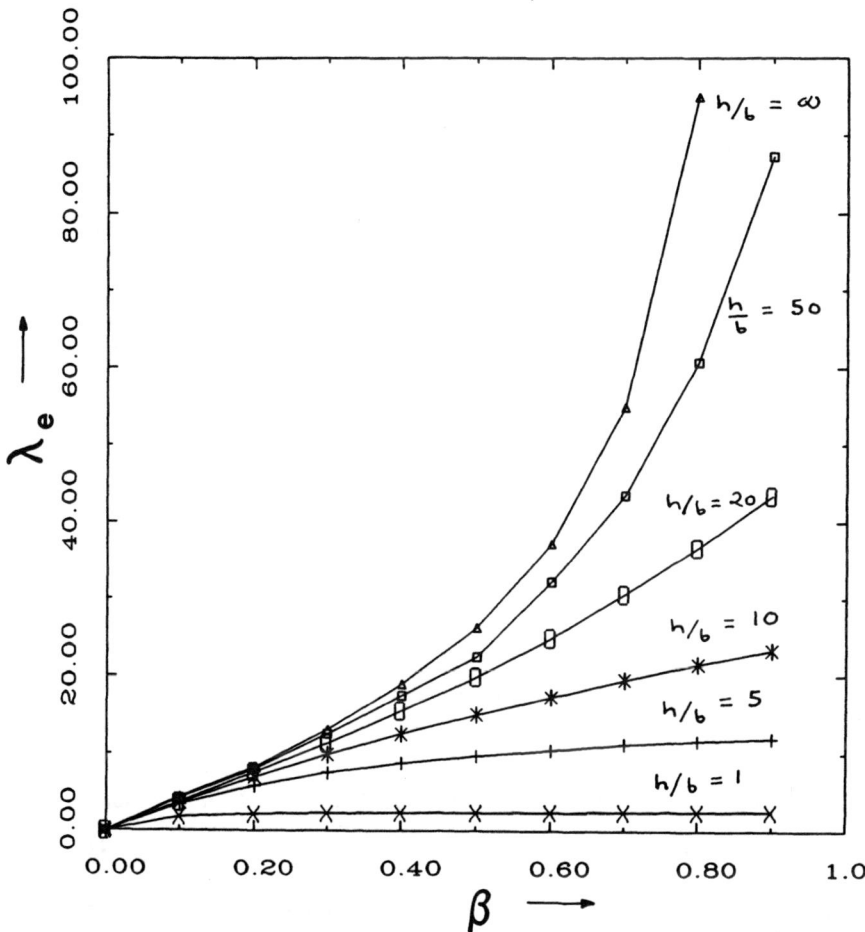

Fig. 4 Equilibrium absolute stand-off distance λ_e as a function of β .

Mader (7) has studied the near interface region of a (0001) $Al_2 O_3-$ (001) Nb thick film couple by lattice-imaging electron microscopy and found a stand-off distance of an edge interface dislocation with $b \parallel <110>$ of six interplanar spacings into the Nb.   While the present analysis is for a screw dislocation, we can   calculate the value of λ_e for the alumina-niobium case   to determine whether the predictions are of the correct order (screw and edge results for λ_e differ by a factor, involving

Poisson's ratio, which is 25 percent for very large h and is more complex and as yet undetermined for the smaller h multiple image case). For the above geometry the appropriate effective value of μ for Nb is C'_{55} given in Eq. (13-43) in ref. (8), while for alumina it is the standard constant C_{44}. With data from refs. (8) and (9), μ_A = 28.7 GPa for Nb and μ_B = 179 GPa for alumina giving β = 0.72. With b = 0.233 nm for Nb, h = ∞ for the thick film case, and γ_A = 0.0338 for the above geometry, we calculate a value of λ_e of 3·6 interplanar spacings in fair agreement with the experimental results.

The results and the above comparison suggest that the stand-off effect is appreciable, in the sense that the stand-off distance exceeds dislocation core dimensions, for β values in the range $\beta \geqslant 0.10$. There are several possible consequences of the stand-off effect. Dislocation injection and motion would be determined by the Peierls stress of the softer material instead of the interfacial value. Pileup distributions and attendant crack nucleation probabilities would be modified at the interface. Trapping of dislocations in a buffer layer would be more difficult if the buffer layer had a larger effective shear modulus then the substrate and would be easier otherwise. These possibilities require detailed calculations to determine their quantitative importance. We are presently extending the anaysis to the case of edge dislocations preparatory to analyzing these more complex problems.

ACKNOWLEDGEMENT

This research was supported by DARPA through ONR Contract Number 0014-86-K-0753 with the University of California, Santa Barbara, with a subagreement with Ohio State University.

REFERENCES

1. Viewpoint Set No. 8, Scripta Metall. 20, 441-488 (1986).

2. J. W. Matthews and A. E. Blakeslee, J. Cryst. Growth 27, 118 (1974).

3. J. W. Matthews and A. E. Blakeslee, J. Cryst. Growth 29, 273 (1975).

4. J. W. Matthews and A. E. Blakeslee, J. Cryst. Growth 32, 265 (1976).

5. J. P. Hirth and A. G. Evans, J. Appl. Phys. 60, 2372 (1986·).

6. S. V. Kamat, J. P. Hirth and B. Carnahan, Scripta Metall. 21, 1587 (1987).

7. W. Mader, Max-Planck Inst. f. Metallforsch., Private communication, March 1987.

8. J. P. Hirth and J. Lothe, Theory of Dislocations (Wiley, New York, 1982).

9. G. Simmons and H. Wang, Single Crystal Elastic Constants and Calculated Aggregate Properties : A Handbook, The M.I.T. Press, 1971

ON THE EXISTENCE OF MULTIPLE EQUILIBRIUM STATES
IN STRAINED-LAYER SUPERLATTICES

WILLIAM C. JOHNSON

Department of Metallurgical Engineering and Materials Science, Carnegie Mellon University,
Pittsburgh, PA 15213-3890.

ABSTRACT

Using recent results from the thermodynamics of stressed solids, two-phase coexistence in a simple binary strained-layer superlattice is examined. We show that for a given temperature and overall composition of the superlattice, there can exist more than one linearly stable, equilibrium thermodynamic state. That is, there may exist several combinations of relative thickness of the phases and corresponding phase compositions that minimize the free energy of the system. The equilibrium state observed experimentally can, therefore, be influenced by the processing path.

INTRODUCTION

The deformation experienced by a heteroepitaxial planar thin-film system deposited on a thick or rigid substrate is essentially determined by the substrate; the lattice parameter in the plane of the film being constrained to match that of the underlying substrate. Such a system may be referred to as "elastically constrained" since the deformation can be viewed as being externally imposed. The thermodynamic state of a phase comprising an elastically constrained system is determined uniquely by its temperature, composition and the state of deformation imposed by the substrate; it is independent of the existence of the other phases. As a consequence, elastically constrained heteroepitaxial systems should obey Gibbs phase rule for phase coexistence, and the tie lines (corresponding to the equilibrium phase compositions) and the field lines (which indicate the limits of single phase stability) should coincide.

In contrast, the state of deformation in a strained-layer superlattice depends strongly on the relative amounts of each phase present, their thickness, difference in lattice parameters and rigidity. This is because the deformation experienced by each phase is not imposed, but is free to assume that value which minimizes the free energy of the system. Thus, the thermodynamic states of each phase are strongly coupled. An immediate consequence is that the tie lines and field lines do not coincide [1-4] and that the Gibbs phase rule, identifying the degrees of freedom in the system, is not applicable [5]. The characteristics of phase coexistence in a two-phase strained-layer superlattice may be expected, therefore, to be fundamentally different from those that obtain when the same two phases comprise an elastically constrained system.

In this paper, we identify some new features pertaining to phase coexistence in strained-layer superlattices (SLS). In particular, we show that there may exist more than one linearly stable equilibrium state that may be obtained by a SLS for a given temperature and overall composition. We assume that the phases are sufficiently thick that regions of each layer behave as bulk material and that a Gibbsian dividing surface may be associated with each interface. Although this precludes consideration of an ultra-thin SLS, some qualitative aspects of the work may still be applicable. We begin by examining the conditions necessary for thermodynamic equilibrium in elastically stressed crystals. These thermodynamic conditions are then used to address the following question, "If two phases are constrained to appear as a SLS, what is the equilibrium composition and relative thickness of each phase when the temperature and overall composition of the system are specified?" The results are expressed in terms of an equation of state for the relative thickness of the layers and an equation for each of the layer compositions.

THERMODYNAMIC EQUILIBRIUM

A strained–layer superlattice occupies an equilibrium thermodynamic state when the conditions necessary for thermal, chemical, mechanical and interfacial equilibrium are satisfied simultaneously. An equilibrium state is linearly stable (metastable) if, upon making a small change (perturbation) in the relative thickness and composition of the layers at constant overall composition, the system decays back to its original state. An equilibrium state is globally stable if the perturbation can be arbitrarily large. If, upon perturbing the system, the system evolves to a new thermodynamic state, the original equilibrium state is termed unstable and is a local energy maximum.

In order to identify some aspects of phase equilibrium unique to SLS, we here restrict ourselves to a simple two–phase binary alloy in which the chemical constituents, A and B, occupy the same lattice site and possess the same partial molar volumes in each of the phases. The superlattice is configured as alternating layers of α and β phase, of thickness t^{α} and t^{β}, respectively. The molar volumes of each phase may be different, resulting in the phases being stressed as shown in Fig. 1. The equilibrium compositions of the layers (phases) in the SLS are denoted c^{α} and c^{β} respectively, and are measured in terms of the mole fraction of component B. The layers are assumed to possess cubic symmetry in the absence of stress with the [001] or [111] crystallographic axis parallel to the x_3 axis and to remain planar and free from defects. Relaxing these assumptions leads to more complicated behavior and, in some cases, new behavior.

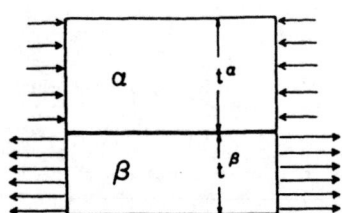

Fig. 1.   When the α and β phases possess different lattice parameters, stresses are engendered that depend on the relative thickness of each phase.

The conditions necessary for thermodynamic equilibrium in two–phase stressed solids are well established [6–8] and are directly applicable to SLS. Chemical equilibrium requires uniformity of the diffusion (chemical) potentials, M, throughout the system and is expressed by [7]

$$M^{\alpha}(\theta,c^{\alpha},\mathsf{T}^{\alpha}) = M^{\beta}(\theta,c^{\beta},\mathsf{T}^{\beta}), \tag{1}$$

where θ is the temperature, assumed uniform, and T is the stress tensor. Since the partial molar volumes of A and B have been assumed to be equal, the diffusion potential can be expressed as the difference in the chemical potentials of the components measured in the absence of stress, $\mu(\theta,c)$, as:

$$M(\theta,c,\mathsf{T}) = \mu_{B}(\theta,c) - \mu_{A}(\theta,c). \tag{2}$$

The diffusion potentials are independent of the state of stress owing to the cubic symmetry and the equality of the partial molar volumes. Mechanical equilibrium is determined from Cauchy's relations, $\nabla \cdot \mathsf{T} = 0$, the continuity of displacement and traction across each layer, and an imposed traction condition along the edge of the superlattice. Neglecting excess forces associated with the interface (interfacial stress), this boundary condition is

$$(1-t)T^{\alpha} + t\,T^{\beta} = 0 \tag{3}$$

where $t = t^\beta/(t^\alpha + t^\beta)$ is a measure of the relative thickness of the two phases and is equal to the volume fraction of β. The stress in each phase is thus:

$$T^\alpha = T^\alpha_{11} = T^\alpha_{22} = Y^\alpha \epsilon t(1+\delta)/(1+\delta t) \tag{4}$$

$$T^\beta = T^\beta_{11} = T^\beta_{22} = -Y^\beta \epsilon (1+\delta)(1-t)/(1+\delta t) \tag{5}$$

and the other stress components are zero. $\epsilon = (a^\beta - a^\alpha)/a^\alpha$ is a measure of the difference in the unstressed lattice parameter of the two phases and $\delta = (Y^\beta - Y^\alpha)/Y^\alpha$ is a measure of the difference in the effective elastic moduli between the phases. The effective elastic moduli for the [100] and [111] film orientations can be expressed in terms of the elastic constants as [9]

$$Y_{100} = (C_{11} - C_{12})(C_{11} + 2C_{12})/C_{11} \tag{6}$$

and

$$Y_{111} = 6C_{44}(C_{11} + 2C_{12})/(C_{11} + 2C_{12} + 4C_{44}). \tag{7}$$

Interfacial equilibrium, associated with the growth and dissolution of the phases, requires that [6-8]

$$\omega^\alpha - \omega^\beta - (E^\alpha_{ij} - E^\beta_{ij})T^\alpha_{ij} = 0 \tag{8}$$

where E is the strain tensor as measured with respect to a common (arbitrary) reference state (not the unstressed state of each phase) and ω is the thermodynamic potential. Equation (8) is a linearized version for planar interfaces of a more general relationship for curved interfaces [8]. For a planar SLS, Eq. (8) becomes

$$\rho_o' [\mu^\alpha_A(\theta, c^\alpha) - \mu^\beta_A(\theta, c^\beta)] + T^\alpha_{ij} E^{\alpha\iota}_{ij}/2 - T^\beta_{ij} E^{\beta\iota}_{ij}/2 = 0 \tag{9}$$

where $E^\iota = E - E^T$ is the strain measured from the stress-free state of each phase, ρ_o' is the density of lattice points in the reference state, and the elastic terms represent the strain energy density of a layer. To uniquely specify the thermodynamic state of the system, there exists a condition on the conservation of mass given by

$$(1-t)c^\alpha + tc^\beta = c^o \tag{10}$$

where c^o is the overall mole fraction of component B in the SLS.

EQUATION OF STATE

If mass diffusion is allowed to occur, the phase compositions and layer thicknesses change so as to minimize the free energy of the system. In the absence of any stress effects, Eqs. (1) and (8) can be solved simultaneously for these equilibrium phase compositions, c^α and c^β, since the thermodynamic state of a phase is then independent of the existence of the other phases and, therefore, the relative thickness of the phases, t. t can then be determined directly from Eq. (10) which is simply the lever law. For the stress-free case, the tie lines (equilibrium compositions of the phases) are equivalent to the field lines (solvi) on the phase diagram.

For the SLS, the equilibrium layer compositions and the relative layer thickness are strongly coupled through Eqs. (1), (8) and (10). This is because the elastic state of a layer depends on t and the presence of the other phase. Consequently, Eqs. (1) and (8) no longer decouple from Eq. (10) and the tie lines and field lines on the phase diagram no longer coincide. This is similar to the behavior observed in bulk coherent alloys [1-3].

c^α, c^β, and t can be determined from the numerical solution of Eqs. (1), (8) and (10). However, some of the physics of the problem can be more easily demonstrated when these transcendental equations are linearized. If the two phases under consideration can coexist in equilibrium in the absence of stress, then Eqs. (1) and (8) can be linearized about the stress-free equilibrium state satisfying:

$$\mu_A^\alpha(\theta, c_o^\alpha) = \mu_A^\beta(\theta, c_o^\beta) \tag{11}$$

$$\mu_B^\alpha(\theta, c_o^\alpha) - \mu_A^\alpha(\theta, c_o^\alpha) = \mu_B^\beta(\theta, c_o^\beta) - \mu_A^\beta(\theta, c_o^\beta) \tag{12}$$

where c_o^α and c_o^β correspond to the stress-free composition of α in equilibrium with β and of β in equilibrium with α, respectively, i.e., the solvi compositions. Expanding Eqs. (1) and (8) to first order in the changes in the equilibrium compositions, c^α and c^β, and substituting the results into Eq. (10) yields the following polynomial, or equation of state, for t (for details of the linearization process see ref. [3]):

$$t(1-t)(A_3 t^3 + A_2 t^2 + A_1 t + A_o) = 0 \tag{13}$$

where

$$A_3 = -2\delta^2 + \delta\xi\lambda^2(1+\delta) \tag{14}$$

$$A_2 = \delta^2(W+1) - 4\delta + \delta\lambda^2(1+\delta) + 2\xi\lambda^2(1+\delta) \tag{15}$$

$$A_1 = 2\delta(W+1) - 2 + \lambda^2(1+\delta)(2-\xi) \tag{16}$$

$$A_o = W+1 - \lambda^2(1+\delta) \tag{17}$$

$$W = 1 + 2(c^o - c_o^\beta)/(c_o^\beta - c_o^\alpha) \tag{18}$$

$$\lambda = \frac{\epsilon}{(c_o^\beta - c_o^\alpha)}\left(\frac{2Y^\alpha}{\chi^\alpha}\right)^{1/2} \tag{19}$$

$$\xi = (\chi^\alpha - \chi^\beta)/\chi^\beta \tag{20}$$

$$\chi = \frac{\rho_o' k\theta}{c_o(1-c_o)}\left[1 + \frac{\partial \ln\gamma_B}{\partial \ln c}\right] \tag{21}$$

The equation of state and the stability of the SLS is determined entirely by the four nondimensional parameters δ, λ, ξ, and W. λ is a measure of the difference in lattice parameters between the two phases and, as such, is a measure of the influence of the elasticity on phase equilibrium. ξ is a measure of the difference in curvatures of the free energies of the phases in the absence of stress, X; it indicates how some of the elastic energy may be relieved by shifting the phase compositions. W is the superlattice composition.

The equilibrium compositions of the layers are:

$$c^{\alpha} = c_o^{\alpha} + (c_o^{\beta} - c_o^{\alpha})[(W+1)/2 - t]/(1+\xi t) \tag{22}$$

$$c^{\beta} = c_o^{\beta} + (c_o^{\beta} - c_o^{\alpha})(1+\xi)[(W+1)/2 - t]/(1+\xi t) \tag{23}$$

DISCUSSION

Equation (13) is the equation of state for the simple SLS under consideration. Each root of the polynomial corresponds to a relative thickness of the layers, t, that satisfies the conditions for thermodynamic equilibrium. The equilibrum state may be either an energy minimum (metastable) or a maximum (unstable). The coefficients of the polynomial and, therefore, the number of equilibrium states, are determined by the three nondimensional materials parameters and the overall composition of the superlattice. The solutions $t = 0$ and $t = 1$ are end-of-the-range extrema, corresponding to the thickness of one of the phases going to zero, and are independent of the overall composition of the superlattice.

When both phases possess the same lattice parameters, $\epsilon = \lambda = 0$, and no stresses are engendered in the superlattice. The only physical solutions ($0 \leq t \leq 1$) to the equation of state are $t = 0, 1$ and $t = (W+1)/2$ which is the lever rule from Eq. (10). The compositions of the layers are $c^{\alpha} = c_o^{\alpha}$ and $c^{\beta} = c_o^{\beta}$. When the layers possess different lattice parameters, additional physical solutions to the equation of state may be obtained. Figure 2 gives an example of how the number of equilibrium states may change as a function of the superlattice composition when the elastic moduli of the two layers are different. The equilibrium relative thickness of the layers, t, is plotted as a function of the superlattice composition, c^o, for a fixed temperature. The heavy solid lines depict the globally stable equilibrium solutions, the fine solid lines the metastable solutions, and the broken lines the unstable equilibrium (energy maxima). c_o^{α} and c_o^{β} delineate the region of two phase coexistence in the absence of stress.

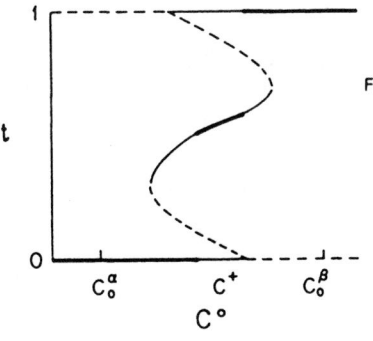

Fig. 2. A stability diagram depicting those values of t that extremize the system energy as a function of alloy composition. Solid lines are minima, broken lines maxima.

When the phases are constrained to appear as a superlattice, both single phase fields are stabilized. As the superlattice composition is increased from c_o^{α} to c_o^{β}, the solution that gives the global energy minimum changes discontinuously from single phase α ($t=0$) to a two phase superlattice structure with $t=0.5$. This solution remains the global minimum over a composition range near c^+ and then changes discontinuously to single phase β. However, both the α and β phases and a two–phase superlattice structure exhibit significant regions of metastability. Since the energy barriers between the global minima may be significant, a system may not always be able to achieve the state of lowest energy.

SUMMARY

We have employed well established results from the thermodynamics of stressed solids to examine phase coexistence in a simple strained–layer superlattice. The following points are emphasized:

1. The thermodynamic description of a SLS is much different from that of an elastically constrained thin–film system. The tie lines (equilibrium phase compositions) and field lines (solvi) do not coincide, the Gibbs phase rule for phase coexistence is inapplicable and phase equilibrium cannot be displayed schematically using a common tangent construction to free energy diagrams.

2. For a given temperature and overall superlattice composition, there may exist more than one linearly stable equilibrium thermodynamic state. Each equilibrium state represents a relative thickness of the layers and corresponding layer composition that satisfies thermal, mechanical, chemical and interfacial equilibrium. These thermodynamic states are not a result of imposing nucleation barriers in that mass is free to redistribute in the system.

3. The number of equilibrium states for a particular superlattice depends on the alloy composition and on a number of materials parameters including the elastic constants, lattice parameters and free energies of each phase in the absence of stress.

ACKNOWLEDGEMENTS

I am grateful to P.W. Voorhees for a number of helpful discussions and to the Division of Materials Research of the National Science Foundation for their support of this work under grant DMR-8620026.

REFERENCES

1. R.O. Williams, Metall. Trans. 11A, 247 (1980).
2. J.W. Cahn and F.C. Larche, Acta Metall. 32, 1915 (1984).
3. W.C. Johnson and P.W. Voorhees, Metall. Trans. 18A, 1213 (1987).
4. W.C. Johnson, J. Appl. Phys., in press.
5. W.C. Johnson, Metall. Trans. 18A, 1093 (1987).
6. P.F.Y. Robin, Amer. Min. 59, 1286 (1974).
7. F.C. Larche and J.W. Cahn, Acta Metall. 26, 1579 (1978).
8. W.C. Johnson and J.I.D. Alexander, J. Appl. Phys. 59, 2735 (1986).
9. J.E. Hillard, in Phase Transformations, edited by H.I. Aaronson (ASM, Metals Park, 1970), Ch. 12.

PART II

Characterization

ATOMIC ARRANGEMENTS IN SHORT PERIOD MO-GE MULTILAYERS DETERMINED BY X-RAY ANOMALOUS SCATTERING AND EXAFS.

LANE WILSON AND ARTHUR BIENENSTOCK
Department of Applied Physics, Stanford University, Stanford, California 94305

ABSTRACT

Studies of atomic arrangements in amorphous and crystalline, short period Mo-Ge multilayers using synchrotron radiation X-ray diffraction (transmission and reflection), supplemented by EXAFS, are described. Differential anomalous scattering and EXAFS were utilized to determine the environment of each species. Intermixing, as well as a BCC epitaxial Ge structure, are among the observed structural characteristics. A wide range of structural variations is present indicating that specification of the layer thicknesses or composition profile alone is insufficient for characterization of the multilayers.

EXPERIMENT AND RESULTS

This paper describes synchrotron radiation X-ray diffraction and EXAFS studies of atomic arrangements in amorphous and crystalline sputtered Mo-Ge multilayers over a wide range of compositions and layer thicknesses. Differential anomalous scattering (DAS) and EXAFS techniques, which determine the environment of each atomic species, were used to help separate the local structure of the individual constituent layers and the interfacial regions of the multilayers. Layer thicknesses were varied to examine the new atomic configurations and bonding patterns that appear when the interfacial regions become a substantial portion of the total system. An appreciable area of each substrate was removed so that three dimensional diffraction data and transmission EXAFS on thin film multilayers were easily obtained. As a result of all these techniques, much more complete structural information was obtained than has been previously possible. This includes information about atomic arrangements in thin amorphous layers, crystallite size anisotropies in thicker films, and intermixing at the layer boundaries; as well as observations of a new Ge epitaxial structure. The studies make apparent the great variety of structural forms in the multilayers and demonstrate that specification of the layer thicknesses is insufficient for characterization of these short period multilayers.

Multilayers were grown on Si (100) substrates by magnetron sputtering with the substrates alternately rotating beneath continuously sputtered Mo and Ge targets[1]. The layer compositions and thicknesses were varied by adjusting deposition rates of the individual targets and substrate rotation speed. Desired individual layer thicknesses were estimated from sputtering rate tests on thick films of Mo and Ge. Rate tests are inherently inaccurate for estimating short period multilayer layer thicknesses because the growth of thicker films differs from that of thin layers dominated by interfacial regions. A technique to give a more accurate measurement of the individual layer thicknesses after deposition was developed by using information about each constituent material from the absorption edge-jumps obtained from transmission EXAFS experiments At the absorption edge of a given element the sharp change in absorption is due to the amount of material of only that particular element. Knowledge of the absorption cross-sections per atom, the number of bi-layers in the sample, and the bi-layer period from low angle x-ray reflections, allowed the determination of the area density of each element in one period of the multilayer. This information was useful as a constraint on estimations of the composition profile in the deposition direction. Overall compositions were determined by this method and individual layer thicknesses were estimated using a bi-layer model with bulk a-Ge (RTN structure) and c-Mo (BCC structure) densities. Fig. 1 shows a matrix of the individual layer thicknesses for the samples examined. The boxes represent the nominal desired layer thicknesses while the circles are the thicknesses if the materials had the bulk structures. The lines are bi-layer thicknesses determined by low angle X-ray reflections. Of interest is the fact that the sums of the bulk structure thicknesses are greater than the measured bi-layer values which supports evidence detailed below that the local structures, particularly that of the Ge component, are more dense than bulk structures. As discussed later, the concept of individual layer thicknesses is ambiguous for short period multilayers where interfacial regions dominate the structure. Exact composition modulation has not been addressed in this paper, rather the local structures that appear as a function of nominal layer thicknesses are examined. The local structures in turn help define the composition modulation. Uncertainties in the individual layer thicknesses do not influence the conclusions about the local structures observed in this work.

The depositions were performed at room temperature with substrates never reaching temperatures above 100°C. Deposition times were of the order of 5 to 10 hours in order to fabricate samples approaching 10um thickness. 10um samples gave strong scattering signals from amorphous structures and also permitted the removal of the substrate over large areas (2cm x 5cm) without breaking the film. Substrates were removed employing techniques developed for the industrial micro-machining of Si wafers[2]. A window in the substrate was etched by immersion of the masked sample in either EDP or KOH for several hours at 100°C. The samples were thus rendered free- standing while retaining a frame of Si which was then inserted into an X-ray diffractometer sample holder. X-ray scattering and EXAFS experiments were carried out at SSRL on the materials diffractometer on wiggler branch line IV-3 under dedicated beam time conditions.

As a starting and reference point for structural investigations, a multilayer consisting of thick crystalline (c-)Mo (21Å) and thick amorphous (a-)Ge (33Å) layers was studied. X-ray scattering from this sample in both transmission and reflection is shown in Fig. 2. Line widths show average Mo crystallite sizes which are at least 250Å in the layer plane but are limited to the Mo layer thickness (21Å) perpendicular to the plane. A Ge edge differential distribution function

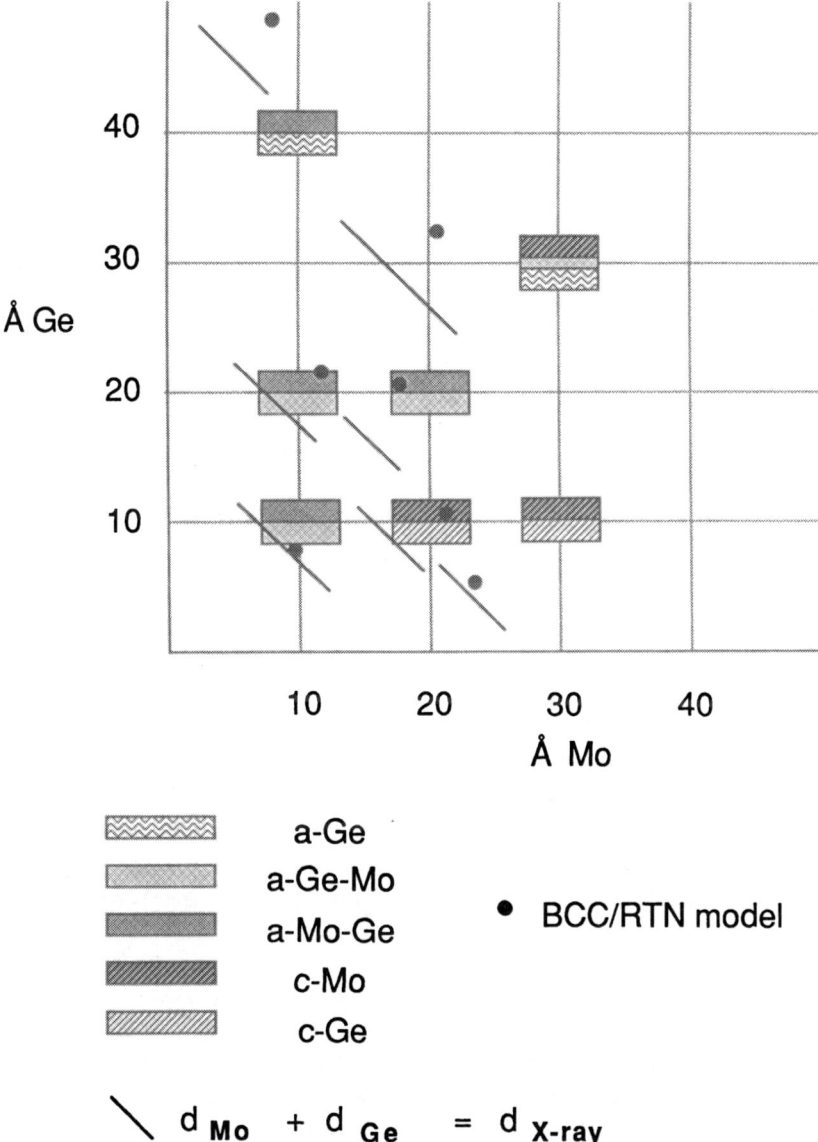

Å Ge

Å Mo

a-Ge
a-Ge-Mo
a-Mo-Ge   ● BCC/RTN model
c-Mo
c-Ge

$d_{Mo} + d_{Ge} = d_{X\text{-ray}}$

Fig. 1. Schematic representation of structure as a
function of individual layer thicknesses.

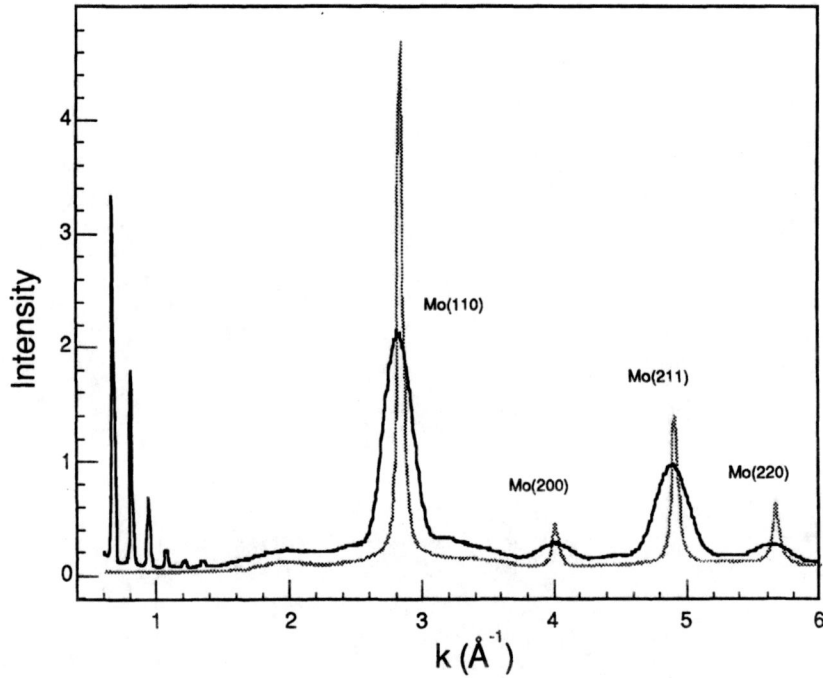

Fig. 2 ------ X-ray scattering intensity as a function of scattering vector (k) perpendicular to the layer plane (dark line) and parallel to the layer plane (light line). Scattering is from a sample of average Mo concentration (χ_{Mo}) =.49 and nominal layer thicknesses 21Å Mo/33Å Ge. Low angle multilayer reflections (k<1.5Å^{-1}) indicate the bi-layer thickness to be 46.6Å. Scattering from an amorphous component, in the Ge layer, is evidenced by the diffuse peaks at k=2.0Å^{-1} and k=3.2Å^{-1}.

(DDF) [3,4] (Fig. 3) allows the removal of the Mo-Mo correlations, leaving information about the Ge layer and interfacial structure, even though the c-Mo contributes heavily to the total scattering (Fig. 2). Comparison of this distribution with that of a-Ge and an a-Mo$_{14}$Ge$_{86}$ alloy[5] (Fig. 3) indicates that the Ge layer is dominated by a-Ge correlations but also suggests that there is an a-Mo-Ge interfacial region. The broadening and shift of the first peak of the DDF outward from the a-Ge value is evidence of Mo correlations, as shown by Kortright[5]. On the other hand, the second and third neighbor peaks of a-Ge are easily seen in the DDF. Comparison with the RDF (not shown), which includes the contribution from the crystalline Mo shows that the Mo-Mo correlations are indeed removed since the peaks at crystalline distances dominate the RDF·at different locations in r than the peaks from the amorphous structure remaining in the DDF.

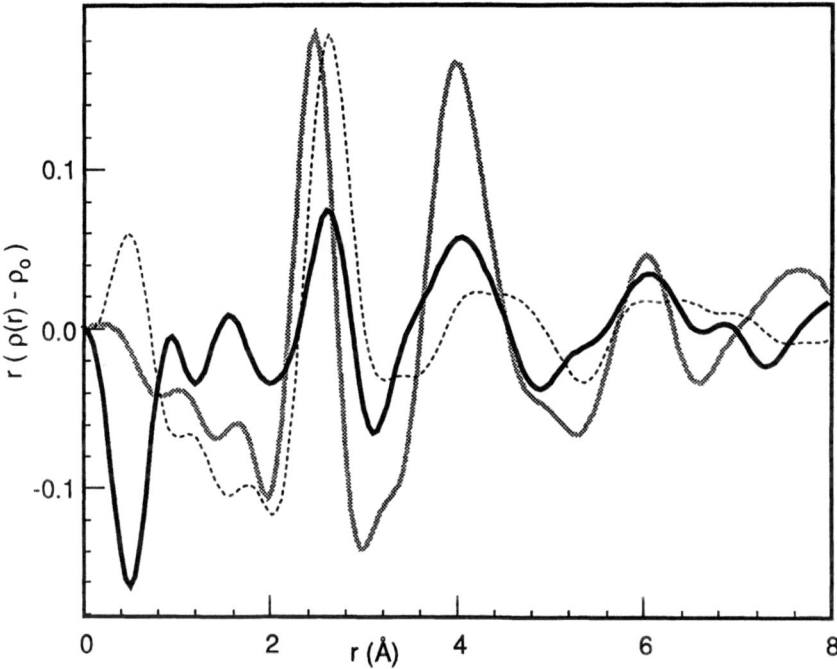

Fig. 3 ------ Ge absorption edge Differential Distribution Function $[r(\rho(r)-\rho_0)]$ for multilayer of average $\chi_{Mo}=.49$ and nominal layer thicknesses 21Å Mo/33Å Ge (dark line), amorphous co-sputtered alloy $Mo_{14}Ge_{86}[5]$ (dashed line), and amorphous sputtered Ge[5] (light line).

In this paper distributions are compared in the form $r(\rho(r)-\rho_0)$ where $\rho(r)$ is the density of atoms about an average atom and ρ_0 is the density averaged over all distances..The form $r(\rho(r)-\rho_0)$ allows a more detailed comparison of structure at high r. Displayed in this manner the region below the first peak is especially sensitive to low frequency error in the scattering data. The error is exacerbated in a DDF since it is the transform of a difference between two scattering data sets. The unphysical ripple at low r is well known to be indicative of errors in the distribution function. However, initial error analysis indicates that errors are primarily limited to the region below the nearest neighbor peak and that any associated error at larger r is unlikely to vitiate any conclusions made in this paper, although some have been softened as a result of possible error. Error analysis will be presented in detail in a later paper.

The structures of thin Ge layers were investigated by examining two samples, each with thin (≤11Å) Ge layers bounded by relatively thick (≥21Å) Mo layers. Fig. 4 shows parts of the scattering data from the two samples, 23ÅMo/5ÅGe and 21ÅMo/11ÅGe. There is no sign of an amorphous multilayer component in either the total scattering (Fig. 4) or the Ge-edge differential structure factor (not shown). The sharpness of the Mo crystalline peaks is evidence that the crystallites have dimensions greater than the layer thicknesses and the asymmetric side bands indicate that the Ge rich layer has assumed a crystalline structure coherent with the BCC Mo[6]. Simple modeling of the side band modulations indicates that the structure consists of strained multilayers. The strain is also evident when comparing transmission and reflection data of the Mo lattice reflections. Peaks from unstrained crystallite orientations have symmetric side bands and occur at the same positions in both geometries. Strained orientations show asymmetric side bands and displacements of the main reflections between the transmission and reflection geometry measurements. EXAFS modeling yielded Ge-Ge distances of 2.78Å for the 5Å Ge sample and 2.73Å for the 11Å Ge sample in agreement with the modeling of the asymmetric side bands. These distances are much longer than the covalent bond lengths of pure Ge (2.44 Å) and are close to the lengths of metallic Mo (2.74Å). The normal covalent nature of Ge-Ge bonds is modified by the thinness of the layers and their boundary of crystalline Mo layers. Although the extent of Mo incorporated into the BCC Ge rich layer has not been completely determined, the Ge EXAFS is best described by Ge-Ge nearest neighbors and the average Ge concentration for the multilayers is well above the 2% solid solubility of Ge in Mo[7] supporting a non-equilibrium Ge phase.

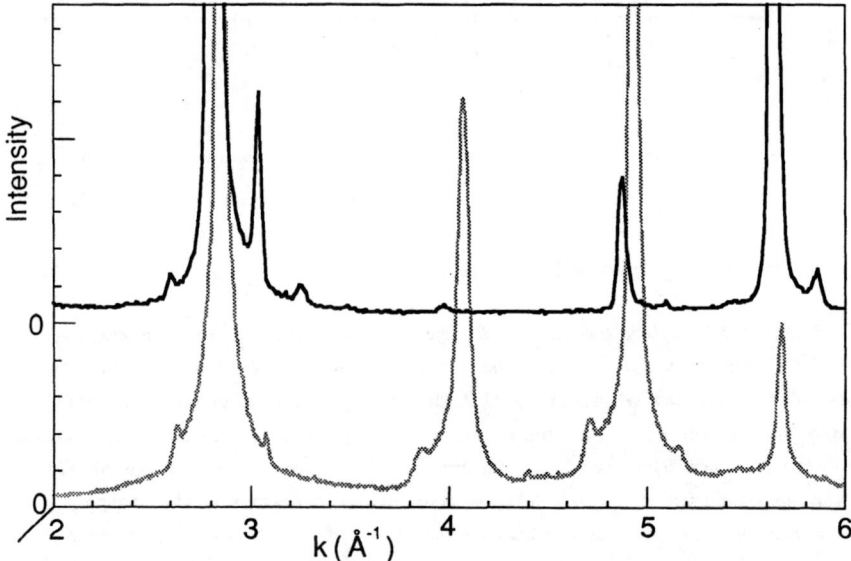

Fig. 4 ------ BCC X-ray reflections and asymmetric satellites from strained epitaxial Mo-Ge multilayers; χ_{Mo}=.87, $d_{(nominal)}$= 23ÅMo/5ÅGe, $d_{(measured)}$= 27.3Å (upper, dark line); and χ_{Mo}=.74, $d_{(nominal)}$= 21ÅMo/11ÅGe, $d_{(measured)}$= 25.6Å (lower, light line).

Texturing occurs with the BCC (110) planes parallel to the layers and is stronger for the sample
with thicker Mo layers as is evident by the greatly diminished (200) and (211) peaks in the
scattering from the sample with 5Å Ge layers.

The structure of thin Mo layers was investigated on a multilayer with layers of 8Å Mo and 49Å
Ge. This sample showed no evidence of crystalline Mo nor of any unique amorphous Mo
structure. Instead, the structure is characterized by a-Mo-Ge alloy material bounded by layers of a-
Ge. The Mo edge DDF is useful in examining the Mo-Ge interfacial region since the scattering
contributions of the a-Ge in the thick Ge layer are removed. The Mo DDF (fig. 5) shows good
agreement with that of an a-$Mo_{42}Ge_{58}$ alloy[5]. The average sample composition of 20% Mo is
periodically modulated between limits of an alloy richer in Mo and a-Ge. The modulation,
consistent with the area density measurements and the amorphous alloy density, is most simply
modelled by a two layer system with an a-Ge layer of 34Å and a 19Å a-$Mo_{42}Ge_{58}$ alloy region.
More detailed analysis, using the anomalous scattering data, is underway.

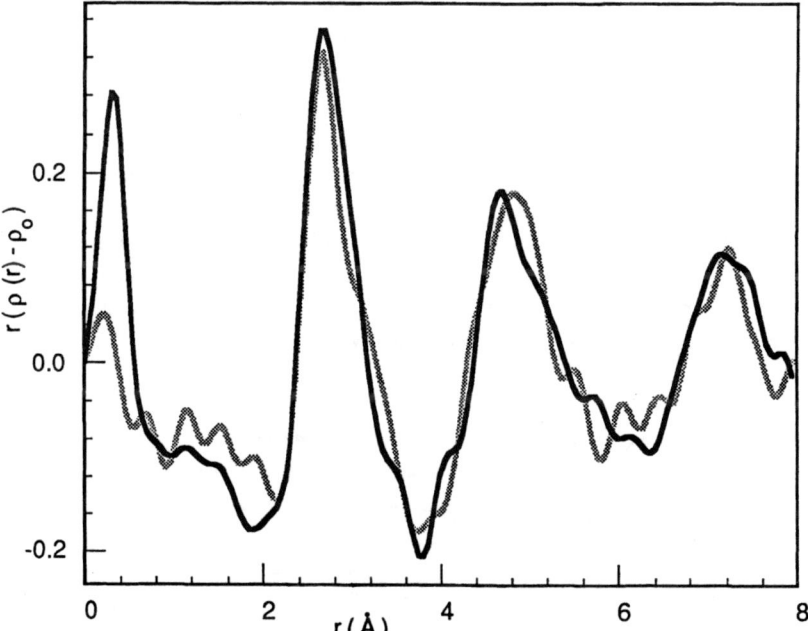

Fig. 5 ------ Mo absorption edge Differential Distribution Function [$r(\rho(r)-\rho_o)$] for
multilayer of average χ_{Mo}=.20, nominal layer thicknesses 8ÅMo/49ÅGe, and
measured bi-layer thickness 50.6Å (dark line); and amorphous co-sputtered alloy
$Mo_{42}Ge_{58}$[5] (light line). Local structure of the Mo layer is in close agreement
with that of the amorphous Mo-Ge alloy with χ_{Mo}=.42 indicating that Ge has
intermixed with the thin Mo layer.

The atomic density for the interfacial alloy used above is 43% more dense than the bulk a-Ge RTN structure and only 6% less dense than the bulk c-Mo BCC structure. The intermixing of the Ge and Mo allows for a more dense structure for a bi-layer. The 53Å spacing of the above simple bi-layer model is closer to the measured 51Å bi-layer spacing than the 57Å predicted by the model using bulk structures. The fact that the measured bi-layer spacing is smaller than that predicted by bulk structures for all of the samples can be explained in a similar way. Because of the unique structures of thin layers and interfacial regions it is difficult to obtain a desired bi-layer spacing for short period multilayers from bulk growth rates alone even if they are well known.

When multilayers of both thin Ge and thin Mo layers are deposited, the structure closely resembles that of a-Mo-Ge alloys although a periodic composition modulation is evidenced by the multilayer low angle Bragg reflections. Two compositions were investigated (65%Mo, 10ÅMo/8ÅGe and 45%Mo, 12ÅMo/21ÅGe). The scattering curves and differential structure factors of both are in good agreement with those from the amorphous alloys of their respective compositions. Radial distribution functions (RDF's) from data taken in transmission and reflection are similar for both compositions (shown in Fig. 6 for the 10ÅMo/8ÅGe sample.) The minor differences are currently being examined in more detail with anomalous scattering. The similarity of the amorphous scattering when probed both in the layer plane and perpendicular to the layers is evidence that the average local structure is isotropic. Anisotropy of average local structure might have been expected to result from the existence of layers or interfacial regions of limited extent on an atomic scale in the layering direction only. This study uncovered no experimental evidence for such anisotropy in the Mo-Ge system. The fact that the distribution functions of the alloys match well the distribution functions of the multilayers supports the conclusion that there are broad compositional ranges where average structure is only slowly changing[5]; thus the multilayers can be structurally homogeneous while compositionally modulated.

A sample of intermediate layer thickness (57%Mo, 18ÅMo/21ÅGe) was also examined. Like the thin layer samples just mentioned, multilayer composition modulation exists but the distribution functions are in close agreement with those of the Mo-Ge alloy of the overall multilayer composition. Thus, increasing the Ge concentration from that of the epitaxial system of 20ÅMo/11ÅGe not only destroys the crystallinity and epitaxy of the Ge but also prevents the Mo from being crystalline. It is evident that the local structure of each layer type is dependent on both its own thickness and the thickness of the neighboring layer. In general, thin layers (≤20Å) of both Mo and Ge exhibit the alloy structure but if Ge is decreased to 50% or less of the Mo thickness the tendency of Mo to crystallize results in an epitaxial structure.

In summary, a wide range of structure is exhibited by the Mo and Ge layers as thicknesses are varied (see fig. 1). Thick layers of Mo (>20Å) exhibit bulk BCC crystalline structure while thick layers of Ge (>30Å) exhibit 4-fold coordinated amorphous RTN structure. An amorphous Mo-Ge

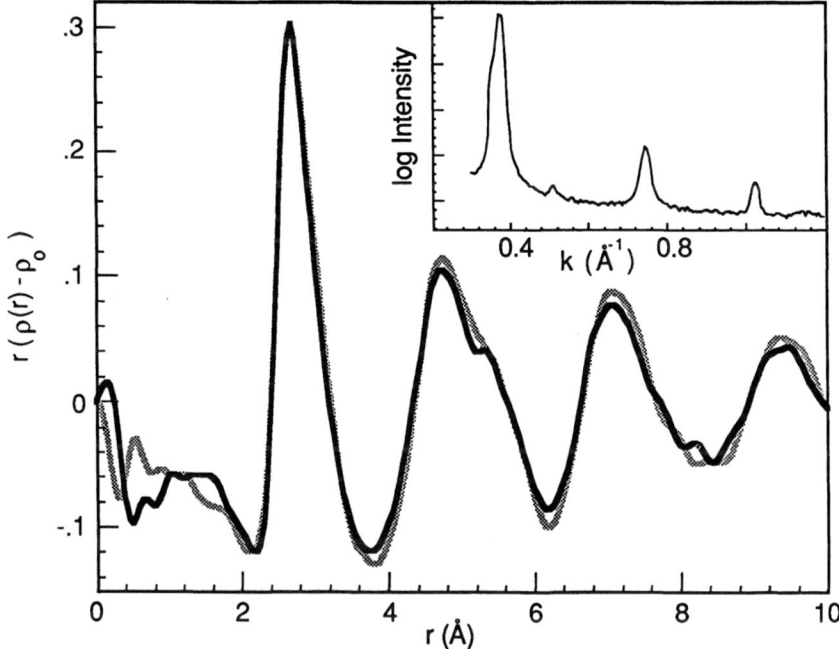

Fig. 6 ------ Radial Distribution Function [r(ρ(r)-ρ$_o$)] from data taken with the
scattering vector (**k**) perpendicular to the layer plane (dark line) and parallel to the
layer plane (light line). Result is from a sample of average χ_{Mo}=.65, nominal layer
thicknesses 10Å Mo/8Å Ge and measured bi-layer thickness 16.9Å. Local structure
of this sample is isotropic although multilayer low angle peaks exist (inset).

alloy interfacial region between the thick layers is suggested by the Ge edge DDF. As the Ge layer
is made thinner (≤10Å) while the Mo remains at least relatively thick (≥20Å), the Ge structure is
transformed into a novel form, a metallic, strained BCC structure coherent with the Mo BCC
structure. If, however, the Mo layer thickness is decreased (≤10Å) while the Ge layers remain
thick, the Mo is no longer crystalline but is incorporated in an interfacial amorphous Mo-Ge alloy
separating layers of a-Ge. If both layers are decreased (Mo ≤20Å, Ge ≤20Å), with the Ge
thickness equal to or greater than the Mo thickness, only an amorphous Mo-Ge alloy with a layered
concentration gradient remains. The amorphous alloys show striking similarities with
homogeneously sputtered samples on the scale of local atomic structure. Indeed, data taken with
the scattering vector in the layer plane and perpendicular to the layer plane indicate the same
distribution function. This demonstrates that the tendency to alloy is strong and bonds are similar
in all directions even though the concentration modulation is strong enough to yield multilayer
Bragg reflections. There is no evidence of an amorphous Mo structure of concentration greater
than 75% in any of the samples.

We are indebted to J. Keem for his involvement in the formulation of this research, to D.B. McWhan for pointing out to us that asymmetric satellite peaks confirm the existence of strained epitaxial layers, and to J.B. Kortright for access to his data on amorphous Mo-Ge alloys. In addition we thank M. Zdeblick for advice in etching Si as well as T.W. Barbee for discussions on fabrication of multilayers and etching procedures. K.F. Ludwig and R.D. Lorentz provided helpful discussions on anomalous scattering and C.A. Kilbourne and M.B. Rice assisted in data collection at SSRL.

This work was partially supported by the NSF-MRL program through the Center for Materials Research at Stanford University. It was partially supported by and performed at SSRL which is supported by the Department of Energy, Office of Basic Energy Sciences; and the National Institutes of Health, Biotechnology Resource Program, Division of Research Resources.

[1] T.W. Barbee and D.L. Keith, in Synthesis and Properties of Metastable Phases, edited by E.S. Machlin and T.J. Rowland, (AIME, 1980), p. 93.

[2] K.E. Petersen, Proceed IEEE 70, 420 (1982).

[3] P.H. Fuoss, P. Eisenberger, W.K. Warburton, and A. Bienenstock, Phys. Rev. Lett. 46, 1537 (1981).

[4] K.F. Ludwig, W.K. Warburton, L. Wilson, and A. Bienenstock, J. Chem. Phys. 87, 604 (1987).

[5] J.B. Kortright and A. Bienenstock, to be published, Phys. Rev. B, 15 Feb.(1988).

[6] D.B. McWhan, in Synthetic Modulated Structures, edited by L.L. Chang and B.C. Giessen (Academic Press, 1985), pp 43-74.

[7] F.A. Shunk, Constitution of Binary Alloys, Second Supplement, (McGraw-Hill Book Co., New York, 1969), p.388.

Structural Perfection of InGaAs/InP Superlattices
Grown By Gas Source Molecular Beam Epitaxy:
A High-Resolution X-Ray Diffraction Study

J. M. Vandenberg, M. B. Panish and R. A. Hamm
AT&T Bell Laboratories, Murray Hill, NJ 07974

ABSTRACT

High-resolution X-ray diffraction (HRXRD) studies have been carried out to determine the structural perfection and periodicity for a number of high-quality InGaAs/InP superlattices grown by gas source molecular beam epitaxy. X-ray scans were carried out with a compact four-crystal monochromator resulting in a resolution of one molecular layer (~3Å), which enables one to observe very small variations in the periodic structure. Sharp and strong higher-order satellite reflections in the XRD profiles were observed indicating smooth interfaces with well-defined modulated structures. Excellent computer simulated fits of the X-ray satellite pattern could be generated based on a kinematical XRD step model which assumes ideally sharp interfaces, and periodic structural parameters such as the strain in the well could be extracted. Our results[3] demonstrate that HRXRD in conjunction with the kinematical step model is a very sensitive method to assess periodic structural modifications in superlattices as a result of the precise growth conditions in the gas source MBE system.

INTRODUCTION

Recent advances in the gas source molecular beam epitaxy (GSMBE) have led to the growth of InGaAs/InP quantum well structures of very high crystal quality.[1] Superlattices (SL's) have been grown with very thin wells, sharp interfaces and well-defined periodic structures along with promising electronic and optical properties for device applications.[2,3] Characteristic for the GSMBE is its capability for excellent control of composition, lattice match and layer thickness, which makes it possible to grow very closely matched SL's as well as strained-layer superlattices (SLS's) having positive or negative strain. In previous work on lattice-matched InGaAs/InP SL's grown by GSMBE, high-resolution X-ray diffraction (HRXRD) has proven to be an effective tool to evaluate their structural integrity.[4,5] Further HRXRD studies have also enabled us to correlate strain in SLS's with the electronic band structure.[6,7] In essence the X-ray studies on GSMBE SL's revealed that kinematical simulation of HRXRD curves enables one to very accurately determine the periodic structure, including small structural modifications in the interfaces, with a precision of one molecular layer. The

Mat. Res. Soc. Symp. Proc. Vol. 103. ©1988 Materials Research Society

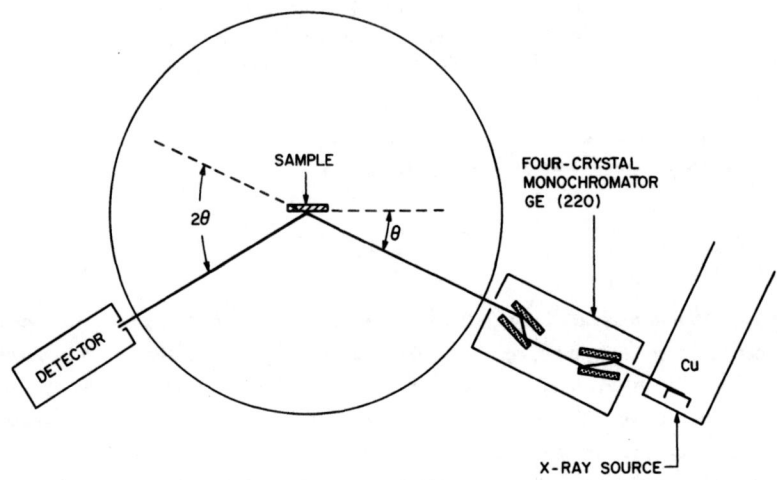

Figure 1.   Bartel's four-crystal monochromator concept.

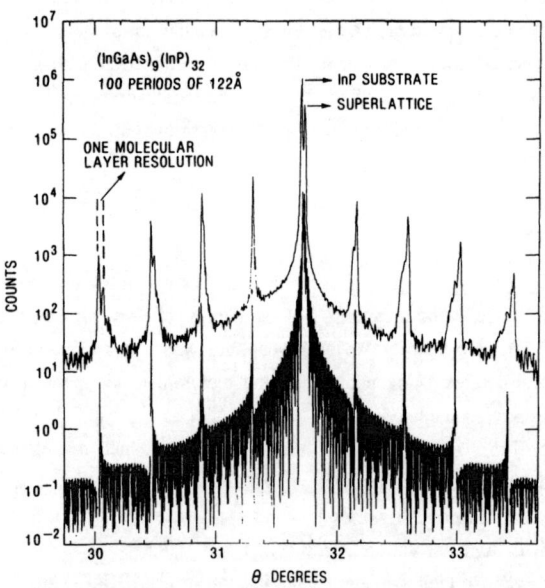

Figure 2.   X-ray scan of the (400) reflection of a closely matched 100-well InGaAs/InP superlattice with simulation of fit in lower part.

high quality of the GSMBE superlattices, in particular their interfacial sharpness, makes them very well suited for precision HRXRD analysis with kinematical simulation based on a step model which assumes alternating layers with ideally sharp interfaces. In the present study HRXRD has been carried out in order to determine the periodic structural parameters of a number of InGaAs/InP superlattices which were grown on [100]InP with various degrees of strain in the InGaAs well.

RESULTS AND DISCUSSION

HRXRD was achieved with a previously described X-ray geometry proposed by Bartels[8] and consisting of a compact four-crystal monochromator, the sample as the fifth crystal and an open-end detector (Fig. 1). The X-ray scan of the (400) reflection is shown in Fig. 2 for a nominally matched $(x = 0.53)$ 100-period 20Å $In_xGa_{1-x}As/100Å$ InP superlattice with a period $\Lambda = 122Å$, calculated from the position of the strongest satellite peaks according to the equation $\dfrac{2\sin\theta_n - 2\sin\theta_{SL}}{\lambda_{CuK_{\alpha1}}}$

$= \pm \dfrac{n}{\Lambda}$ where n represent the order number of the satellites around the main SL peak. The experimental scan shows splitting of the satellite peaks with differences in periodicity Λ of ~3Å, which corresponds to one molecular layer. This indicates that variations in the periodic structures as small as one molecular layer can be observed in HRXRD analysis. The lower trace in Fig. 2 shows the simulation of the (400) scan for the strongest satellite structure, using a kinematical step model. This model[9,10] permits calculation of the diffracted amplitude F of the structural periodicity of the superlattice along the [100] growth direction. The variable input parameters of the step model are the number of molecular layers N_W and N_B in the InGaAs well (W) and InP barrier (B) respectively, and their corresponding lattice spacings d_W and d_B. For alternating wells and barriers the diffracted amplitude F in the vicinity of the (400) reflection is given by:[9]

$$F(\vec{h}) = \sum_{N_{SL}} \exp 2\pi i h N_{SL} N \left[f_W \sum_{n=0}^{N_w-1} \exp[2\pi i h(n+1/2)d_w/d_{SL}] + \right.$$
$$\left. + f_B \exp[2\pi i h N_W d_W/d_{SL}] \times \sum_{n=0}^{N_B-1} \exp[2\pi i h(n+1/2)d_B/d_{SL}] \right] \qquad (1)$$

where f_W and f_B are the total scattering factors for the InGaAs well and the InP barrier respectively, calculated at the (400) diffraction angle θ, and N_{SL} is the number of periods Λ in the superlattice. An excellent fit was obtained with $N_W = 9$, $N_B = 32$ and $d_W = 1.4676Å$, assuming $d_B = 1.4671Å$ $(=a_{InP}/4)$ which demonstrates one can analyze these SL's with one molecular layer precision. Once the variable parameters are extracted from the step model, the thickness of the well and barrier can then be calculated from the equations $t_W = 2 d_W N_W$ and $t_B = 2 d_B N_B$, and the corresponding lattice parameters from $a_\perp^W = 4 d_w$ and $a_\perp^B = 4 d_B$. The strain in the well is then given by the equation $\varepsilon_\perp^W = \dfrac{a_\perp^W - a_\perp^B}{a_\perp^B}$, where $a_\perp^B = a_{InP} = 5.8687\text{Å}$. For the 100-period SL InGaAs/InP in Fig. 2 this strain is calculated to be $\varepsilon_\perp^W = + 0.03\%$.

Figure 3 shows the (400) X-ray scan of a perfectly matched Sn-doped 50-period 75Å InGaAs/68Å InP superlattice with extremely sharp single satellite peaks. This structure is an unusual one in which the quantum wells were doped with Sn and there was a very heavily ($\sim10^{14}$) doped InGaAs cap layer. This X-ray scan indicates that this SL was grown with a very stable beam flux resulting in one well-defined periodic structure of 141Å. The fact that the overall satellite intensities are fairly symmetric (+n vs. −n) around the main superlattice peak, for the 50-period SL (Fig. 3) as well as the 100-period SL (Fig. 2), gives evidence that there is very little overall strain between the well and the barrier. However the scan of Fig. 3 shows a noticeable asymmetry in the ±2° and ±4° order satellite intensities. A much better fit for those satellites could be obtained with a step model which assumes the presence of an

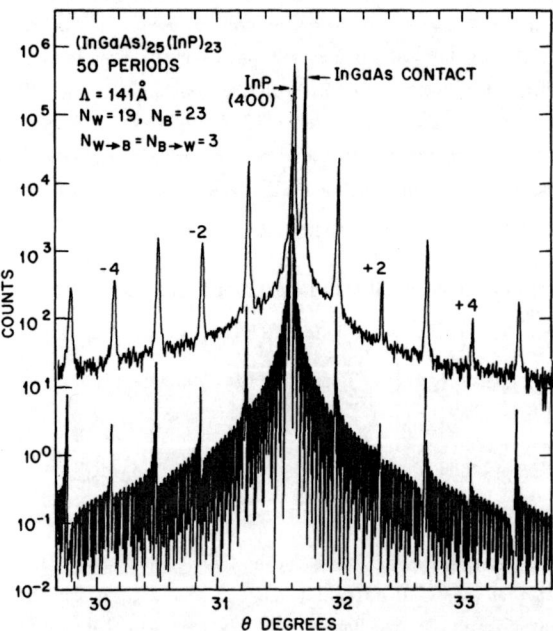

Figure 3. X-ray scan of the (400) reflection of a matched n-doped 50-well InGaAs/InP superlattice with the simulation in the lower part.

additional strained interfacial region, with a thickness of three molecular layers, on each side of each InGaAs well. This was achieved by adding two more terms to the diffracted amplitude F in equation (1), one for the well-to-barrier (W $\rightarrow$ B) and one for the barrier-to-well (B $\rightarrow$ W) interface. By applying this previously described modified step model, a good fit for 2° and 4° order satellites was now obtained for the number of molecular layers $N_{W \rightarrow B} = N_{B \rightarrow W} = 3$ and a corresponding strain $\varepsilon_{\perp}^{W \rightarrow B} = \varepsilon_{\perp}^{B \rightarrow W} = +0.11\%$ (Fig. 3 and Table I). The strain modulation may have been induced by the Sn-doping during growth of the InGaAs well and suggests accumulation of Sn at the well inferfaces. The traces in Fig. 2 and 3 are two illustrative examples where HRXRD analysis provides detail of structure that may be related to the GSMBE growth conditions.

Kinematical simulation of HRXRD curves can also be successfully applied to strained-layer superlattices. This method appears to be very useful in order to quantitatively determine the strain with the aim of exploring the relationship between strain and the optical and electronic properties of the quantum wells.[6,7] With the thickness of the well small enough to accommodate the mismatch strain coherently, it is possible to grow either positively or negatively strained $In_xGa_{1-x}As/InP$ SLS's by GSMBE spanning a wide range of composition x. In the X-ray scan of a 10-period 75Å $In_{.43}Ga_{.57}As/450Å$ InP SLS (Fig. 4a) the asymmetric shape of the satellite structures with stronger intensities for positive order numbers (+n) indicates a negative strain in the well. The presence of very sharp satellites demonstrate that high-quality SLS's can be grown with great precision. This is further confirmed by the excellent agreement between the simulated and experimental scan. The structural parameters extracted from the kinematical step model are $N_W = 26$, $N_B = 154$ and the strain in

Table I. Structural parameters of InGaAs/InP superlattices grown by GSMBE.

	Λ	$\Delta a_{\perp}/a_{\perp}$	N_W	N_B	$\varepsilon_{\perp}^{W}$	t_W	t_B
	(Å)	(%)			(%)	(Å)	(Å)
100-well	122	−0.09	9	32	+0.03	27	94
50-well	141	0	19	23	0	65	76
			$N_{W \rightarrow B} =$		$\varepsilon_{\perp}^{W \rightarrow B} =$		
			$N_{B \rightarrow W} = 3$		$\varepsilon_{\perp}^{B \rightarrow W} = +0.11$		
10-well ($x_{In}=0.43$)	528	−0.2	26	154	−1.4	75	452
10-well ($x_{In}=0.13$)	316	−0.3	7	101	−5.0	20	296

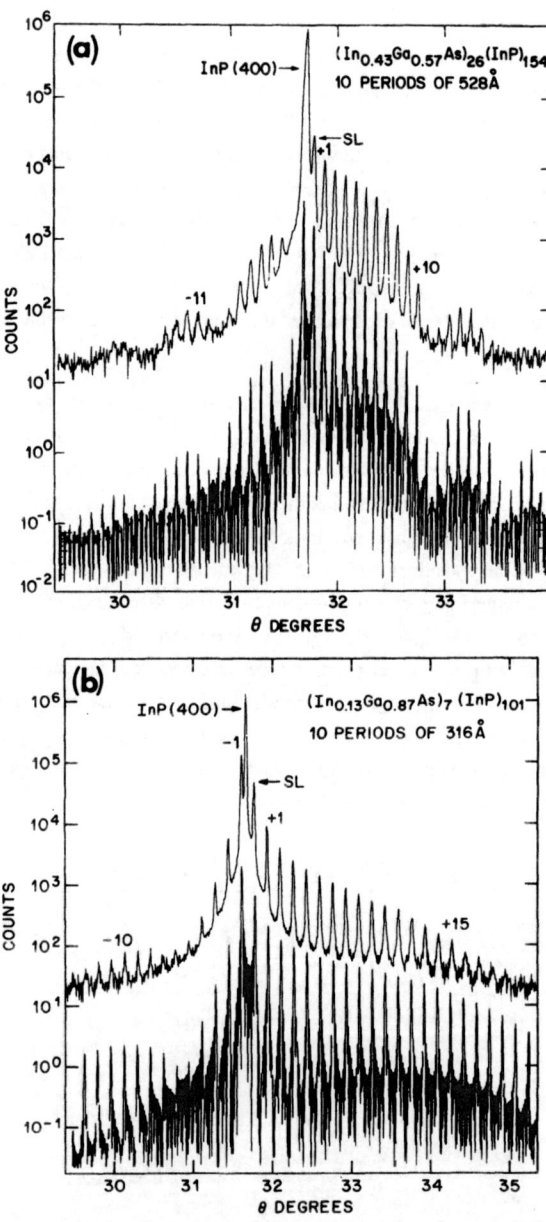

Figure 4.   X-ray scan of the (400) reflection of a 10-well strained-layer $In_xGa_{1-x}As/InP$ superlattice part: (a) $x_{In} = 0.43$, (b) $x_{In} = 0.13$; simulation of fit in bottom part.

the well $\varepsilon_\perp^W = -1.4\%$ (Fig. 4 and Table I).

This kind of structural perfection was also observed in the HRXRD of a highly strained 10-period 20Å $In_{.13}Ga_{.87}As/300Å$ InP SLS (Fig. 4b). The very strong asymmetry in the satellite pattern reveals an even larger negative strain ($\varepsilon_L^W = -5.0\%$) as a result of the much lower In concentration in the well. Despite the large lattice mismatch (Table I), this SLS retains its structural integrity, as judged by the sharpness and intensity of the satellites and the excellent simulated fit.

CONCLUSION

Our results demonstrate that HRXRD curves of GSMBE superlattices can be computer simulated in great detail by using a kinematical step model which assumes a series of alternating layers with ideally sharp interfaces. The successful application of this X-ray method is strongly enhanced by the fact that GSMBE superlattices can be grown with extreme precision and a large degree of structural perfection resulting in a very accurate assessment of the periodic structure of closely-matched as well as strained-layer superlattices.

REFERENCES

1. M. B. Panish, H. Temkin and S. Sumski, J. Vac. Sci. Tech. B3: 657 (1985).

2. H. Temkin, M. B. Panish and S. N. G. Chu, Appl. Phys. Lett. 49: 859 (1986).

3. H. Temkin, D. Gershoni and M. B. Panish, Appl. Phys. Lett. 50: 1776 (1987).

4. J. M. Vandenberg, S. N. G. Chu, R. A. Hamm, M. B. Panish and H. Temkin, Appl. Phys. Lett. 49: 1305 (1986).

5. J. M. Vandenberg, R. A. Hamm, M. B. Panish and H. Temkin, J. Appl. Phys. 62, 1278 (1987).

6. D. Gershoni, J. M. Vandenberg, R. A. Hamm, H. Temkin and M. B. Panish, Phys. Rev. B, 36: 1320 (1987).

7. D. Gershoni, H. Temkin, J. M. Vandenberg, S. N. G. Chu, R. A. Hamm and M. B. Panish, unpublished.

8. W. J. Bartels, J. Vac. Sci. Technol. B1: 338 (1983).

9. A. Segmüller and A. E. Blakeslee, J. Appl. Cryst. 6: 19 (1973).

10. D. B. McWhan, M. Gurvitch, J. M. Rowell and L. R. Walker, J. Appl. Phys. 54: 3886 (1983).

ATOMIC LEVEL INVESTIGATIONS OF W-Si MULTILAYERS BY HIGH
RESOLUTION TEM AND X RAY SCATTERING

S.R. Nutt* and J.E. Keem**
*Brown University, Providence, RI 02910
**Ovonic Synthetic Materials Co., Inc., 1788 Northwood Drive,
Troy, MI 48084

ABSTRACT

We have prepared multilayer films of W-Si with bilayer
repeat spacing from approximately 1.5 nm to 9 nm and performed
high resolution electron microscopy and low angle x-ray
scattering on them. Average composition estimates as inferred
from deposition conditions, x ray scattering and electron
microscopy are compared. Determinations of the individual layer
thickness ratios by electron microscopy and x ray scattering
vary significantly from expectations as the bilayer thickness
approaches 1.5 nm. Layer intermixing to increase as the bilayer
thickness decreases. Composition profiles as inferred from the
Cuk x ray profile are compared to those inferred from the high
resolution electron micrographs. Visual observations from
electron microscopy are presented indicating that the interface
roughness is rapidly damped in the W-Si multilayer system.
Estimates of the layer uniformity are made from the high
resolution images.

INTRODUCTION

Synthetic multilayered structures [1,2] with bi-layer
thicknesses from 1.5 nm to 20 nm are used extensively as
quasi-Bragg reflector coatings on optical elements functioning
in the soft x ray and extreme ultraviolet regions of the
electromagnetic spectrum [3]. Similar structures are used in
the investigation of atomic diffusion on very short length and
time scales [4] and in the study of magnetism [5] and electronic
transport [6]. In general, the x ray reflecting structures are
composed of alternating non-crystalline or partially crystalline
layers of a high Z and low Z materials. The studies of
diffusion, magnetism and electronic properties are less
restricted in the amount of crystallinity and Z contrast.
The average composition profile and its variations are the
essential physical parameters describing multilayer structures.
These parameters have been effectively studied by high
resolution transmission electron microscopy [7,8] and x ray
scattering [1,2,3]. Estimates of the composition profile and
its variations are obtained by interpretation of the
photographic plates obtained from high resolution electron
microscopy [8] and are based primarily on observations of very
restricted volumes of the multilayer. The x ray scattering data
are collected from almost the entire volume of the multilayer
and estimates of the composition profile and its variations are
generally best obtained from fitting the low angle Cuk x ray
scattering profile to the optical multilayer model using the
profile and interface thickness as variables [9].
In this work both techniques have been used to estimate the
composition profile and its variations for four multilayers of
W-Si. The comparison was made on very regular multilayer

Mat. Res. Soc. Symp. Proc. Vol. 103. ©1988 Materials Research Society

specimens as well as those chosen to have certain types of
structural irregularities like slow bi-layer thickness
variation, oscillation of bi-layer thickness and diffuse
interfaces. These features in variations of the composition
profile are identified by both techniques. Significant
discrepancies are observed in the quantification of the average
profile and its variations, especially in the estimates for
bi-layer thickness near 1.5 nm. Evidence supports the results
of the x-ray scattering as being more reliable in the small
bi-layer limit. One additional multilayer was investigated by
high resolution electron microscopy alone to obtain evidence for
the conjecture that irregularities in the layers are smoothed
away during the synthesis process.

EXPERIMENTAL TECHNIQUE

The W-Si multilayers were prepared by magnetron sputtering
on the (001) surface of super polished, integrated circuit grade
silicon wafers. The target power and substrate rotation period
were varied in order to obtain different thicknesses of the
individual layers. A large number of multilayers were culled to
find those with striking structural irregularities.
Cross-sectional transmission electron microscopy specimens
were prepared following the method described by Bravman and
Sinclair [10]. Final thinning was done by ion beam milling.
Total milling time was one to two hours. To minimize the
possibility of structural damage to the multilayers the
specimens were mounted on a liquid nitrogen cooled milling
holder.
High resolution images were obtained on a JEM 400EX
electron microscope operated at 400kv. This microscope has a
point to point resolution of approximately 0.17 nm. The images
were made close to optimum defocus conditions to obtain accurate
determinations of the layer thickness [11]. Slightly lower
magnification images were obtained on 100kv electron
microscope. Spurious variations and reductions of layer
thicknesses may be observed if the specimen droops or is
otherwise not correctly oriented to the electron beam. Special
care was taken to align the specimens so that the growth
direction was exactly perpendicular to the electron beam. This
was achieved by aligning the diffraction patterns from the
silicon substrates assuming that the multilayer growth direction
was parallel to the Si (001) direction [8].
The x ray diffraction measurements were made on a Norelco
Powder diffractometer using Cuk (0.154 nm wavelength)
radiation. The region of total internal reflection and critical
angle as well as the first order reflections were measured using
a matched pair of 1/30 degree divergence and receiving slits.
This allows accurate determination of the critical angle and
measurement of fine structure on the first order diffraction
peak. Higher order reflections, including the Bragg angle of
the highest discernible order of reflection, were measured with
broader (1 degree) beam dimensions. Those reflections which
were suppressed in intensity with respect to their neighboring
orders were also noted.

RESULTS

XRO#1214-16

Figure 1. shows a high resolution image of the first four bi-layers deposited on the silicon substrate. The first layer is W and is unusual because it shows incipient crystallization. Subsequent W and Si layers are completely amorphous. Modulation from 100% W to 100% Si occurs in each bi-layer. The interfaces between the layers are smooth and slightly diffuse, approximately 0.5 nm to 0.6 nm, or about two atomic diameters, in width. Examination of the entire film shows a very slow monotonic decrease in the bi-layer thickness. As was mentioned above, special care was taken to align the specimen perpendicular to the electron beam to minimize the possibility of observation of foreshortening of the layer thicknesses due to sample curvature. The estimated bi-layer thickness near the substrate is 9.2 nm and the layer thickness ratio, W/Si was 2/5.

The critical angle θ_c was determined to be 0.319 degrees. The highest order of reflection observed was the (0017) at approximately 17 degrees 2θ. The (005), (0010) ands (014) reflections were systematically suppressed in this specimen. The fwhm of the first order peak was approximately 0.16 degrees 2θ. The first order peak showed evidence of the asymmetry associated with diffraction from dynamical systems. The higher order reflections showed significant broadening, but with no observation of splitting.

XRO#1130-1

The high resolution image in Figure 2 shows the first few layers deposited on the wafer substrate. The first layer is unusually thick, and probably contains native oxides of silicon and/or absorbed contaminants. All subsequent Si layers appear to be the same thickness but thinner than the first. The first W layer forms a sharp interface and shows clear evidence of incipient crystallization. Subsequent W layers are completely amorphous. The interfaces formed after the first W-Si interface although smooth, are not as sharp as the first interface. As in the 9.2 nm multilayer, complete composition modulation is observed in this specimen. The interfaces appear to be about 0.5 nm, or approximately two monolayers wide. There is very little difference between the images of the W-Si and the Si-W interfaces in this multilayer. The bi-layer spacing was 5.15 nm and the layer thickness ratio, W/Si was 9/14.

The critical angle θ_c was determined to be 0.342 degrees. The highest order of reflection observed was the 010 at approximately 15 degrees 2θ. The 004 and 008 reflections were systematically suppressed in this specimen. The fwhm of the first order peak was approximately 0.15 degrees 2θ. The first order peak showed evidence of the asymmetry associated with diffraction from dynamical systems. None of the reflections showed significant distortions.

XRO#545-4

This specimen, shown in Figure 3, exhibits unusual roughness in the first deposited layers. These irregularities are rapidly damped (or smoothed) out in the following 3 - 4 layers. The roughness was apparently a localized phenomenon and was only observed in some regions of the specimen. The bi-layer spacing was 2.9 nm and the layer thickness ratio, W/Si was 5/6.

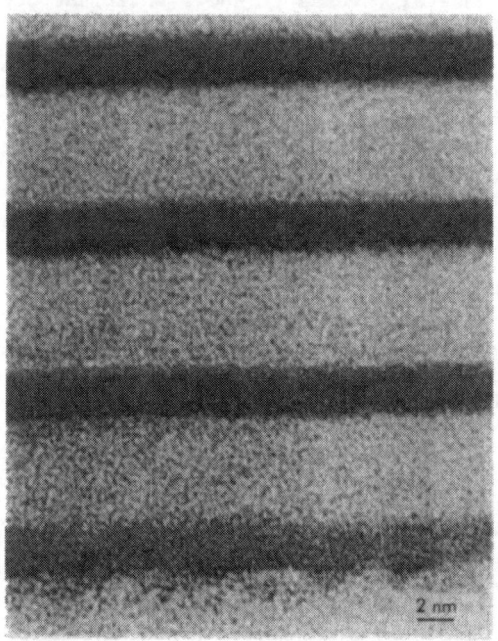

Figure 1.

Figure 2.

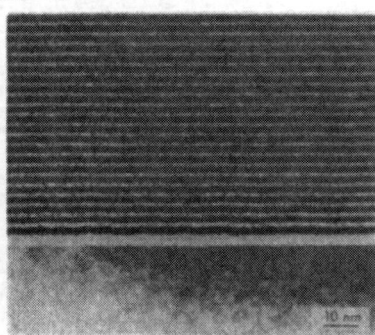

Figure 3.

XRO#702-16

The image in Figure 4 clearly indicates a repeating
variation in the bi-layer spacing of the multilayer. Note the
three broad dark bands at the top of the micrograph. A portion
of the lower part of the image has been cut out, inverted, and
overlaid on the top half of the image. The three arrows show
points of mismatching and matching, highlighting the local
variations in bi-layer thickness. The repeat spacing of this
larger period is approximately 20 bi-layer spacings.
Densitometer traces were performed across the negative parallel
to the growth direction. The variation in bi-layer spacing was
roughly sinusoidal in nature and appeared to regularly exceed
the nominal bi-layer spacing of 1.87 nm but rarely fell below
this value. It appeared that this systematic bi-layer thickness
variation was caused by increased thickness of the W layers. In
regions of constant bi-layer thickness it had a value of 1.87 nm
and the layer thickness ratio, W/Si was 3/2.
    The critical angle θ_c was determined to be 0.327
degrees. The highest order of reflection observed was the (004)
at approximately 17.5 degrees 2θ. There were no
systematically suppressed orders observed in this specimen. The
fwhm of the first order peak was approximately 0.08 degrees
2θ. The first order peak was symmetrically shaped and had
symmetrically located side bands spaced at about 0.12 degrees
2θ.

XRO#682-10

The shortest period multilayer structure examined consists
of layers only a few atomic diameters thick. The electron
micrograph of this structure is shown in Figure 5. Unlike the
other specimens, the composition profile for this specimen may
not reach pure W or pure Si, but oscillate between W-rich and
Si-rich compositions in a sinusoidal manner. No crystallinity
is observed, and Si layers are barely continuous because of W
intermixing. The first two layers in this structure are
perturbed from the nominal value to approximately twice
nominal. The apparent bi-layer thickness is 1.48 nm and the
apparent layer thickness ratio, W/Si is 7/3. The diffuse nature
of the interface boundaries make this a difficult parameter to
estimate.
    The critical angle θ_c was determined to be 0.361
degrees. The highest order reflection observed was the (003)
reflection at approximately 17.5 degrees 2θ. No reflections
were observed to be systematically suppressed. The first order
peak was symmetric. The fwhm of the reflection was less than
0.05 degrees 2θ. Unidentifiable structure on the high angle
side of the first order peak was observed.

 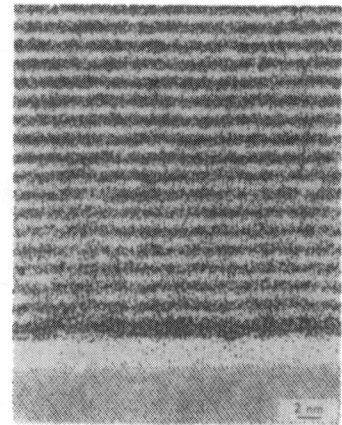

Figure 4.   Figure 5.

DISCUSSION

As illustrated in Table I, the qualitative features of the composition profiles as determined by either technique are completely consistent (see modeling results in ref 12).

Table I - Comparison of Qualitative Features of Composition Profile Determinations

Sample	TEM	X Ray
1214-16	1. Slowly varying bi-layer spacing	1. Dynamical scattering profile maintained. 2. Broadened high order diffraction lines.
1130-1	1. High degree of perfection	1. Dynamical scattering profile for low order lines. 2. Higher order lines only weakly perturbed
702-16	1. Extra periodicity (longer than fundamental period)	1. Sidebands on diffraction lines observed [12].
625-16	1. Interdiffused layers - some spacing irregularity. 2. Bi-layer spacing approaching interface thickness.	1. Loss of dynamical scattering profile. 2. Appearance of unidentifiable low intensity structures in the scattering profile.

Using the relationship between the critical angle and the
real part of the dielectric constant of the material, the
optical constants from the literature for W and Si at Cuk one
can obtain the ratio of W to Si in the multilayer. From this
the relative thicknesses of the W and Si layers can be
calculated [13]. The systematically suppressed orders may also
be used to estimate the W/Si thickness ratio using the
relationship derived from fourier analysis [9,13]. Finally, the
d spacing (bi-layer spacing) may be determined from the Bragg
equation applied to the higher orders of reflection of the
multilayers [12]. The scattering angle at which Bragg
reflection vanishes can be used to estimate the smallest length
in the system which contributes to Bragg scattering. We use
this length as an estimate for the interfacial roughness.

Table II is a comparison of the estimates of the bi-layer
spacing, layer thickness ratios and the interfacial roughness
made by electron microscopy and x ray scattering. The x ray
scattering estimates are made by using Bragg's Law for the
bi-layer thickness estimates and critical angle measurements and
the order suppression method to obtain the other estimates. The
more ambitious project of fitting the x ray traces to the
Optical Multilayer Model with interfaces [9] and then extracting
a calculated composition profile to compare to the electron
microscopy was a failure, details of which will be reported in
the future.

Table II - Comparison of Composition Profile Parameter
Estimates

Sample		1214-16	1130-1	702-16	682-10
Bi-layer	TEM	9.20 nm	5.15 nm	1.87 nm	1.48 nm
thickness	X ray	9.03 nm	5.7 nm	2.02 nm	1.50 nm
W thickness	TEM	40%	39%	60%	70%
% bi-layer	Crit.	17%	24%	21%	29%
	Order	16%	20%	NA	NA
Si thickness	TEM	60%	60%	40%	30%
% bi-layer	Crit.	83%	76%	79%	71%
	Order	84%	80%	NA	NA
Interface	TEM	0.55 nm	0.5 nm	NA	NA
thickness	Order	0.52 nm	0.59 nm	0.5 nm	0.5 nm

In terms of the profile parameters which we have access to,
the bi-layer thickness estimates by either x ray or electron
microscopy are in good agreement. There are significant
differences in estimates of the individual layer thicknesses by
these two techniques. (This lack of agreement between the coarse
estimates of the profile parameters dooms more sophisticated
comparisons using the Optical Multilayer Model with interfaces.)
However, there is good agreement in the estimates of the
interface thickness.

CONCLUSIONS

1. Optical Multilayer Models with a single interface can not quantitatively describe the x ray scattering profiles from W-Si multilayer structures

2. High resolution electron microscopy does not provide accurate information about composition profiles of small bi-layer spacing structures.

3. Good agreement between estimates of bilayer thicknesses directs attention to investigation of the interface profiles as key to achieving success at fundamental chaacterization of these multilayer materials.

REFERENCES

1. E. Spiller, AIP Conf. Proc. 75, 12 (1981).

2. T.W. Barbee, Proc. Soc. Photo-opt. Insturm. Eng. 563, 2 (1985).

3. G.F. Marshall, editor, Applications of Thin Film Multilayered Structures to Figured X Ray Optics, Proc. SPIE 563 (1985), papers and references therein.

4. L. Greer and F. Spaepen, Synthetic Modulated Structure Materials, edited by L.L. Chang and B.C. Giessen (Academic Press, New York, 1985) p. 419.

5. M.B. Stearns, J. Appl. Phys. 55,1729 (1984).

6. C.M. Falco and I.M. Schuller, Synthetic Modulated Structure Materials, edited by L.L. Chang and B.C. Giessen (Academic Press, New York, 1985) and references therin.

7. Y. Lepetre and G. Rasigni, Optics Lett. 9, 433 (1984).

8. Amanda K. Petford-Long, Mary Beth Stearns C.H. Chang, S.R. Nutt, D.G. Stearns, N.M. Ceglio, and A.M. Hawryluk, J. Appl. Phys. 61, 1422 (1987).

9. A.M. Kadin and J.E. Keem, Scripta Met., 20, 443 (1986).

10. J.C. Bravman and R. Sinclair, J. Elec. Mic. Tech. 1, 53 (1984).

11. C.S. Baxter and W.M. Stobbs, Ultramicroscopy 16, 213 (1985).

12. J.L. Wood, N. J. Grupido, K.L. Hart, S.A. Flessa, A.M. Kadin, J.E. Keem, and D.H. Ferris, Applications of Thin-Film Multilayered Structures to Figured X Ray Optics, Proc. SPIE, 563, p. 238 (1985).

13. J.P. Chauvineau, J. Corno, D. Decanini, L. Nevot, and B. Pardo, Applications of Thin-Film Multilayered Structures to Figured X Ray Optics, Proc. SPIE, 563, p. 245 (1985).

TUNGSTEN-CARBON MULTILAYER SYSTEM STUDIED WITH X-RAY SCATTERING

J.B. KORTRIGHT AND J.D. DENLINGER
Center for X-ray Optics, Lawrence Berkeley Laboratory, University of
California, Berkeley, California 94720

ABSTRACT

The tungsten-carbon multilayer system has been studied using x-ray
scattering techniques as a function of multilayer period and annealing.
Grazing incidence x-ray scattering shows that interatomic arrangements in the
W-rich layers depend strongly on their proximity to the interface with the C-
rich layers. Amorphous interface layers are stabilized by intermixing of W
and C during deposition into local structures similar to those in the tungsten
carbides. Further intermixing and structural relaxation in the amorphous
state occurs on annealing these structures. This interatomic structural
interpretation is consistent with trends in the observed x-ray optical
properties.

INTRODUCTION

The tungsten-carbon multilayer system was one of the first to demonstrate
utility as a robust Bragg reflector of x-rays [1], and continues to find new
and varied applications as an x-ray optical element. This utility results
from the ability to form extremely thin layers of these constituent materials
with significant uniformity and stability. Detailed knowledge of the
interatomic arrangements of these ultra-thin layered systems, and thermal
stability of these interatomic arrangements, has been lacking.

This paper reports on a study undertaken to learn details of these
interatomic arrangements and their thermal stability, and to correlate these
interatomic arrangements with the multilayers' x-ray optical performance. In
addition to this structural interest motivated by x-ray optical properties,
thin tungsten and carbon films are of interest for protective thin-film
coating applications. X-ray scattering techniques formed the basis of this
experimental study. Multilayer Bragg scattering was measured as was the in-
plane scattering from interatomic correlations in the grazing-incidence
scattering geometry.

EXPERIMENTAL

Multilayer samples were prepared by magnetron sputtering onto substrates
floating in temperature and rotating in turn over two elemental targets. One
multilayer period was deposited during a single rotation of the substrates.
Substrates were highly polished optical flats 1 inch in diameter. The samples
discussed here were deposited to have different multilayer periods d, with the
same nominal relative amounts of W and C, by keeping the W and C sputtering
rates fixed and varying only the rotation velocity to change d for the
different samples. Based on the sputtering rates of the individual elements,
the W layers would make up roughly 0.4 of the period of the various
multilayers, a typical relative thickness for x-ray optical applications. The
total thickness of multilayer samples ranged from 0.5 to 1.5 μm. One
multilayer was annealed at 10^{-6} torr for 18 hours at 420°C.

X-ray measurements were of several types. A two-crystal, Bragg geometry
utilizing Cu $K\alpha_1$ radiation was used to measure the absolute reflectivity
spectrum of low-angle multilayer Bragg peaks with the multilayer in the
position of the second crystal. Large-period samples showed as many as 19
orders of multilayer Bragg scattering. The grazing-incidence scattering (GIS)

geometry was adopted to measure scattering at a larger range of scattering vector, resulting from interatomic correlations. GIS measurements were made at the Stanford Synchrotron Radiation Laboratory on beamline 6 utilizing 1.2398 Å radiation with an incidence angle with respect to the sample surface of about 0.7°, which is slightly above the critical angle for total external reflection. The GIS geometry and these small incidence angles allow for measurement of scattering from the thin films of interest without measuring scattering from the substrate. One sample was studied using the Seeman-Bohlin geometry.

RESULTS AND DISCUSSION

X-ray Optical Performance

By measuring several orders of Bragg reflectivity for each sample, multilayer periods of as-prepared samples were determined as 108.8 Å, 59.4 Å, 37.0 Å, 19.0 Å and 8.3 Å. First order multilayer Bragg peaks for the samples are collectively shown in Figure 1. The incident beam and total reflection regions are shown in addition to the first order peak for the sample with period d = 108.8 Å. Two general trends of W/C multilayer x-ray optical properties are demonstrated by these data.

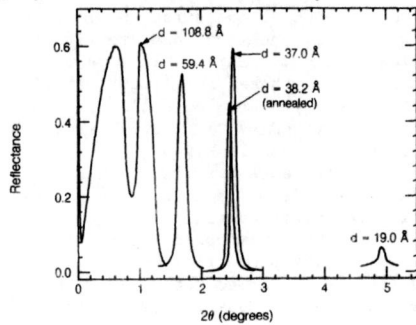

Figure 1. First order Bragg peaks multilayers with different periods.

First, significant peak reflectance values drop sharply as multilayer periods decrease below roughly 20-25 Å. Reflectance values in excess of 0.5 are typically observed for samples with periods larger than these values. The peak reflectance for the 19.0 Å period multilayer in this case is roughly 0.06 [2]. Not shown in Fig. 1 is the first order peak from a sample with period 8.3 Å. For this sample the first order peak reflectivity was of order 10^{-5}. For practical purposes, composition modulation in this sample is negligible. Calculations of multilayer first order of peak reflectance, assuming ideal models with layers consisting of homogeneous, elemental materials with compositionally sharp and smooth interfaces, predict that first order reflectance values remain high with decreasing period. The results presented below are consistent with the hypothesis that decreasing reflectivity with period is associated with intermixing of discrete atoms at the interfaces.

Another trend demonstrated in Fig. 1 is the observation that the period of W/C multilayers expands on annealing. This phenomenon appears to be a general feature in this system [3,4]. In the present case, the sample with period 37.0 Å expanded during annealing to d = 38.2 Å, causing the first order peak to shift by a significant fraction of its width. In this case the multilayer peak reflectivity shows a decrease of 23 percent on annealing. Decreasing peak reflectances are not always observed on annealing this multilayer system under similar conditions; sometimes peak reflectivity remains essentially unchanged or even increases [3,5]. Annealing temperatures of roughly 600°C must be reached before peak reflectances of W/C multilayer systems show catastrophic decreases, signalling gross departures from the layered microstructure of the as-deposited samples.

Grazing-incidence Scattering

Grazing-incidence x-ray scattering was used to study the same multilayer samples to understand interatomic arrangements in these systems and to relate this understanding with trends in multilayer optical performance with period and annealing. In this geometry [6], the scattering vector is predominantly in the plane of the multilayer thin film, so that scattering from in-plane interatomic correlations is measured. The atomic scattering factor describing the scattering amplitude is roughly an order of magnitude greater for W than for C atoms. Thus, W-W correlations scatter with roughly an order of magnitude greater intensity than W-C correlations, which in turn scatter with greater intensity than C-C correlations. Radial distributions obtained from the measured intensity thus have contributions primarily from correlations involving W, and we learn primarily of the W environment in the W-rich layers of these samples with this technique.

Figure 2a shows the GIS intensity for this set of multilayers. These scattering patterns show that for samples with d ≤ 60 Å, interatomic arrangements appear to be entirely amorphous. The sample with d = 108.8 Å shows polycrystalline diffraction peaks resulting from W in its equilibrium, body-centered cubic phase. The observed polycrystalline peaks, together with Seeman-Bohlin geometry diffraction results from this sample, indicate that the bcc W is not highly textured. Utilizing the Scherrer equation for particle-size broadening [7], an estimate of the grain size of the bcc W in the plane of the W-rich layers of 170 Å is obtained from the GIS data. Broadening of these peaks in the Seeman-Bohlin geometry, where the scattering vector is directed more along the sample normal than in the GIS geometry, results in an estimate of the grain size normal to the layers in rough agreement with the nominal thickness of the W-rich layers in this sample. The absence of crystalline peaks associated with C suggests that the C-rich layers of all

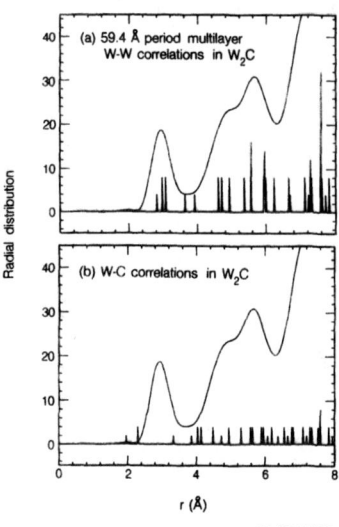
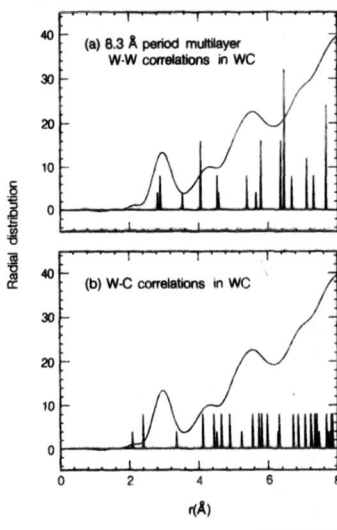

Figure 2. Grazing-incidence scattered intensity from in-plane interatomic correlations in multilayers with different periods is shown in (a). The vertical line at k = 2.8 Å^{-1} is positioned at the bcc W 110 peak as a guide to the eye. Radial distribution functions for multilayers with amorphous interatomic arrangements and different periods are shown in (b).

samples are amorphous.

Several inferences about the nature of the interfaces between W- and C-rich layers in this multilayer system can be drawn by observing the trends in the scattering patterns in Fig. 2a with period. In addition to the polycrystalline bcc W in the W-rich layers, a diffuse amorphous scattering pattern makes a significant contribution to the scattered intensity in the 108.8 Å period sample. Considering that the W-rich layers of smaller period samples appear entirely amorphous while the 108.8 Å period sample exhibits both amorphous and polycrystalline components, we conclude that the polycrystalline bcc W exists at the center of the thin W-rich layers and that the interface regions between W- and C-rich layers exhibits the amorphous scattering pattern. Evidently, amorphous interatomic arrangements are stabilized at the sputter-deposited interfaces between W and C, and bcc W does not form until some critical thickness is exceeded. Given that bcc W appears at a period between 59.4 and 108.8 Å, and that the W-rich layers are nominally 0.4 of the period, this critical thickness for bcc W formation must be at least 10-15 Å. The interfaces thus strongly influence local atomic arrangements over significant distances in the W/C multilayer system. Possible differences in W-on-C and C-on-W interfaces cannot be distinguished from these data, though could be significant.

Subtle changes are seen in the amorphous scattering patterns for samples with varying period in Fig. 2a. The amorphous interface regions of the W-rich layers of the d - 108.8 Å sample have an amorphous scattering profile most closely matching the d - 59.4 Å sample. Samples with decreasing periods show systematic changes in peak positions and shapes. Interpretation of these subtle differences is more easily accomplished by studying the radial distribution functions obtained by Fourier transformation of the normalized and background-corrected scattered intensity patterns.

Figure 2b shows radial distribution functions (RDFs) obtained from the amorphous scattering patterns in Fig. 2a. Trends in peak shape and position are observed with period and annealing, especially in the intermediate-range order between roughly 4 and 6 Å. Interpretation of the changing interatomic

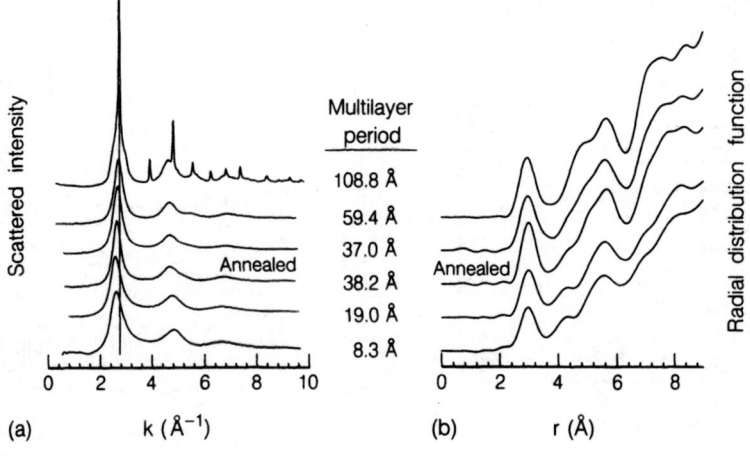

(a) $k\ (\mathring{A}^{-1})$

(b) $r\ (\mathring{A})$

XBL 873-8947A

Figure 3. Comparison of RDF for d - 8.3 Å sample with radial correlations in the compound WC.

Figure 4. Comparison of RDF for d - 59.4 Å sample with radial correlations in the compound W_2C.

arrangements for samples with d ≤ 60 Å is aided by comparison with radial distributions of W-W and W-C correlations in the tungsten carbides W_2C and WC. These comparisons are, in part, made in Figures 3 and 4, where the RDFs of the d = 8.3 Å and d = 59.4 Å samples are compared with the carbides WC and W_2C respectively. In each figure, (a) shows the radial distribution of W-W correlations and (b) shows that of W-C correlations in the carbides compared to the RDF of the multilayer sample. In these comparisons, the W-W and W-C correlations for the carbides give the number of atoms at a given distance. In the RDFs, W-W correlations are weighted more heavily than W-C correlations, as previously discussed.

Figure 3 shows the RDF for the 8.3 Å period sample compared to the stronger scattering correlations in WC. A small shoulder in the RDF at 2.2 Å coincides with the short W-C first neighbor distances in WC. The peak in the RDF at 4.2 Å corresponds closely to a strong second neighbor W-W feature in WC. Because of the coincidence of these and other features for both the amorphous sample and the compound WC, and because there is closer similarity of these features to correlations in WC than in W_2C, we conclude that the short-range interatomic arrangements in the 8.3 Å period sample are, on average, similar to those in WC. This is consistent with the known small composition modulation of this sample, and is direct evidence for significant intermixing between W and C.

Figure 4 shows the RDF for the 59.4 Å period sample compared to the correlations in W_2C. The features in the RDF are similar to those of typical transition metal-metalloid amorphous alloys, and are better described by the correlations in W_2C than those in WC. Note in particular the close similarity of the first distances and the shape of second neighbor distribution. The close match of the RDF and the correlations in W_2C suggests that intermixing of W and C into local structures as in this carbide provides a reasonable first order description of the W-rich layers in this sample.

Given this interpretation of the 8.3 and 59.4 Å period RDFs, the trends in Fig. 2 are interpreted as resulting from intermixing of W and C in amorphous interface layers with local structures similar to those in the carbides. As the multilayer period decreases, and with annealing, local atomic arrangements in the W-rich layers tend toward those in the more C-rich carbide WC. The similarity of the amorphous part of the d = 108.8 Å scattering pattern with that of the d = 59.4 Å sample suggests that the amorphous interfaces for the large period sample are stabilized by intermixing similar to that in W_2C. The 37.0 Å period sample clearly has features intermediate between those identified for the 59.4 and 8.3 Å period samples. Presumably, more W-rich amorphous arrangements exist in the center of the W-rich layer, with local structures more similar to those in WC in an interface region between the C-rich layers. The d = 19.0 Å RDF is nearly identical to that of the d = 8.3 Å sample, which we take as a model for a nearly homogeneous amorphous mixture of W and C. Thus, significant intermixing in the 19.0 Å period sample is correlated with the decreased x-ray optical performance of this sample.

Annealing causes the 37.0 Å period RDF to change towards that of the 19.0 Å period sample. These structural rearrangements on annealing in the amorphous state thus result in further intermixing of W and C towards local structures like those in WC. The first peak in the RDF of the annealed sample is narrower than that of the unannealed sample, and features at larger r appear more distinct in the annealed sample. This suggests that structural relaxation into more well-ordered local arrangements like those in the carbides is a part of the rearrangements accompanying annealing. The observation of atomic rearrangements on annealing is evidence that the observed intermixing at interfaces of as-deposited samples occurs during the deposition process, rather than after deposition.

CONCLUSIONS

In summary, this atomic-scale structural understanding of the W/C multilayer system helps to explain the changing x-ray optical performance of these structures with period and annealing, and provides insight on the deposition process and stability of this system. GIS techniques have provided direct information about interatomic arrangements in amorphous W-rich layers of W/C multilayer samples of interest as x-ray optical elements. Interatomic arrangements are seen to be highly correlated with the spacing between the interfaces, which is proportional to the multilayer period. Multilayers of large period have polycrystalline bcc W at the center of the W-rich layers and amorphous interfaces with the C-rich layers, while lower period samples have W-rich layers which are entirely amorphous. RDFs suggest that these amorphous arrangements are stabilized by intermixing of W and C during deposition. This intermixing results in local atomic arrangements like those in the tungsten carbides, which may help to account for the room-temperature thermal stability of the ultra-thin metastable layers in these small-period structures. Further intermixing and structural relaxation in an amorphous state are observed on annealing.

ACKNOWLEDGEMENTS

This work was supported by Department of Energy contract DE-AC03-76SF00098. One author (JD) was supported by an Air Force Office of Scientific Research contract with the Regents of the University of California for performance at the Lawrence Berkeley Laboratory, which is operated under the above DOE contract. We thank P. Plag for assistance in sample preparation and P.H. Fuoss for lending experimental apparatus for part of this work. The work reported herein was partially done at SSRL, which is supported by the DOE, Office of Basic Energy Sciences; and the National Institutes of Health, Biotechnology Resource Program, Division of Research Resources.

REFERENCES

1. T.W. Barbee, Jr., in Low Energy X-ray Diagnostics--1981, edited by D.T. Attwood and B.L. Henke (AIP Conf. Proc. 75, AIP New York, 1981), p. 131.

2. The authors have fabricated other W/C multilayers with period roughly 19 Å having first order peak reflectance as great as 0.32. Even so, the general trend of rapidly decreasing reflectivity with period below 20 Å remains.

3. T.W. Barbee, Jr., in X-ray Microscopy, edited by G. Schmahl and D. Rudolph (Springer-Verlag, Berlin, 1984), p. 144.

4. Y. Lepêtre, E. Ziegler, I.K. Schuller and R. Rivoira, J. Appl. Phys. 64, 2301 (1986).

5. E. Ziegler, Y. Lepêtre, I.K. Schuller and E. Spiller, Appl. Phys. Let. 48, 1354 (1986).

6. J.B. Kortright and A. Fischer-Colbrie, J. Appl. Phys. 61, 1130 (1987).

7. B.E. Warren, X-ray Diffraction (Addison-Wesley Publishing Co.. Reading, MA, 1969), p. 251.

GLANCING ANGLE EXAFS STUDIES OF TUNGSTEN-CARBON MULTILAYERS

G. M. LAMBLE, S. M. HEALD*, D. E. SAYERS, AND E. ZIEGLER+
North Carolina State University, Raleigh, NC 27607
*Brookhaven National Laboratory, Department of Applied Science, Upton, NY 11973
+Materials Science and Technology Division, Argonne National Laboratory, Argonne, IL 60439.

ABSTRACT

Results are presented from glancing angle EXAFS studies of three tungsten-carbon multilayer systems, of different W/C thickness ratios, by monitoring the fine structure above the tungsten L3-edge. The purpose of the investigations was to determine the structural changes occurring in the multilayer as a result of moderate annealing. Surprisingly, dramatic changes are observed in the structural environment of the W atoms as a result of heating to temperatures as low as 350°C for 3 hours. It is found that, when the W layer is sufficiently thick, and the W/C layer thickness favorable, W_2C is preferentially formed and crystallization is extensive.

The experiment demonstrates the suitablility of the EXAFS technique for the study of multilayer systems since, 1) contribution to the signal from the interface is significant and 2) the phenomenon is not dependent on the existence of long range order.

INTRODUCTION

The fabrication of multilayered structures from pairs of materials with a large difference in refractive index has provided an extension of the low energy limit for monochromatized X-radiation beyond that attainable with natural crystals. The creation of these multilayers with a "tailored" d-spacing of sufficient quality for them to function as optical elements requires very stringent experimental procedure, there are additional problems associated with the behavior of the multilayer itself. A better understanding of these chemical and physical properties is required in order to optimize its performance.

There is understandable concern about the thermal stability of W/C multilayers which undergo dramatic structural changes after heat treatments. In an x-ray diffraction experiment Takagi et al. [1] demonstrated that structural changes occurring, after annealing various multilayer samples to 730°C for several hours, depend largely on the thickness ratio of W/C. In particular, when the tungsten layer is very thin (10-20 Å) an anomalously large expansion of the d-spacing is observed at 730°C, and analysis of the resulting samples shows the components to be crystallized in some form or other. Some controversy exists as to the identification of the annealed products [1-3] and also as to whether the expansion is a result [4] or cause [2] of agglomeration of the tungsten atoms. Since while the multilayer structure is destroyed at this temperature, changes in the d-spacing are observed long before this occurs.

We present the results from glancing angle EXAFS studies of various W/C multilayer samples; monitoring the fine structure above the tungsten L3 edge. The experimental technique is a particularly useful probe of the interface region where substantial chemical and physical effects might be expected. Contribution to the signal from the interface is enhanced by the presence of many layers [5]. The technique is additionally advantageous in that, unlike diffraction methods, the EXAFS phenomenon is not dependent on long range order, while it can provide information on the extent of the latter. Measurements were taken before and after an annealing treatment of

350°C for 3 hours. This is a significantly milder heat treatment than that required for multilayer destruction, the objective being to observe the structural changes occurring in an intermediate temperature range.

EXPERIMENT

The samples were prepared on float glass or fused silica substrates using a multisource hot filament magnetically enhanced triode sputtering technique. The multilayer compositions being: A) 43.0 Å W/39.6 Å C B) 65.0 Å W/17.6 Å C and C) 37.4 Å W/131.0 Å C.

The glancing angle experiments were performed at the National Synchrotron Light Source (NSLS) on beamline X-11A and at the Cornell High Energy Synchrotron Source (CHESS) on beamline A3 and C2. The multilayer samples wre mounted on a stepping motor controlled angle stage with a step resolution of 0.3 arc seconds. Collimation of the incident beam was by a fine slit (50-100μm). The incident, reflected and fluorescence signals were monitored by ionization chambers. The fluorescence chamber being a large area detector situated to collect a large solid angle of   steradians.

The EXAFS was measured in the fluorescence signal while the reflected signal was used to optimize the alligment. The data were taken at an incident angle of 50 mR which was large enough to avoid amplitud distortions [6]. Bulk materials, to be used as references, included WC, W_2C and W metal. Data from these standard materials were taken in conventional transmission mode.

RESULTS AND DISCUSSIONS

Sample A; 43.0 Å W/39.6 Å C

In this case the layer thickness of each component is approximately equal. The background subtracted EXAFS spectra of the sample before and after annealing are shown in Fig. 1(a). with the corresponding Fourier transforms in (b). The first real space component apparent in these and all other transforms is a non-structural feature due to the scattering of the photoelectron off the central atom potential barrier [7]. These peaks are frequently observed in EXAFS involving heavy absorber atoms and appear at a distance roughly equal to the absorbing atom radius. It should also be pointed out that the structural features in the transforms are shown without phaseshift corrections and thus appear at slightly smaller distances than the true bond lengths.

Carbon neighbors are expected to contribute to the region between 1 and 2 Å. Two components are observed in the unannealed spectrum with greatest intensity at around 2 Å, a slightly smaller peak appearing at 1.5 Å. The carbon contribution in the heat treated sample appears as a single feature at around 1.7 Å, with a shoulder at around 1.5 Å. Additionally, an overall increase in intensity by a factor of 2 occurs as a consequence of annealing. However, the most outstanding difference observed is the very large increase in the signal due to backscattering tungsten atoms, manifested in the Fourier transform at around 3 Å, which indicates that substantial ordering has occurred within the sample. Figure 2(a) shows the spectrum from the annealed sample with a transmission EXAFS spectrum of bulk W_2C, along with the Fourier transform in (b). A striking similarity is evident. EXAFS data from the other carbide standard, WC, is shown along with that of W_2C in Figs. 3(a) and (b) in order to illustrate that sufficient dissimilarity exists in the spectra to allow simple distinction between the two carbides and thus the unambiguous identity of the annealed sample A. Crystallinity extends beyond 6 A as is demonstrated in Fig. 4 by the reproducibility of the more distant features in the Fourier transforms of the

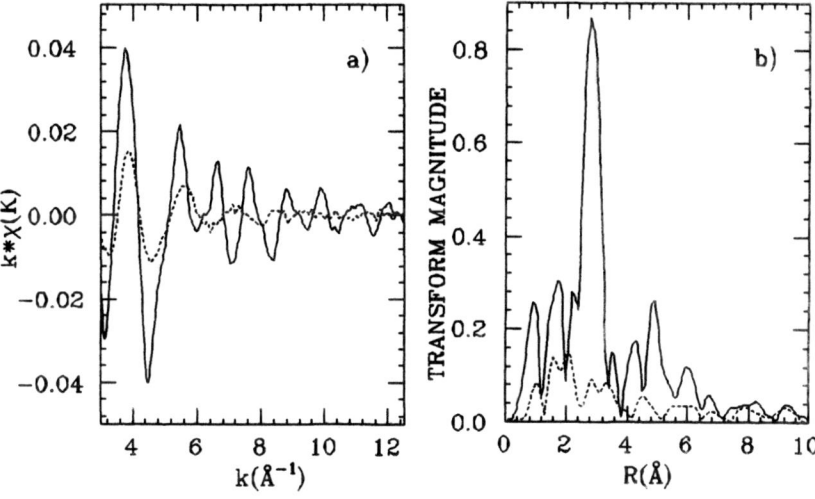

Fig. 1(a). EXAFS of sample A before (full line) and after (broken line) annealing to 350°C. (b) Fourier transforms of (a).

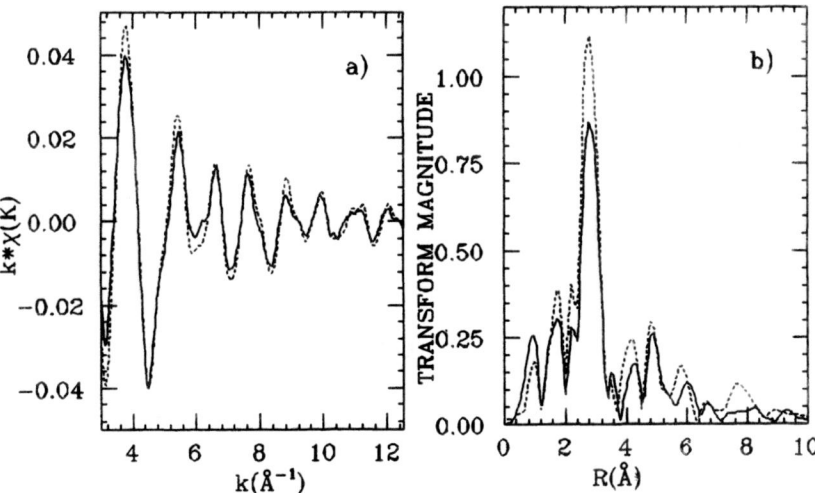

Fig. 2(a). Comparison of EXAFS data from sample A (full line) and W_2C (broken line). (b) Fourier transforms of (a).

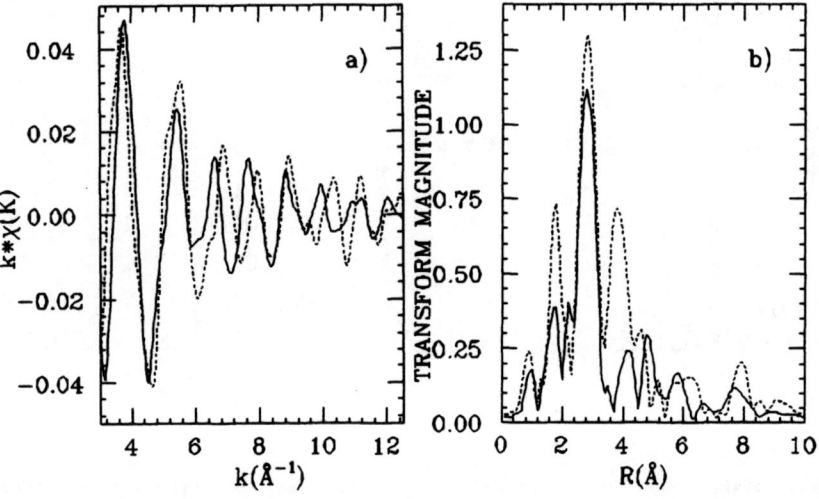

Fig. 3(a). Superimposition of EXAFS from W_2C (full line) and WC with (full line) and WC with Fourier transforms in (b).

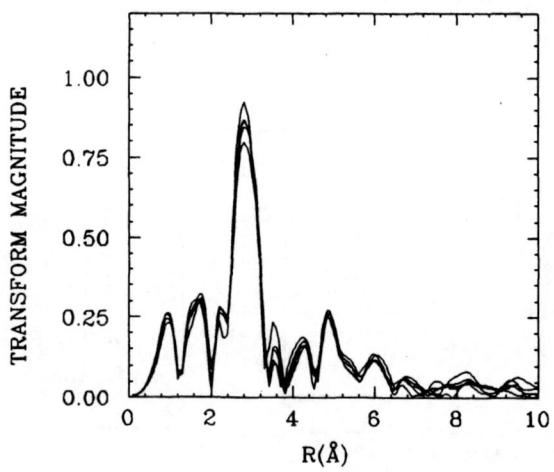

Fig. 4. Superimposed Fourier transforms of six separate EXAFS scans from the annealed sample A.

EXAFS from six separate data runs. Thus, the formation of W_2C as a consequence of the heat treatment is extensive. Preliminary quantitative analysis shows that more than 90% of the W atoms are in the form of W_2C.

The precise identification of the low R peak for the unannealed sample is more difficult, mainly because of its small itensity. Nevertheless, it is quite likely that it is due to a trace oxide contaminant [8]. The second peak is due to carbon and its frequency is very close to that of WC. However, crystallinity is absent in the sample, so we attribute the peak to being representative of the tungsten-carbon bonds within the amorphous layers. Although the experiments are unable to provide information about concentration gradients, amplitude considerations indicate that the average number of nearest neighbor C atoms around a given W atom is approximately equal to 1 (in W_2C this coordination number is 3).

Our observations reflect, in essence, those of Takagi et al. [1], though our annealing conditions were not nearly as severe. After heating a sample, of composition 40 Å W/10 Å C, to 730°C for 4 hours, they found that crystalline W_2C was dominant.

Sample B; 65.0 Å W/17.6 Å C

The EXAFS from the annealed and unannealed samples is shown in Fig. 5(a) with the Fourier transforms in (b). Observations in terms of changes in the low R peaks are similar to those of sample A except that the peak assigned to trace oxide contamination is missing in the untreated sample. Though slightly puzzling, this may be due to a smaller presence of oxygen in the atmosphere during preparation. At the average coordination number of C to W is 0.6. After annealing, the carbon peak has shifted to a shorter distance which corresponds to that of bulk W_2C, with an accompanying increase in intensity.

A notable difference observed by comparing the unannealed samples A and B is that in this case there is a substantial contribution from the tungsten atoms. By comparison to the standard, this component can be identified as a bulk tungsten feature, which might be expected in view of the greater W layer thickness. On annealing, the center of gravity of this feature shifts to become coincident with the W-W feature in W_2C though some intensity remains in the bulk tungsten position. It appears that around 80% of the tungsten is in the form of W_2C, the remainder being of crystalline form.

Again our results are not incompatible with those in Ref. 1., insofar as W_2C is dominant, though we are additionally observing crystalline W in the annealed sample. However, our layer composition in this case is somewhat different. In particular, the W layer is thicker. Additionally, the difference in temperature regime of the heat treatments should be borne in mind.

Sample C; 37.4 Å W/131.0 Å C

The behavior of this sample is distinctly different from samples A and B insofar as very little structural change is observed on annealing. The data from the sample before and after annealing are presented in Fig. 6(a) and (b). Two peaks appear in the low R range of the unannealed spectrum in the sample positions as those of sample A; the coordination of W to C also appears to be the same as in sample A. After annealing the intensity shifts to approximately the position of the W-C distance in W_2C, though there is probably still some contribution, within this feature, from the oxide contaminant. It is clear that there is negligible formation of crystalline W_2C, though there appears to be generally a stronger interaction between the W and C atoms indicated by the shortening of the bond. In this sense the behaviour is similar to the previous two samples.

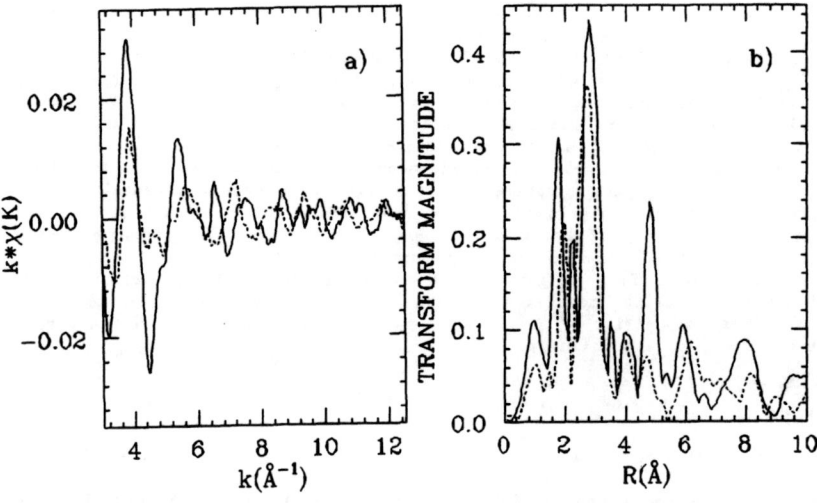

Fig. 5(a). EXAFS of annealed (full line) and unannealed (broken line) sample B, with corresponding Fourier transforms in (b).

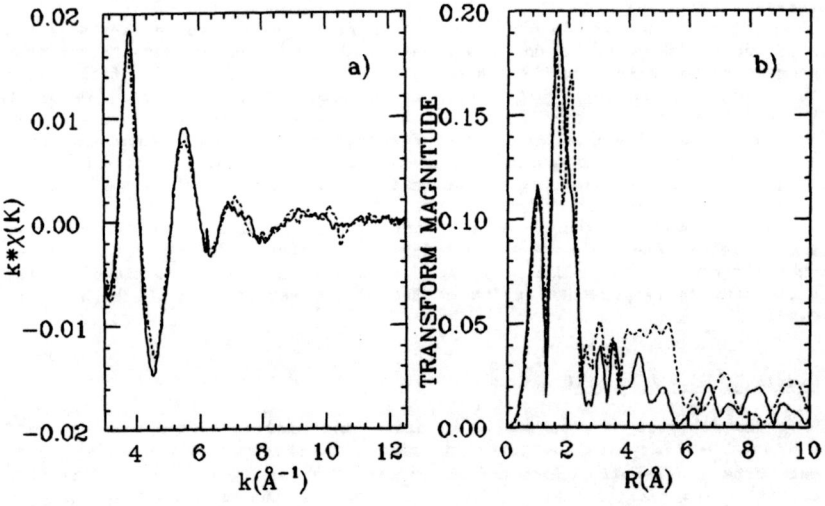

transforms are in (b).

The composition of this sample (in terms of initial W/C d-spacing ratio) bears a resemblance to another of Takagi et al. (10 Å W/40 Å C) and there is again a marked similarity in the results. They observed little change in the single broad diffraction peak of the unannealed sample (other than a general sharpening of the feature) even after heating to 730°C for 7 hours.

SUMMARY AND CONCLUSIONS

The study presented here demonstrates that the EXAFS technique is able to provide a wealth of information about multilayer behavior, to complement that produced from diffraction methods. For instance, the independence of the technique on long range order permits simple, direct observation of amorphous structures. This is demonstrated in the observations of the unannealed W/C multilayers.

Dramatic changes in the W/C multilayer structure occur at temperatures as low as 350°C. We find that when the W thickness is sufficiently small, W_2C is preferentially formed. The picture emerging is in good agreement with that of Takagi et al. [1] whose studies showed that the structures formed after annealing were strongly dependent on the W/C thickness ratio. A notable difference in the studies reported here is that the annealing conditions were much less severe. It is somewhat surprising, and of considerable significance, that such dramatic structural changes are accompanied by only minor alterations in the reflectivity behavior. A more rigorous data analysis of these systems is currently in progress and will be reported elsewhere.

ACKNOWLEDGEMENTS

This work was supported by the Materials Science Division of the U.S. Department of Energy under Contract Nos. DE-AS05-80-3410742 and DE-AC02-76CH00016 and by support from Argonne National Laboratory Advanced Photon Source R and D funds.

REFERENCES

1. Y. Takagi, S. A. Flessa, K. L. Hart, D. A. Pawlik, A. M. Kadin, J. L. Wood, J. E. Keem, and J. E. Tyler, Proc. SPIE 563, 66 (1985).
2. E. Ziegler, Y. Lepetre and K. Schuller, Appl. Phys. Lett. 48, 1354 (1986).
3. Y. Lepetre, E. Ziegler, and K. Schuller, J. Appl. Phys. 60, 2101 (1986).
4. Y. Takagi and A. I. Bienenstock, SSRL Activity Report p. 184 (1986).
5. See for instance, E. A. Stern and S. M. Heald, Handbook of Synchrotron Radiation, E. E. Koch, Editor, Vol. 1b, p. 955, North-Holland, Amsterdam, 1983.
6. S. M. Heald, J. M. Tranquada, B. M. Clemens, and J. P. Stec, J. Phys. C8, 1061 (1986).
7. B. W. Holland, R. Pettifer, J. B. Pendry, and J. Bordas, J. Phys. C11, 631 (1978).
8. D. Roux, A. Rolland, P. Renucci, and J. P. Petrakian, Appl. Surf. Sci. 28, 93 (1987).

X-RAY SCATTERING (MODELING AND EXPERIMENT) OF $In_xGa_{1-x}As$/GaAs MULTIPLE QUANTUM WELLS

Jichai Jeong, J. C. Lee, M. A. Shahid[*], T. E. Schlesinger, and A. G. Milnes

Department of Electrical and Computer Engineering
* Department of Metallurgical Engineering and Materials Science
Carnegie Mellon University
Pittsburgh, PA 15213

Abstract

X-ray diffraction, transmission electron microscopy (TEM), and photoluminescence measurements have been made on strained $In_xGa_{1-x}As$/GaAs quantum well structures. The well widths measured from TEM are 187, 115 and 69 Å for an interrupted growth, and 218, 126, 60 Å for a non-interrupted growth. In the measured x-ray diffraction patterns, the Pendellosung fringes due to GaAs barriers are modulated by a broad weak peak mostly coming from the thickest $In_xGa_{1-x}As$ well layer and is fairly symmetric for the non-interrupted sample. For the interrupted quantum well, the x-ray diffraction pattern is less symmetric, since there is further modulation by another broader and weaker peak. This results show that the In content in the $In_xGa_{1-x}As$ well layers are not well controlled for the interrupted quantum well. Using actual thickness measured from TEM, x-ray diffraction patterns are calculated and good agreement is obtained between the measured and the calculated x-ray diffraction patterns. The three strained $In_xGa_{1-x}As$/GaAs quantum wells grown without interruption produce high intensity and narrow full-width at half-maximum (FWHM) of 2.9 meV of the photoluminescence peak. The photoluminescence peaks for the interrupted quantum well are relatively broad and asymmetric, and have lower intensities, indicating that better quality $In_xGa_{1-x}As$/GaAs quantum wells can be grown without interruption.

I. Introduction

X-ray characterization of hetero-epilayers is a non-destructive technique which has been used to characterized heterostructures[1] and superlattices[2,3]. This technique provides information on lattice constant, thickness, and degree of crystal perfection. Pendellosung fringes due to thin layer effects have been obtained in 0.78 μm $In_xGa_{1-x}As$ grown on InP by vapor-phase epitaxy[4], and provide a way of estimating layer thickness[5].

In this paper, x-ray studies are reported for a strained $In_xGa_{1-x}As$/GaAs structure having three quantum wells grown with and without growth interruption of 30 sec. at each interface. For heterointerfaces between $Al_xGa_{1-x}As$ and GaAs it has been demonstrated that interface roughness can be reduced to within one or two monolayers by interruption[6]. Significantly different x-ray diffraction patterns for the interrupted and the non-interrupted samples in our work were observed. To explain these patterns, we have modeled the x-ray diffraction patterns and have been able to obtain satisfactory agreement with the experimental results. TEM and photoluminescence measurements have also been made on the same samples.

Mat. Res. Soc. Symp. Proc. Vol. 103. ©1988 Materials Research Society

II. Sample preparations

The samples were grown in a Perkin Elmer Model 400 MBE system on Si-doped (10^{18} cm^{-3}) (100) GaAs substrates at a temperature of 500 °C. An undoped GaAs buffer layer about 0.77 μm thick was grown at growth rate of 0.77 μm/hr on the substrate, followed by three strained In$_x$Ga$_{1-x}$As/GaAs quantum wells at 500 °C in increasing order of thickness. The growths were made under As stabilized condition (As/Ga=10). The well thicknesses from the calibration of the growth rate calibration were estimated to be 61 Å, 121 Å, and 183 Å and each quantum well was separated by 513 Å thick GaAs barriers. The thickness of the GaAs top layer was 1280 Å. Fig. 1 shows the sample structures and layer thicknesses measured from TEM and estimated from growth time and rate. The well thicknesses calculated from growth rate are in good agreement with those measured from TEM. Another sample was grown under similar conditions except for growth interruptions of 30 sec. at every interface by closing the Ga and In shutters, or the In shutters while the As shutter was always kept open. The first GaAs barrier layer was 374 Å instead of the planned 513 Å because of a timing error.

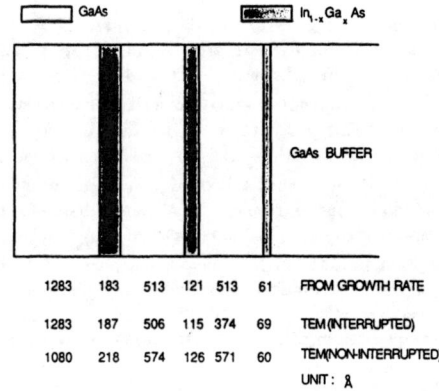

1283	183	513	121	513	61	FROM GROWTH RATE
1283	187	506	115	374	69	TEM (INTERRUPTED)
1080	218	574	126	571	60	TEM(NON-INTERRUPTED)
						UNIT : Å

Figure 1: Sample structures and layer thicknesses of three strained In$_x$Ga$_{1-x}$As/GaAs quantum wells (x=0.28) for calculations.

III. Model

The x-ray diffraction patterns can be calculated from the Takagi-Taupin theory[7]:

$$\frac{dX}{dZ} = \frac{i\pi}{\lambda \sin\theta_o} \left\{ cY_H X^2 + [2Y_o - \alpha_H(Z)] X + cY_H \right\} \tag{1}$$

where X is the diffracting amplitude normalized by incident x-ray beam amplitude, Z is the depth below the surface of the crystal, θ_o is the Bragg angle, c, is the polarization factor(1 for σ and cos θ for π polarization), subscript H represents the (400) reflection and subscript o the (000) reflection. Y and $\alpha_H(Z)$ are given by

$$Y = -\frac{\lambda^2 r_e F}{\pi a^3} \tag{2}$$

$$\alpha_H(Z) = -2\lambda (\theta - \theta_o) \cos\theta_o / d \tag{3}$$

where λ is the wavelength, a is the lattice constant, $r_e (=2.818 \times 10^{-5}$ Å$)$ is the classical electron radius, F is structure factor, θ is the incident angle of beam, and d is the spacing of the parallel (400) atomic planes.

The solution of Eq. (1) can be obtained by integration with appropriate boundary conditions[8]. The model assumes that the interfaces are abrupt and flat, and the layers are perfect. Both CuKα_1 and CuKα_2 wavelengths were included and the incidence beam angle step was 0.001 ° in the calculations. The lattice constant of In$_x$Ga$_{1-x}$As was calculated from Vegard's rule for a linear relationship between composition and lattice constant. Tetragonal distortion of the In$_x$Ga$_{1-x}$As lattice was not taken into consideration in the calculations. The structure factors for GaAs and In$_x$Ga$_{1-x}$As were calculated from the bases on their unit cell. For single layer structures, x-ray diffraction patterns were calculated with the boundary condition that X is equal to zero at the bottom surface. Using X at the bottom layer, a new X at the top layer can be calculated. Therefore, this technique can be applied to obtain x-ray diffraction patterns in multilayer structures.

IV. Experimental and calculated results

A. X-ray diffraction patterns for the non-interrupted quantum well

Besides the strong GaAs peaks due to CuKα_1 and CuKα_2, weak ripples are observed between 32° and 33° diffraction angles in Fig. 2.

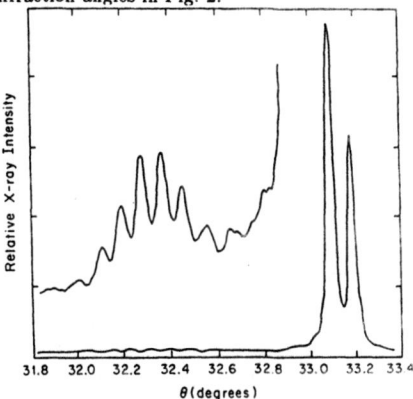

Figure 2: Measured x-ray diffraction patterns from the three strained In$_x$Ga$_{1-x}$As/GaAs quantum wells without growth interruption.

These Pendellosung fringes are modulated by a broad weak peak located at about 32.3° that is quite symmetric. This peak should be related to an epilayer that has a larger lattice constant than GaAs. Therefore, the thickest In$_x$Ga$_{1-x}$As well layer is mostly responsible for this modulation.

With three layers in the model and layer thicknesses estimated from the growth rate, a 1 μm GaAs buffer, a 183 Å $In_{0.28}Ga_{0.72}As$ well, and a 513 Å GaAs barrier layer, the calculated results show patterns similar to those in Fig. 2. To obtain more closely matched patterns, the model was applied to a structure of five layers based on the TEM measurements for the non-interrupted growth, namely: 1 μm GaAs, 126 Å $In_{0.28}Ga_{0.72}As$, 571 Å GaAs, 218 Å $In_{0.28}Ga_{0.72}As$, and 574 Å GaAs with the results shown in Fig. 3. There is quite good agreement between Fig. 2 and Fig. 3.

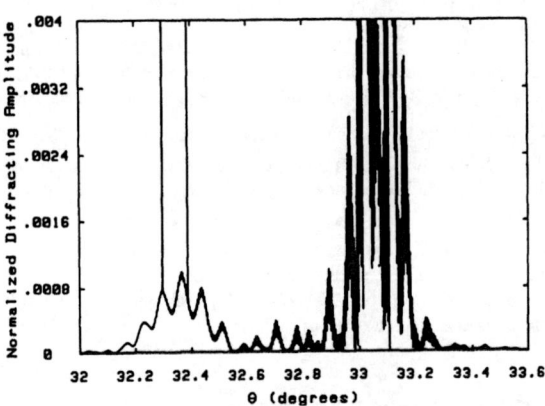

Figure 3: Calculated x-ray diffraction patterns from five layers: 1μm GaAs buffer, a 126 Å $In_{0.28}Ga_{0.72}As$, 571 Å GaAs, 218 Å $In_{0.28}Ga_{0.72}As$, and 574 Å GaAs.

Two spikes at the Bragg angles of the two incident wavelengths appear due to the discrete incident beam angle steps of 0.001° used in the calculations. These spikes did not show up with 0.005° steps. We can estimate the In composition to be about 28 percent in the well layers by comparing measured and calculated patterns. Compositions in quantum wells have usually been calculated from x-ray diffraction patterns of a thick layer grown under the same growth conditions as the quantum well structure[9]. However, the present study suggests that compositions in quantum wells can be estimated quite well from experimental x-ray diffraction patterns if the thickness of the well layer is above 100 Å.

B. X-ray diffraction patterns for the interrupted quantum well

In the x-ray diffraction patterns of the interrupted quantum well, the modulation of Pendellosung fringes is fairly similar to the diffraction pattern in Fig. 2 but not quite as symmetric. From the modulation of the Pendellosung fringes, it appears that two peaks are overlapped with slightly different diffraction angles, and this implies that this portion of the x-ray diffraction pattern is coming from two $In_xGa_{1-x}As$ well layers with slightly different In composition. The peak at the slightly lower diffraction angle is a little bit stronger and therefore comes from a thicker $In_xGa_{1-x}As$ well layer of slightly higher In composition. Another possible explanation is that the pattern irregularity may be associated with compositional grading within each $In_xGa_{1-x}As$ well layer or interface roughness between the $In_xGa_{1-x}As$ and GaAs layers. Compositional grading can probably be neglected, because the growth time of each well layer was relatively short and this effect was not observed in the structure with continuous growth.

Fig. 4 shows the calculated x-ray diffraction patterns for a five layer structure including two $In_xGa_{1-x}As$ well layers with In composition of 24 and 28 %.

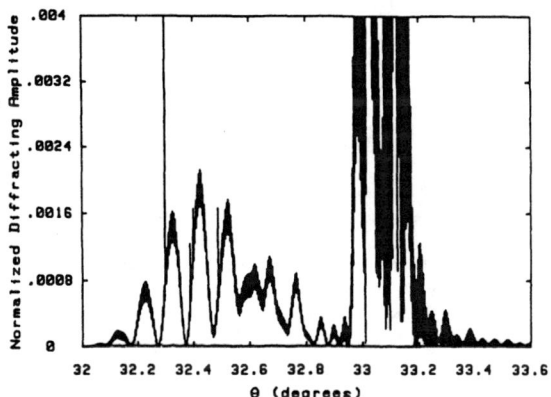

Figure 4: Calculated x-ray diffraction patterns for five layers:
$1\mu m$ GaAs, 115 $\text{Å}In_{0.24}Ga_{0.76}As$, 374 ÅGaAs,
187 $\text{Å}In_{0.28}Ga_{0.72}As$, 506 ÅGaAs.

In the experimental data (not shown here), the diffraction pattern in the diffraction angle range between 32° and 32.4° is weaker and the pattern in the diffraction angle range between 32.4° and 32.8° is moved up. The Pendellosung fringes in the diffraction angle between 32.4° and 32.8° have small peak to valley ratios. Somewhat similar behavior is seen in the x-ray diffraction measurements. The difference in angular spacing of the Pendellosung fringes between Fig. 3 and Fig. 4 is due to the different thickness of the GaAs barriers.

C. Photoluminescence measurements for two kinds of quantum wells

The photoluminescence spectra for the interrupted and the non-interrupted $In_xGa_{1-x}As$/GaAs quantum wells are shown in Fig. 5. The semi-log plot of photoluminescence peak amplitudes reveals more information on the lower portion of the peaks. The emission peaks of photoluminescence in the three quantum wells with growth interruption occur at higher wavelengths than those in the non-interrupted quantum wells. This was also reported in the interrupted $Al_xGa_{1-x}As$/GaAs quantum well study[10]. In our case the peak shift is related to In composition changes rather than well thickness variation. The full widths at half maxima (FWHM's) in the non-interrupted quantum wells from the thinnest to the thickest well are 3.7, 2.9 and 4.1 meV, compared with the corresponding values of 8.6, 3.9, and 4.7 meV for the interrupted quantum wells. The photoluminescence peaks of the three quantum wells without growth interruption are relatively more symmetric than for the interrupted quantum wells. In the interrupted quantum wells, the longer wavelength region of each photoluminescence peak is relatively stronger. This is possibly related to the incorporation of a shallow acceptor impurity or to In sublimation at the surface during the relatively long interruption in the In-based alloys[11]. There is a possibility that the surface mobility of the In would yield smoother interfaces. However, from the FWHM's of the photoluminescence peaks in the interrupted quantum wells, we infer that the interfaces between the $In_xGa_{1-x}As$ and GaAs are rough, compared with the interfaces of the non-interrupted quantum wells.

114

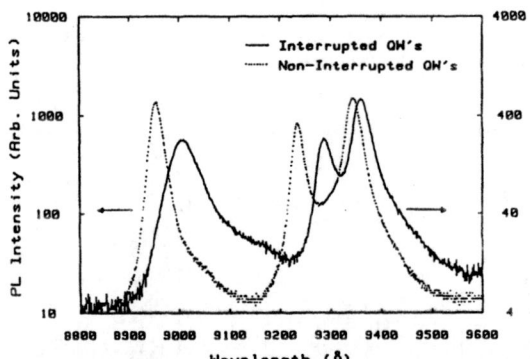

Figure 5: Photoluminescence spectra for the interrupted and non-interrupted quantum wells measured at 4.2 K. FWHM's of the interrupted one are 3.7, 2.9 and 4.1 meV and those of the non-interrupted one are 8.6, 3.9 and 4.7 meV, for the thinnest to the thickest well.

V. Conclusions

From calculations of x-ray diffraction patterns for the interrupted and the non-interrupted quantum wells, good agreement has been obtained between calculated and measured results and it appears that $In_xGa_{1-x}As$ well layers grown with interruption have In composition differences of as much as 3 %. Presumably this is due to time dependent flux or temperature changes in the emission ovens when the shutters are manipulated. Photoluminescence results show that high quality strained $In_xGa_{1-x}As$/GaAs quantum wells can be grown without growth interruption.

Acknowledgments

This work was supported in part by NSF Grant ECS-8521139 and made use of the facilities of the NSF funded Center for the Science of Materials. One of us (TES) would like to acknowledge the support of an IBM Faculty Development Award.

References

1. K. Kamigaki, H. Sakashita, H. Kato, M. Nakayama, N. Sano, and H. Terauchi, Appl. Phys. Lett. **49**, 1071(1986)
2. J. Kervarec, M. Baudet, J. Caulet, D. Auvray, J. Y. Emery, and A. Regreny, J. Appl. Cryst. **17**, 196(1984)
3. M. Quillec, L. Goldstein, G. Le Roux, J. Burgeat, and J. Primot, J. Appl. Phys. **55**, 2904(1984)
4. A. T. Macrander and K. E. Strege, J. Appl. Phys. **59**, 442(1986)
5. W. T. Stacy and M. M. Janssen, J. Cryst. Growth **27**, 282(1974)
6. M. Tanaka and H. Sakaki, J. Cryst. Growth **81**, 153(1987)
7. M. A. G. Halliwell, M. H. Lyons, and H. J. Hill, J. Cryst. Growth **68**, 523(1984)
8. J. Jeong, T. E. Schlesinger and A. G. Milnes, accepted for J. Crystal Growth
9. N. G. Anderson, W. D. Laidig, R. M. Kolbas, and Y. C. Lo, J. Appl. Phys. **60**, 2361(1986)
10. C. W. Tu, R. C. Miller, B. A. Willson, P. M. Petroff, T. D. Harris, R. F. Kopf, S. K. Sputz, and M. G. Lamont, J. Cryst. Growth **81**, 159(1987)
11. J.-L. Lievin and C. G. Fonstad, Appl. Phys. Lett. **51**, 1173(1987)

Fe-W SUPERMIRRORS FOR POLARIZING NEUTRONS

C. F. MAJKRZAK*, D. A. NEUMANN*, J. R. D. COPLEY** and R. P. DINARDO***

*Institute for Materials Science and Engineering (IMSE), National Bureau of Standards (NBS), Gaithersburg, MD 20899.

**Department of Physics and Astronomy, University of Maryland, College Park, MD 20742, and IMSE, NBS, Gaithersburg, MD 20899.

***Instrumentation Division, Brookhaven National Laboratory, Upton, NY 11973.

ABSTRACT

Thin film bilayers of ferromagnetic Fe and nonmagnetic W have been deposited by sputtering on flat glass substrates according to a sequence of gradually varying bilayer thicknesses which in effect extends the critical angle for external mirror reflection for neutrons of one spin state when the Fe is magnetized in the plane of the film. The measured reflectivity of this Fe-W multilayer system is compared with that of other supermirror polarizers consisting of different materials and layer sequences.

INTRODUCTION

Polarized neutrons play an important role in scattering studies of condensed matter [1]. Consequently, an efficient means of polarizing a neutron beam is very valuable. In addition to conventional, bulk single-crystal polarizers such as the Heusler alloy Cu_2MnAl, polarizing mirrors are often used for this purpose. This is possible because for most materials the refractive index for neutrons is less than unity so that external reflection occurs from flat surfaces in air or vacuum at glancing angles up to a critical angle of the order of 0.1 degree/Å neutron wavelength. Furthermore, because the neutron possesses a magnetic dipole moment, its propagation through ferromagnetic materials is determined by a birefringent refractive index. Thus for a ferromagnetic material which is magnetized parallel to its reflecting surface, the critical angle depends on the neutron spin state so that only one spin state is effectively reflected between two distinct critical angles. For many applications it is advantageous to reflect one spin state over as wide a range of angles as possible with a high efficiency. To this end polarizing supermirrors, first suggested by Mezei [2] and Turchin [3], have been developed in recent years [4,5] which in effect extend the critical angle of reflection for one spin state by constructive interference of the neutron wave in a thin-film, multi-layered structure consisting of a ferromagnetic and nonmagnetic material. Reviews of these devices and their applications are given in references 6-8. Nevertheless, it has not yet been possible to produce polarizing supermirrors with high reflectivity beyond about twice the critical angle of ordinary Ni (0.1 deg/Å). The purpose of the present work has been: 1) to make a supermirror with a very high polarizing efficiency; and 2) to attempt to extend the critical angle for one spin state beyond what has been previously achieved.

SUPERMIRROR ALGORITHMS

There are a number of algorithms for generating a sequence of layers which have been shown to produce a supermirror effect. The algorithm we have chosen is based on a semi-empirical approach [9]. Use is made of the fact that N bilayers (of two different materials) of thickness D will give rise to a well-defined peak in reciprocal space at $Q = 2\pi/D$ where $Q = 4\pi\sin\theta/\lambda$ and where θ is the glancing angle of reflection and λ is the neutron wavelength. In the kinematic limit where extinction is negligible, the peak reflectivity is proportional to N^2D^4 and the width of the reflection ΔQ is proportional to Q/N. By superimposing sets of N_i bilayers of thickness D_i in such a way that the peak reflectivity remains constant and the individual peaks intersect at their respective half-heights (for example), a simple supermirror sequence of bilayer numbers and thicknesses can be obtained. To determine whether in principle such a sequence gives the desired effect when interference occurs (as must happen for reflectivities approaching unity), a calculation analogous to that used in the solution of thin film optical interference problems is appropriate.

This amounts to solving the Schrodinger equation for a neutron plane wave incident on and propagating through a layered but continuous medium [10]. For neutrons the refractive index n is given by $n = [1 - \lambda^2 N(b\pm p)/\pi]^{1/2}$ where N is the number of atoms per unit volume, and b and p are the nuclear and magnetic scattering lengths, respectively, for a given material. Boundary conditions are imposed at each interface and the reflectivity and transmission coefficients subsequently evaluated. Random as well as systematic variations in bilayer thickness can be readily incorporated. The results of such a calculation, for an Fe-W supermirror (on a glass substrate) consisting of the sequence of bilayer numbers and thicknesses given in Table I, is shown in Figure 1. The critical angle or Q for the "up" spin state was chosen to be about twice that for ordinary Ni. The ratio of spin-up to spin-down reflectivity (defined as the flipping ratio R) at $Q = 0.04$ Å^{-1} is about 500 and is due to the well-matched scattering densities, Nb for W and N(b$\pm$p) for Fe.

TABLE I

D (Å)	$2\pi/D$ (Å^{-1})	# BILAYERS
250	0.025	5
216		7
192		8
175		10
161		11
151		13
142	0.044	15

Sequence of numbers and thicknesses of Fe-W bilayers generated by the supermirror algorithm described in the text.

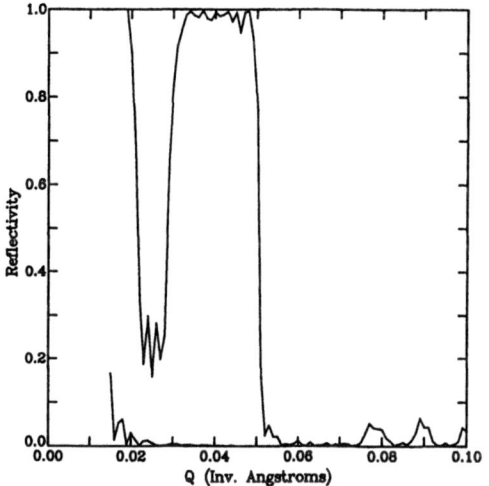

FIGURE 1   Calculation of the reflectivity for the "up" and "down" spin
states as a function of Q = $4\pi\sin\theta/\lambda$ for an Fe-W supermirror
on a glass substrate with the sequence of bilayers given in
Table I. The "up" spin state corresponds to the solid line
with the higher reflectivity.

SPUTTER DEPOSITION SYSTEM

The deposition system, located at Brookhaven National Laboratory,
consists of a large, central stainless steel box containing two planar, 8-
inch diameter targets mounted with their planes vertical. Substrates up to
2 ft. long by 4 inches wide can be accommodated on a carriage which
translates back and forth across the two targets into cylindrical extensions
of the central vacuum chamber. Using a liquid nitrogen cold trap and a
turbomolecular pump, a vacuum of about 2×10^{-7} Torr can be achieved prior
to the introduction of the high purity Ar sputtering gas. Sputtering
pressures are typically 5μ and deposition rates are of the order of 100
Å/min. The Fe is sputtered by application of a 13.56 MHz radio frequency
potential whereas the W is sputtered using a magnetron cathode and an
applied DC potential. Details regarding the hardware components of the
system are given in Ref. 11. Quartz crystal oscillators are used as a rough
measure of the deposition rate but an accurate calibration of the rate for
each material under a given set of conditions (Ar pressure, power, etc.) is
obtained by making a set of single-D-spacing multilayers and measuring the
positions of the Bragg diffraction peaks with neutrons. The widths of these
multilayer diffraction peaks, the observation of higher-order diffraction
harmonics, and the reproducibility achieved in a number of independent
depositions is evidence that the deposition rates are constant. Layer
thicknesses were consequently varied during the supermirror deposition by
adjusting only the translational speed of the carriage (via a programmable
microprocessor).

REFLECTIVITY AND FLIPPING RATIO MEASUREMENTS

The reflectivities of several Fe-W supermirrors made with the sequence of bilayers given in Table I, on a float glass substrate 3" x 18" x 1/4" thick, were measured on a spectrometer at the National Bureau of Standards Reactor (NBSR). A Be-filtered beam was monochromated by a pyrolytic graphite crystal [(002) reflection] and collimated by a pair of 0.007" wide slits. The neutron wavelength λ was approximately 4Å, $\delta\lambda/\lambda \approx$ 3%, and the beam divergence incident on the sample was about 1 minute of arc. By using an absorbing Cd mask attached to the downstream end of the glass substrate to block the incident beam for $Q \lesssim 0.005$ Å^{-1}, it was possible to leave a relatively wide aperture in front of the neutron detector. Measurements of the reflectivity as a function of Q were performed by scanning the glancing angle of incidence θ of the neutron beam relative to the mirror surface while simultaneously moving the detector in steps of 2θ. Because of the relatively large detector aperture, the precise positioning of the detector was not critical. Special care was taken in aligning the spectrometer to ensure that the beam center coincided both with the center of the reflecting mirror surface and with its axis of rotation.

The Fe-W supermirrors were placed in a uniform magnetic field of the order of 200 Oe in order to saturate the magnetization of the Fe layers in the plane of the substrate. Under these conditions only the "up" spin state is effectively reflected at angles corresponding to reflection by the supermirror coating. The incident beam intensity was measured with $2\theta - 0.0$ and the supermirror removed from the spectrometer. The reflectivity R is then given by the ratio of the reflected intensity to one-half the unpolarized incident beam intensity. No corrections were made for the relatively small wavelength spread of the incident beam, of the order of 3%. Our results for the reflectivity of the up spin state as a function of Q are shown in Figure 2. The reflectivity has a gap between the region of normal

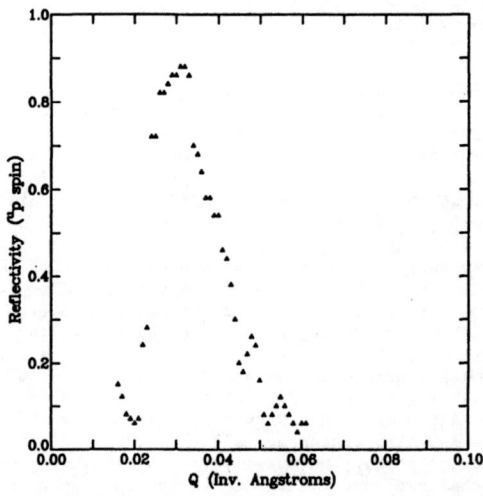

FIGURE 2   Measured reflectivity for the "up" spin state of an Fe-W supermirror with layers deposited on a glass substrate according to the sequence given in Table I.

mirror reflection and that region where the reflectivity is due principally to the supermirror layer sequence as expected for the particular set of bilayers deposited. For the thicker bilayer sets which contribute at lower values of Q, the reflectivity is relatively good. For higher values of Q, the reflectivity is significantly reduced. Scanning electron micrographs taken at Oak Ridge National Laboratory [12] and the National Bureau of Standards [13] on cross sections of similar superlattice structures show a progressive roughening of the layers. Furthermore, Fe-W supermirror coatings consisting of more than about 70 or 80 bilayers buckle and peel from the glass substrate. All of these observations indicate a roughening of the layers presumably due to thermal expansion coefficient differences and lattice mismatch between the two different bilayer materials.

Scharpf [4,6] has made Fe-Ag supermirrors which have 90 % and somewhat better reflectivities, and Co-Ti supermirrors with significantly reduced reflectivites, out to about twice the critical angle of Ni. For further comparison, the reflectivity measured by us for a sputtered $Fe_{0.85}Si_{0.15}$ - Si supermirror manufactured by J. Bradshaw and coworkers at the Opto-Line Co. of Andover, MA, is shown in Fig. 3. A different algorithm was used in this case and the layers were deposited on a single-crystal Si substrate [14]. Nevertheless, in all cases, attempts to extend the critical angle out to three times that of Ni with high reflectivities have failed.

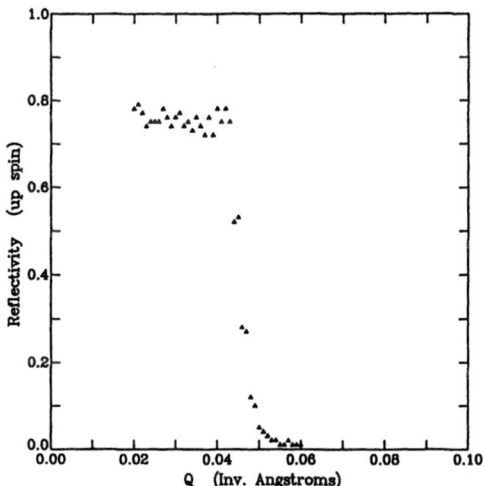

FIGURE 3   Measured reflectivity for the "up" spin state of an $Fe_{0.85}Si_{0.15}$ - Si supermirror fabricated by the Opto-Line Co. on a single crystal Si(111) substrate as discussed in the text.

Measurements of the flipping ratio R for the Fe-W supermirrors were performed at the High Flux Beam Reactor at Brookhaven National Laboratory using a spectrometer geometry similar to that described above but with a polarizing Heusler crystal analyser and flat-coil spin flipper following the supermirror sample. Flipping ratios of about 50 were measured. Because this is the limit of the sensitivity of the particular instrumental configuration, the flipping ratios could in fact be significantly higher.

Work is now in progress to overcome the layer roughness problem that is believed to be the cause of the reduced reflectivity. Deposition conditions favoring amorphous growth and reduced stress are being sought along with alternative techniques including electrodeposition.

ACKNOWLEDGEMENTS

The authors would like to acknowledge valuable discussions with J. Bradshaw, J. B. Hayter, H. Mook and A. Saxena. Work at Brookhaven was supported by the DOE under contract DE-AC02-72CH00016.

REFERENCES

1. R. M. Moon, T. Riste, and W. C. Koehler, Phys. Rev. 181, 920 (1969).
2. F. Mezei, Comm. Phys. 1, 81 (1976).
3. V. F. Turchin, Deposited Paper, At. Energy 22, No. 2 (1967).
4. O. Scharpf, AIP Conf. Proc. 89, 182 (1981).
5. T. Ebisawa, N. Achiwa, S. Yamada, T. Akiyoshi, and S. Okamoto, J. Nucl. Sci. Tech. 16, 647 (1979).
6. R. Pynn, Rev. Sci. Instrm. 55, 837 (1984).
7. C. F. Majkrzak, Applied Optics 23, 3524 (1984).
8. C. F. Majkrzak et al., Physica B, to be published
9. C. F. Majkrzak and L. Passell, Brookhaven National Laboratory Report No. 34343, February, 1984 (Upton, NY 11973) and C. F. Majkrzak and L. Passell, Acta Cryst. A41, 41 (1985).
10. P. Croce and B. Pardo, Nouv. Rev. Opt. Appl. 1, 229 (1970).
11. R. P. DiNardo, P. Z. Takacs, C. F. Majkrzak, P. M. Stefan, Brookhaven National Laboratory Report No. 36742, June, 1985 (Upton, NY 11973).
12. H. Mook, private communication.
13. D. Lashmore, private communication.
14. J. Bradshaw, private communication.

TECHNIQUES FOR CHARACTERISING ARTIFICIAL LAYER STRUCTURES
USING TRANSMISSION ELECTRON MICROSCOPY

W.M. STOBBS
Cambridge University, Dept. of Materials Science and Metallurgy, Pembroke
Street, Cambridge CB2 3QZ, United Kingdom.

ABSTRACT

T.E.M. methods are described for the quantitative characterisation of
the compositional and structural changes at interfaces and in homo- and
hetero-phase multilayer structures. Many of the newer approaches described
including the Fresnel and Centre Stop Dark Field Imaging Methods were
developed specifically for such characterisations. The range of
applications of each of the techniques is assessed as is the importance of
delineating the limiting effects of inelastic and inelastic/elastic
multiple scattering.

INTRODUCTION

That a modern transmission electron microscope can allow the formation
of images exhibiting fringe periodicities well below the atomic spacings
present in the material examined can lead to the falsely optimistic
approach suggesting that all that is needed for the characterisation of a
man made nanometre scale multilayer structure is the large amount of money
required to purchase an instrument of sufficient quality. Such optimism
should be dispelled by a brief examination of an image of gold at the cube
normal, showing the 0.1nm fringe spacings associated with the non-linear
interference of {200} beams: the spacing of the periodic potential causing
the elastic scattered beams which combine to form the image is of course
0.2nm. This single example demonstrates that the use of the term
"structure imaging" to describe this sort of imaging technique is at the
very least simplistic and, indeed, usually grossly misleading. Actually
there are many other techniques at the disposal of the microscopist for the
quantitative analysis of a multilayer structure and in any case high
resolution "interference images" can be analysed fairly rigorously to
provide an interpretation of the structure examined which, if not usually
unique, can at the very least provide very useful supportive detail for
information obtained by other methods. This is fortunate: the demands of
the electronics physicist in the design of quantum engineered multilayers
of III-V, II-VI or alloyed IV-IV compound semiconductors put increasing
pressures on the grower and the characteriser alike; so too do the needs
of the new generation of high energy sychrotron sources for multilayers
which, according to the optical theorists, should have a smoothness and
regularity of fractional (!) atomic dimensions for use as high efficiency
monochromators and mirrors. Equally there is still considerable interest
in the anomalous properties of low wavelength (circa 3nm) metallic multilayers; the anomalous changes in the physical behaviour of such systems
will not be understood without detailed structure characterisations
obtained with sufficient accuracy to enable the measurement of the
modifications in retained elastic strain as the wave-length is changed from
values at which properties are interpretable conventionally to those at which
they are not. It is pressures of this sort which have led to the
development of a whole new range of electron microscopical techniques for
the quantitative analysis of the localised changes in structure which can
occur at an interface. Conventional methods employing energy dispersive
spectroscopy or energy loss analysis are, for example, wholly inadequate
for the analysis of the sharpness of a compositional irregularity at an

interface with the atomic level accuracy which problems of the type alluded to above demand. As we will see, the new techniques which have been found allowing such analyses to be made have opened up, in parallel, new avenues of research in more conventional and older fields such as the characterisation of grain boundary segregation and the modelling of interface reaction behaviour as occurs during oxidation.

In fact the diversity of the types of boundary which have been studied by TEM is as wide ranging as are the methods now available for their characterisation. The choice of the best technique to use, as well as an understanding of the limitations of each, can often thus be as important as the recognition of the specific interface characteristic of most relevance in understanding the material property being investigated. Elsewhere [1] I have already discussed the way in which it is sometimes not even clear whether the local heterogeneity of structure or chemistry in question can properly be described as being delineated at the atomic scale by a discrete plane. Such problems are apparent in the study of amorphous structure, spinodal decomposition, short range order and the characterisation of icosahedral structure. Man made multilayers can show periodic changes of both composition and lattice parameter with wave forms of anything from simple sine to square wave character. Thus even problems involving the specification of the class of interface which is to be characterised can be of equal relevance in the study of a multilayer. A fundamental difficulty in this connection is that a simple rigid body displacement at a boundary is necessarily associated with a local change in scattering potential which can be very difficult to distinguish uniquely from that caused by a local change in composition.

In this necessarily brief examination of the methods open to the microscopist for the characterisation of multilayers I will draw heavily on the work done in my own group over the last few years and will not attempt a broad review. Other recent descriptions of relevant characterisation techniques for heterostructures [2] industrial semiconductor systems [3] and multilayers generally [4] may be found elsewhere.

I will firstly consider what can now be done for single interfaces and then go on to describe how the techniques developed in this context can often be used for the characterisation of metallic and semiconductor multilayers. Arguably it often seems to be the justification of the expenditure required for a modern HREM or small probe STEM which dominates the choice of method applied for boundary characterisation. If the real problem to be solved is kept in mind, the relatively qualitative application of very standard methods can usually reap substantial benefits provided an appropriate specimen first can be made, though this in itself can be far from trivial [5].

INTERFACE CHARACTERISATION METHODS

There are three individual measurements which need to be made for single interfaces. These are:
1) The quantification of any rigid body coincident site lattice (C.S.L.) displacement for a homophase boundary.
2) The measurement of the concentration of any boundary phase segregant.
3) The characterisation of the local relaxation structures associated with the presence of any intrinsic boundary defects.

Here I will examine in turn the approach best taken for each of the above measurements:

The Rigid Body Displacement

This might seem the simplest of the three measurements to make accurately, given the advances made over the last few years in high resolution electron microscopy at the sub 0.2nm level. With such a microscope the scattering due to a laterally large weak phase object with spacings of say 0.25nm will be well represented in the interference image where, for such an object, the local contrast modulations seen can be related directly (through the contrast transfer behaviour of the microscope) to the projected potential of the object. For even marginally thicker specimens a full chemical characterisation using measurements of modulations in intensity is more dubious given that it is now recognised [6] there can be contributions exhibiting high resolution detail due to multiple inelastic/ elastic scattering. However, this need not worry us yet. What is of more immediate concern is that, even for a weak phase object, a direct relationship between the contrast detail and the projected potential is not maintained locally near to a boundary if the interplanar spacing is changed there, even if this is an increase (as will usually be the case). The problem is related to the Fourier representation of a square well: in order that the true planar positions at and near to the boundary plane are properly represented in the image the microscope must transfer information about the second Fourier components at least, and even for the best instruments now available this is usually impracticable. In actual fact, even in the conventionally accepted view, the direct interpretation of a small boundary displacement using the comparison of modelled simulations with experimental image series is still more difficult, simply because few real boundaries will approximate sufficiently accurately to a weak phase object. This means that when independent methods are found to determine the various different parameters affecting an image series (defocus, specimen thickness, beam tilt, lens characteristics etc.) a unique solution is unlikely. Given further the recent theoretical [6] and experimental [7] appraisals of the effects which can be associated with inelastically scattered electrons contributing atomic level detail to an image, and necessarily differentially at a defect, any direct appraisal of boundary structure using current simulation methods must be viewed with at best considerable caution.

Fortunately, at least for a limited range of boundaries such as the $\Sigma 3$, which exhibit a common reflection across them when appropriately oriented, there is no need to rely on a complex image simulation at all! The fringe positions within about 2nm of the boundary position will still be non-uniform and very dependent upon the imaging conditions. However, as long as these conditions are identical on either side of the boundary, then whatever the relationship of the fringes to the atomic planes relatively far from the boundary this will be maintained on either side of it; the image shift in registry will thus represent accurately the C.S.L. displacement. Such a shift can be measured to much higher accuracy than the microscope resolution: the accuracy attainable is related to the overall size of the uniformly scattering regions whose relative positions are required. This approach has been used to yield C.S.L. displacement values at twin boundaries in copper and in gold to an accuracy about a hundred times better than the microscope "resolution" [8,9] and systematic errors caused by, for example, foil bending can also be assessed [9]. It should be noted that, unlike conventional high resolution methods, the above approach is similar to that involving the assessment of moire displacements [10, 11] in that it does not require that the interface exhibit a uniform projection or even be vertical. Nonetheless the methods described have severe limitations for more general boundaries as discussed elsewhere[11], though the requirement of a common, or at least parallel, reflection across the interface is often met in a multilayer structure.

The Boundary Composition

As has been noted a rigid body displacement must be associated with a change in the scattering potential which can easily be misinterpreted as due to a local change in composition. However, if the rigid body displacement is known, then this can be included in high resolution image simulations and the local boundary composition can then be adjusted in a computed model to provide a match between the simulations and an experimental image series. Vitek has recently assessed the influence of segregation on the diffraction of a homophase boundary [12]. He compared the effects of a bulk displacement and segration while noting the difficulties in distinguishing them. This, in turn, emphasises the importance of using independent methods for the measurement of these two parameters. When a high resolution dark field Fresnel method was used to assess the displacement at a Σ3 twin boundary in gold the inconsistency of the result obtained with that previously indicated by the fringe shift technique [9], allowed the inference that segregation was the cause. Refinements of the Fresnel matches using the known displacement data would actually allow the quantification of the change in scattering potential caused by segregation if this were needed but this would require that the problem be worth solving!

Interestingly, and rather fortunately, the Fresnel approach does not require high resolution imaging: the Fresnel effects seen at the edge of a foil, where there is of course a discrete change in the scattering potential, are very familiar in the correction of astigmetism and the assessment of beam coherence. However, it has been demonstrated that the intensity of fringed images of a boundary, as a function of defocus, can be related to the magnitude of the localised change in the elastic scattering potential [14] and thus to a composition change. In our past assessments of the technique it has proved surprisingly accurate and, given care, has yielded results apparently limited to about ±5% only by the accuracy of the foil thickness measurements required (and obtained using the weak beam technique [15]). Current work [16] would however suggest that differential inelastic scattering by the segregated element (or layered material) can cause systematic errors in even the thinnest foils. These would obviously be expected to be different for different segregations, and can be difficult to limit but are nonetheless assessable and seldom exceed 30%. On the positive side, the approach is increasingly sensitive the more localised is the change in potential (as is clear on the basis of a Fourier analysis of the problem). Thus it is ideally complementary to the use of EELS quantitatively. Of course EELS still has to be used to determine the nature of the element present when the quantity of this element as segregated is assessed using the Fresnel method. This of course emphasises the inapplicability of the Fresnel method if the concentration of more than one element varies at the interface to be analysed. We are still in the process of refining the technique, and even with uncertainties in the absolute determination of a concentration change as mentioned above it is becoming clear that the local form of a composition change can be measured to near atomic level accuracy. This requires using the sensitivity of the details of the Fresnel contrast to the Fourier analysed shape of the projected composition change. As we will see this development of the method is particularly useful in the characterisation of multilayers, but important applications have been found for single interfaces in, for example, the measurement of the width of SiO_x layers (to atomic plane accuracy) on oxidised silicon surfaces [17,18].

Localised Relaxation Structures at a Boundary

It will be clear from the above discussion that if there are problems in the use of conventional high resolution methods for the assessment of a rigid body displacement then the situation is far worse when it comes to the measurement of the localised relaxation structure of, for example, an intrinsic boundary defect. Nevertheless several authors have tackled this type of problem, and many beautiful micrographs have been obtained. In recent work it is significant that useful information can sometimes be obtained about the relative energy of different geometrically possible defects without having to rely on details of simulations [19]. Krakow et al [20,21] have, however, suggested that, with care, the detailed nature of a theoretical boundary defect can be related to the local image form, after digital image processing. The uniqueness of image fits of this type has been questioned [eg. 22] but they have nonetheless clearly been achieved. That this is so is less surprising when it is remembered that the localised image distortions which tend to arise relative to the true atomic projections when the image is formed, tend further to be emphasised when the image is Fourier filtered. These are those which tend to even up local variations in the spacings, and it is of course just such displacements which would be expected to lower the energy of on unrelaxed boundary structure [23]. Various assessments have been made of the inaccuracies which can result from too direct an interpretation of atomistic images of localised displacements [eg. 24]. Although it is clear that Fourier filtering will exacerbate the problem, most computer modelled simulations tend to be limited by the wrap-around effects inherent in periodic continuation so the conclusions drawn [24] could well be somewhat optimistic. Further work needs to be done in this area particularly given the problems noted above relating to inelastic scattering [6].

This application of high resolution electron microscopy is, at least for defects which can be viewed in projection, immensely rewarding. It is nonetheless, as we have seen, open to question whether or not the solutions obtained to date are unique and data obtained using the weak beam technique [15] could well be less prone to misinterpretation. Attempts have thus been made to use weak beam images quantitatively in a "mapping approach" [24] comparing the contrast effects of the unknown displacement field directly with those exhibited by a dislocation of known displacement field. The full analysis of the technique [25] is, however, discouraging in that simplistic interpretation is rarely viable because of dynamical effects. These, though not dominant at high deviation parameters, can be sufficiently different for a boundary and a matrix defect at the relatively lower deviation parameters needed to view the former. This usually means that image simulations are still required.

MULTILAYER CHARACTERISATION METHODS

Are there further complications in taking over the techniques described above for the characterisation of single interfaces in the assessment of hetero- and homostructural multilayers? These, as we shall see, depend both on the nature of the multilayer to be examined and on the level of detail required. In many instances, as is usual in the examination of trial multilayer mirrors, all that is required is the degree of local smoothness at unit cell dimensions for materials of (in this case necessarily!) very different scattering behaviour, and this can be accomplished easily (see eg. [26]). Nevertheless even images of materials of this type can be grossly misleading: often layering is "visible" even when strongly inclined, at apparent spacings which, as a function of the diffraction conditions, can then be either bigger or smaller than the true spacing [27]. The hall marks of such misinterpretable images are apparent "dislocations" in the layering.

The more interesting systems, at least in terms of the problems involved in their characterisation, tend to be of two general types; those which are, or are nearly, lattice matched (eg. GaAs - $Al_x Ga_{1-x}$ As) and those which exhibit both compositional and lattice spacing variations (eg. GaAs - $In_x Ga_{1-x}$ As or Cu - $Ni_x Pd_{1-x}$). The characteristics of such systems, as required for the quantitative modelling of their properties are normally:

    1) The amplitude and form of the compositional modulation
    2) The layer thickness and its vicinality
    3) The form of any interface steps

Consider first systems such as GaAs - $Al_x Ga_{1-x}$ As. The methods required have been discussed from an industrial viewpoint elsewhere [3] and will only be summarised here. Much of the approach needed can be inferred from the methods described above for single interfaces. Actually, however, the compositional modulation in this system can be analysed by a technique which is of special relevance to III-V alloy systems for which the contrast in an 002 reflection is, on a kinematic basis, sensitively dependent upon the Ga:Al ratio [eg. 28]. For accurate work this method is complicated by the effects of inelastic scattering, but the systematic errors involved can at least be characterised to allow the use of a "fudge factor" approach [29]. Other techniques for composition measurements in this industrially relevant system include the characterisation of thickness fringe profiles for a cleaved wedge of known thickness [30] (though the method described requires modification to include absorption) and a refinement of the CBED technique [31]. However, none seem more accurate than the dark field approach. It is also unclear in the CBED technique how far the information obtained is laterally localised.

An interesting application of the Fresnel technique [14] lies in the measurement of the abruptness of the compositional change in multilayers such as these. We have recently applied the approach to the assessment of this parameter in a heterostructure with variably spaced fine layers of supposedly equal Al content and abruptness. We were able to demonstrate that, at each interface, composition changes occured for at least a monolayer further than the roughness or layer vicinality would allow [32]. The layer vicinality and thickness can of course be readily assessed to near monolayer accuracy using conventional dark field and high resolution methods, but it is important in this context to obtain images at a series of grossly different specimen orientations, in each case using the 002 reflection approximating to the growth direction [29,33].

The problem of determining the interface step density in a multilayer must be distinguished from that of examining the _form_ of the local displacement field at the steps. The former is easy to deal with and benefits from the use of Fourier filtered high resolution images [eg. 34] whereas the latter problem is difficult and does not so benefit, for reasons given earlier. For the AlGaAs system we have recently been applying centre stop dark field methods [35] in an attempt to improve the contrast differences between the layers while retaining the required resolution in order to characterise a step. (See figure 1) Unfortunately it appears [7] that inelastic contributions to the image severely complicate the interpretation as is demonstrated by the simulations shown in figure 2. For more strained systems it is unlikely that the problem of analysing the form of a step displacement field will be solved in any satisfactorily unique manner before high resolution energy filtered imaging becomes available [6]. The image analysis of an interface step structure thus remains the unsolved problem in multilayer characterisation.

Consider now the characterisation of multilayer systems with modulation of both composition and lattice parameter: an interesting and alarming example is provided by the GaAs - $In_x Ga_{1-x}$ As system. This can exhibit the reverse contrast between the layers to that expected for known compositions. While this anomaly has now been characterised [36], its origin is not understood quantitatively and does not lie in surface stress

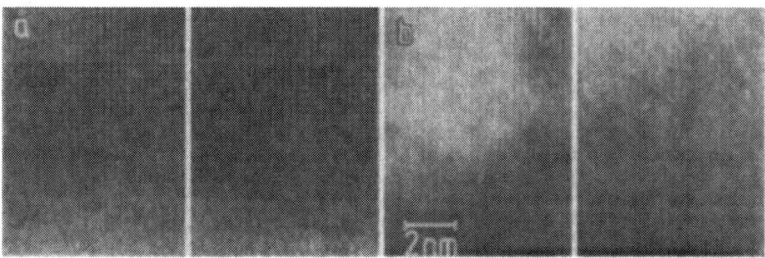

Figure 1:(a) two conventional H.R. images of an $Al_{0.3}Ga_{0.7}As/GaAs$ multilayer of approximately 2.5nm wavelength. The images were taken at a specimen tilt of about $3°$ (about the horizontal layer normal) from the cube direction. The layering is hardly visible even in the left of the two images which is well out of focus to emphasise Fresnel effects. (b) A pair of images taken with a centre stop obscuring the 000 beam. Note that the presence of 0.28nm fringes is surprising given that 200/200 interference would be expected to yield fringes of 0.14nm spacing. We also see that the layering is no more visible than before despite simulations incorporating only elastic scattering indicating that the layering should now exhibit much higher contrast than in bright field (See [7] for further details).

Figure 2: "Centre-Stop" dark field simulations of (on the left) GaAs and (on the right) $Al_{0.3}Ga_{0.7}$. As for a specimen thickness of 41nm and a defocus of 80nm. The parameters used were as appropriate for the Cambridge H.R.E.M. at 500kV. It is assumed that 75% of the electrons contributing to the image do so elastically. In (a) half of the 25% of electrons which are inelastically scattered (mainly into the plasmon loss centred at 20eV) are taken to be further elastically scattered, this contribution being added incoherently in a first order model (see [6]). The simulation in (b) is for the "hollow cone" of the other half of the 25% of the inelastically scattered electrons which are taken to have been scattered into a cone at an angle to evade the centre stop aperture. It is the elastic scattering of these electrons which yields a 0.28nm fringe spacing which is rather insensitive to defocus because of the hollow cone geometry as well as the large (8eV) energy spread about the plasmon loss value. The simulation in (c) is the sum of (a) and (b) and it is only by including both types of inelastic contribution that the characteristics of figure 1b are achieved. For further details see [7].

128

relief although it is certainly affected by inelastic scattering. It should be emphasised that surface relaxation effects can be very important in the structural assessment of a wide variety of images of strained multi-layer systems [37] unless only very long extinction distance reflections are used [eg. 38].

While the full structural and compositional characterisation of strained multilayer systems remains extremely difficult, it is encouraging that, in one case at least, this has been achieved for a metallic system which, for a variety of reasons, should be more difficult to assess than most semiconductor multilayers. The system examined consisted of alternate layers of Cn an Ni_xPd_{1-x} at various wavelengths above and below those for which anomalous properties are reported for systems of this type (eg. [39]). In broad terms the TEM techniques employed were simple developments of those described above for the characterisation of interfaces. High resolution techniques were used, not in an attempt to determine precise atomic positions, but instead to assess the average spacings both perpendicular and parallel to the growth direction of the layering as a function of the wavelength. Even in this context the improved accuracy achievable using non-axial methods [40] proved essential for the strain measurements. (See figure 3). Furthermore, modelling the fringe spacing variations observed (as exemplified in figure 4) required an assessment of the form of the compositional modulation (as achieved by the sputter growth technique used (eg. [41])) and this required the full application of the Fresnel method. (See figure 5).

Hence the degree of interlayer mixing was determined to very nearly atomic spacing accuracy [42]. The results of this investigation have been reported for sometime [43] and we will only summarise them here. It was found that whereas for long wavelengths (4.4 nm) the ratios of the planar spacings in and perpendicular to the [001] growth direction ($^-$0.99) (measured to ± 0.2%) could be explained on a conventional elasticity approach, those for a lower (1.6 nm) wavelength ($^-$ 1.025) could not. The result appears to require that a novel form of "phase transition" occurs as the wavelength is decreased. We emphasise that in our opinion the results are not due to loss of epitaxy or to the rich variety of image artefacts present which were quantitatively assessed before coming to this conclusion.

Figure 3: A non-axial high resolution image of an artificially layered Cu/Ni-Pd multilayer. Approximately four periods of the layering of wavelength circa 4.4nm can be seen. Despite the axial point resolution of the microscope used being better than 0.18nm variations in the lattice spacing are only reproduced measurably in a non-axial imaging mode (see [40]).

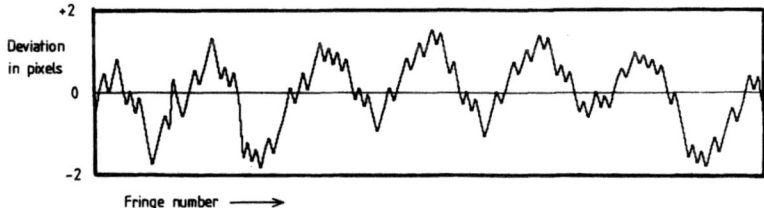

Figure 4: Deviations of the measured lattice fringe positions from the line of least squares fit using positions taken from a digitised region of figure 3 of six wavelength extent. 1 pixel corresponds to approximately 5% of the fringe spacing. It should be noted that the true lattice strains present have to be determined by the comparison of simulations of models given different strains but the composition profile as determined from the lower resolution Fresnel images as shown in figure 5. (See [42] and [43]).

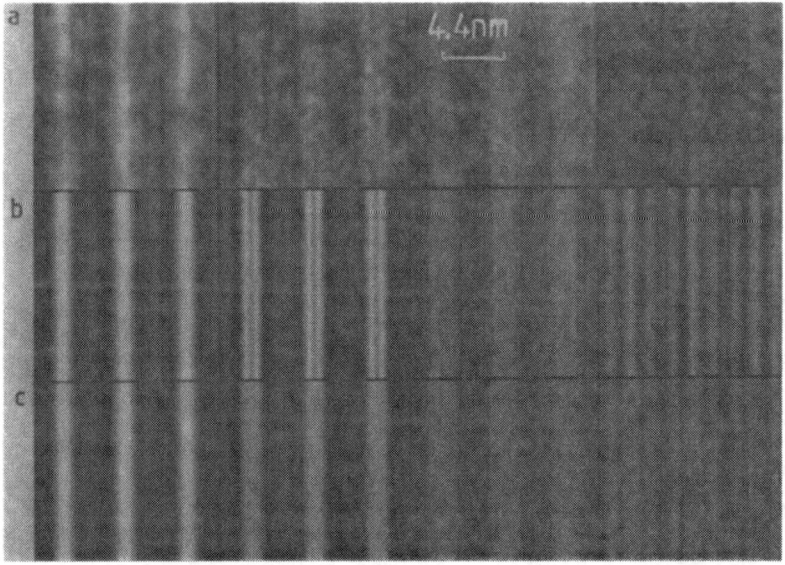

Figure 5: Four of a through focal series of 32 images for the 4.4nm wavelength film, (as shown in figure 3,) are shown at (a). The defoci at 100kV (using a Philips 4000ST with a low convengence LaB_6 source are from left to right: -570nm, -220nm, -20nm and +200nm. The simulations shown in (b) exhibited the best match (for a specimen thickness of 22nm) and required intermixing over 2 atomic planes. The simulations shown in (c) are included for comparison and are for the identical conditions as those in (b) but now intermixing has been allowed over 4 atomic planes. Compare in particular in the second image from the left the double weak fringe in (a) and (b) and the incorrect single fringe in (c).

DISCUSSION

In this review of the newer TEM techniques which can be applied to the characterisation of a multilayer I have concentrated on those which can allow the fullest structural and compositional analysis. Frequently, such analyses can take months rather than days to complete and should be attempted only if some new physics is involved (as appears to be the case for metallic multilayers) or for the very careful modelling of some of the structurally sensitive properties of well made semiconductor heterostructures. It should not be forgotten that many of the simpler measurements of layer thickness and quality for a multilayer can nonetheless be completed in a matter of hours and here the use of a TEM is competitive with most other methods. It is, however, destructive, but even so a variety of new approaches to reflection electron microscopy, in particular for the assessment of steps [33], make non destructive assessment in more general terms using an electron microscope a real possibility.

ACKNOWLEDGEMENTS

I am grateful to Professor D. Hull for the provision of laboratory facilities, to the S.E.R.C. and to a variety of industries including British Telecom, G.E.C., I.B.M., Johnson Matthey, Philips and S.T.L. for support in connection with different aspects of the work described. I would also like to thank C.S. Baxter, E.G. Bithell (nee Britton), C.B. Boothroyd, P.E. Donovan, S.B. Newcomb, F.M. Ross and E.J. Williams for their help in the development of many of the techniques described.

REFERENCES

1.   W.M. Stobbs in Electron Microscopy in Materials Science, Proceedings of the Erice School, edited by A. Howie and W. Valdre (Plenum Press, New York, 1988).
2.   W.M. Stobbs, in The Physics and Fabrication of Microstructures and Microdevices, edited by M.J. Kelly and C. Weisbuch (Springer Proceedings in Physics 13, Springer Verlag, Berlin, 1986) p. 136.
3.   E.G. Britton, K.B. Alexander, W.M. Stobbs, M.J. Kelly and T.M. Kerr, GEC Journal of Research 5, 31 (1987).
4.   W.M. Stobbs, in IIIrd M.S.S. Conference Montpelier, J. de Physique; in press, (1987).
5.   S.B. Newcomb, C.B. Boothroyd and W.M. Stobbs, J. Miscrosc. 140, 195 1985).
6.   W.M. Stobbs and W.O. Saxton, J. Microsc. in press, (1988).
7.   C.B. Boothroyd and W.M. Stobbs in E.M.A.G., IOPCS, edited by L.M. Brown (Adam Hilger, Bristol, 1987); submitted to Ultramicroscopy.
8.   J. Wood, W.M. Stobbs and D.J. Smith, Philos. Mag. A50, 375 (1984).
9.   W.M. Stobbs, G.J. Wood and D.J. Smith, Ultramicrosc. 14, 145 (1985).
10. J.W. Matthews and W.M. Stobbs, Philos. Mag. 36, 373 (1977).
11. R.C. Ecob and W.M. Stobbs, J. Microsc. 129, 275 (1983).
12. J.M. Vitek, Ultramicrosc. 22, 197 (1987).
13. C.B. Boothroyd, A.P. Crawley and W.M. Stobbs, Philos. Mag. A54, 633 (1986).
14. J.W. Ness, W.M. Stobbs and T.F. Page, Philos. Mag. A54, 679 (1986).
15. D.J.H. Cockagne, I.L.F. Ray and M.J. Whelan, Philos. Mag. 20, 1265 (1969).
16. F.M. Ross and W.M. Stobbs, in Proceedings of A.E.M., edited by G. Lorimer (Institute of Metals, London 1988).
17. F.M. Ross and W.M. Stobbs, Surf. Int. Anal. (1988).

18. F.M. Ross and W.M. Stobbs, presented at MRS Boston Fall Meeting 1987.
19. K.M. Mertile and D.J. Smith, Ultramicrosc. 22, 57 (1987).
20. W. Krakow, J.T. Wetzel and D.A. Smith, Philos. Mag. A53, 739 (1986).
21. W. Krakow, and D.A. Smith, Ultramicrosc. 22, 47 (1987).
22. W. Wunderlich and M. Ruhle, in XIth Proc ICEM, edited by T. Imura, S. Maruse and J. Suzuki (Jap. Soc. E.M. Tokyo, 1986) p. 1335.
23. A.P. Sutton and V. Vitek, Phil. Trans. R. Soc. (London) A309, 1 (1983).
24. P.E. Donovan and W.M. Stobbs, J. Microsc. 130, 361 (1983).
25. P.E. Donovan and W.M. Stobbs, Ultramicrosc. 23, 119 (1987).
26. K.B. Alexander, C.S. Baxter, J.E. Evetts, R.E. Somekh and W.M. Stobbs in RAL Workshop on Advanced Technology Reflectors for Space instrumentation, edited by M. Grande (Rutherford Appleton Laboratory, 1987) p. 154.
27. C.S. Baxter and W.M. Stobbs, Appl. Phys. Lett. 48, 1202 (1986).
28. P.M. Petroff, J. Vac. Sci. Technol. 14, 974 (1977).
29. E.G. Britton, Ph.D. Thesis, University of Cambridge (1987).
30. H. Kakibayashi and F. Nagata, Jap. J. Appl. Phys. 25, 1644 (1986).
31. D.J. Eaglesham and C.J. Humphreys, in XIth Proc ICEM, edited by T. Imura, S. Maruse and J. Suzuki (Jap. Soc. E.M. Tokyo, 1986) p. 209.
32. F.M. Ross, E.G. Britton and W.M. Stobbs in Proc of A.E.M., edited by G. Lorimer (Institute of Metals, London 1988).
33. C.B. Boothroyd, E.G. Britton, F.M. Ross, C.S. Baxter, K.B. Alexander and W.M. Stobbs, in Microscopy of Semiconducting Materials, I.O.P.C.S. 87, edited by A.J. Cullis (Adam Hilger, Bristol, 1987) p. 15, and p.195.
34. A.F. de Jong, W. Coene and M. Bender, in Microscopy of Semiconducting Materials, edited by A.T. Cullis (IOPCS Adam Hilger, Bristol, 1987) p.9.
35. W.O. Saxton, K.M. Knowles and W.M. Stobbs, in Proc EMAG I.O.P.C.S. 87, edited by G.J. Tatlock (Adam Hilger, Bristol, 1985) p. 75.
36. C.S. Baxter, W.M. Stobbs, in Proc of A.E.M., edited by G. Lorimer (Institute of Metals, London, 1988).
37. M.M.J. Treacy and J.M. Gibson, J. Vac. Sci. Technol. B4, 1458 (1986).
38. K. Sato and W.M. Stobbs, in Proc E.M.A.G., I.O.P.C.S., edited by L.M. Brown (Adam Hilger, Bristol, 1988).
39. G.E. Henein and J.E. Hilliard, J. Appl. Phys. 54, 728 (1983).
40. D.J. Hall, P.G. Self and W.M. Stobbs, J. Microsc. 130, 215 (1983).
41. R.E. Somekh and C.S. Baxter, J. Cryst. Growth 76, 119 (1986).
42. C.S. Baxter and W.M. Stobbs, in E.M.A.G., I.O.P.C.S. 78, edited by G.J. Tatlock (Adam Hilger, Bristol, 1985) p. 387.
43. C.S. Baxter and W.M. Stobbs, Nature 322, 814 (1986).

EVIDENCE FOR INSTABILITY IN $Pb_{1-x}Eu_xTe$ ALLOYS.

L. Salamanca-Young[+], M. Wuttig[+], D.L. Partin[$] and J. Heremans[$]

[+] Chemical and Nuclear Engineering Department, University of Maryland, College Park, MD 20742

[$] General Motors Research Laboratories, Warren, MI 48090-9055

ABSTRACT

We have used transmission electron microscopy to study the structure of $Pb_{1-x}Eu_xTe$ alloys grown by molecular beam epitaxy. We have observed ordered solid solutions of the $Pb_{1-x}Eu_xTe$ alloys as well as spinodal decomposition for $0.35 \leq x < 0.75$. The spinodal decomposition corresponds to a modulation of both the composition and the lattice spacing of the $Pb_{1-x}Eu_xTe$ alloy. These modulated structures have periodicities of ~18Å along the <111> and <110> directions and indicate that the solid solution of $Pb_{1-x}Eu_xTe$ is unstable in this range of compositions.

INTRODUCTION

Coherent heterostructures have received much attention due to their technical importance in electronic devices such as lasers, transistors and detectors. The devices operate because the energy gap of adjacent layers is different. The difference in the energy gap is usually associated with a difference of the lattice parameters which conflicts with the requirement of coherency. The coherence between the layers is required because the number of defects at the interface between the layers is considerably reduced when the layers are coherent. As a result, the performance of the devices fabricated with coherent heterostructures is greatly improved. Thus, the key to the development of high performance heterostructures consists of finding the proper compromise between the difference of the energy gap on one hand and the structural match of the layers on the other hand. The band gap ΔE and lattice parameter a of PbTe/$Pb_{1-x}Eu_xSe_yTe_{1-y}$ heterostructures, for example, are adjusted by varying x and y in the $Pb_{1-x}Eu_xSe_yTe_{1-y}$ layers. It is known that x increases both ΔE and a whereas y decreases a and has a negligible effect on ΔE.[1] Not all of the phase diagrams underlying the solid solution $Pb_{1-x}Eu_xSe_yTe_{1-y}$ are completely known. It is known, however, that PbTe, EuTe and PbEu are immiscible. Recent high resolution transmission electron microscopy (TEM) studies have shown spinodally decomposed $Pb_{1-x}Eu_xTe$ layers indicating that $Pb_{1-x}Eu_xTe$ is thermodynamically unstable for a range of compositions.[2] Similar observations have been reported for III-V semiconductor compounds.[3,4] Thus, the conflict between band gap and lattice parameter adjustment and thermodynamical stability of the different layers forming a heterostructure is generic. The $Pb_{1-x}Eu_xTe$ system is of particular importance because of its use in the fabrication of devices based on the PbTe/EuTe system.

It is now apparent that the coherency stresses can stabilize disordered solid solutions [5] or in extreme cases stabilize ordered compounds which do not appear in the equilibrium phase diagram of the bulk material [6-9]. The excess free energy of an epitaxially stabilized solid solution can be written as [7]

$$\delta F = K_c(1-K_e/K_c)c(1-c) + kT[c\ln(c) + (1-c)\ln(1-c)] \qquad (1)$$

where K_c represents the solute–solute, solvent–solvent and solute–solvent interactions, and K_e is a measure of the stabilizing epitaxial strain energy. The quantities k, T and c denote Boltzman's constant, the temperature and the concentration, respectively.

It is possible to distinguish between three kinds of epitaxial alloys. Those with essentially no lattice mismatch are characterized by $K_e/K_c \ll 1$. In this case the properties and range of stability of the epitaxial alloy do not differ significantly from the known bulk properties. The extreme case of large mismatches is characterized by a ratio $K_e/K_c \gg 1$. This is the situation where new strain stabilized epitaxial structures are formed.[6,8,9] The intermediate case, $K_e/K_c \approx 1$, corresponds to epitaxial structures which are partially strain stabilized so that disordered or ordered solid solutions are marginally stable.[10] Potential instabilities with respect to long or short wave compositional instabilities exist at finite temperatures in these alloys. This intermediate case is of central interest in this research as we present studies on the stability, meta-stability and instabilities of partially strain stabilized epitaxial structures.

Instabilities of a solid solution are found for a range of compositions where the base elements forming the alloy are immiscible. Spinodal decomposition is the result of fluctuations in the composition of a solid solution of average composition within the miscibility gap. The modulations in the composition are along the primary crystal directions and have relatively large wavelengths.[5] Spinodal decomposition has been observed mostly in metallic alloys and recently in ternary and quaternary III-V semiconductor alloys.[11] The loss of stability occurs when the difference between the free energy of a homogeneous solid solution and that of a solid solution with an infinitesimal concentration fluctuation is less than zero. As a consequence, a spontaneous increase in the amplitude of the fluctuation takes place. In the range of compositions within the miscibility gap a homogeneous solid solution might be stable at temperatures above a critical temperature T_c. On cooling from above T_c, the system becomes unstable to even small fluctuations in the composition at temperatures below T_c.[5] If strain is associated with the composition modulation T_c is lower (elastic spinodal) than when there is no strain (chemical spinodal).[12] In a binary system, the region for instability is defined by the locus of

$(\partial^2 F/\partial c^2)_{\sigma,T} < 0$ where F is the free energy, c the composition, σ the stress and T the temperature.

Several ternary [13,14] and quaternary [3,13,15] epitaxially grown III-V semiconductor alloys have been observed to be unstable below a critical temperature T_c. This instability could be detrimental to devices fabricated with compositions within the miscibility gap if they are to operate at temperatures below the critical temperature. It has also been observed that in some instances the strain between the substrate and the film can stabilize the solid solution of the epitaxial alloy over a larger temperature range if the growth is performed at a temperature above the critical temperature T_c.[16]

EXPERIMENTAL DETAILS

$Pb_{1-x}Eu_xTe$ alloys were grown on the (111) surface of BaF_2 substrates using molecular beam epitaxy. In some instances a buffer layer of PbTe was grown first in order to study the effect of the lattice mismatch between the $Pb_{1-x}Eu_xTe$ film and the substrate. The growth temperatures were ~300°C. Samples were prepared with several values of the relative concentration of Eu, x, in the range 0.2-0.75. The samples for electron microscopy were prepared first by mechanical grinding and subsequent ion milling at low temperatures. We obtained information on the structure of the samples from both electron diffraction patterns and high resolution lattice imaging.

RESULTS AND DISCUSSION

Both, electron diffraction patterns as well as high resolution lattice images obtained from the $Pb_{1-x}Eu_xTe$ alloys showed that the alloys are unstable for 0.35<x<0.75. For example, the lattice image and electron diffraction pattern shown in Fig. 1 show a modulation of ~21 Å along the <1$\bar{1}$0> direction perpendicular to the <111> growth direction. This modulation is observed in the electron diffraction pattern by the appearance of diffuse satellites around the crystalline Bragg reflections and along the <1$\bar{1}$0> direction and corresponds to spinodal decomposition as explained below. It is important to note that the intensities of the satellite peaks n and -n (n is the order of the satellite) are not the same (see inset to Fig. 1). This is in agreement with a modulated structure where there is a sinusoidal modulation of both scattering (c) and interplanar spacing (d) of the form

$$c = c_0 + A\cos(\beta \cdot r) \qquad (2)$$

$$d = d_0 + A\eta\cos(\beta \cdot r) \qquad (3)$$

where A and $2\pi/\beta$ are the amplitude and wavelength of the modulation, respectively, c_0 and d_0 are the average composition and spacing, respectively, and η is the linear expansion per unit composition change. In this case the intensity of the nth satellite is given by [17]

$$|F(\pm n)| \propto |\mp \gamma J_{n-1}(h\eta QA/d_o) + fJ_n(h\eta QA/d_o) \mp \gamma J_{n+1}(h\eta QA/d_o)| \quad (4)$$

where $J_n(x)$ is the nth Bessel function, γ is a constant, h is the crystalline Bragg reflection, and $Q=2\pi/\beta d_o$.

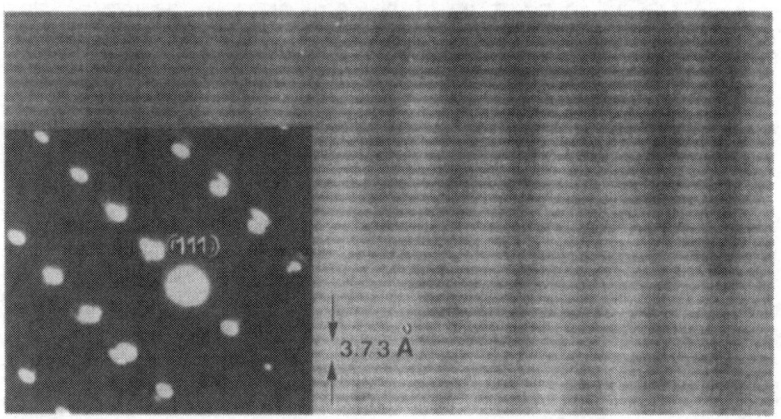

Figure 1. $(11\bar{2})$ lattice image of a $Pb_{0.59}EuTe$ alloy showing spinodal decomposition along the $<1\bar{1}0>$ direction with a modulation wavelength of ~21Å. The inset shows the corresponding electron diffraction pattern.

In the case where only one kind of modulation is present, symmetric satellites (with almost equal intensities) are obtained.[17] For the alloy shown in Fig. 1, the scattering modulation corresponds to a modulation in the composition of the Pb and Eu atoms in the lattice along the $<1\bar{1}0>$ direction. The modulation in the lattice spacing is a result of the modulation in the composition since Eu is bigger than Pb and the interplanar spacings of EuTe are larger than the corresponding ones of PbTe. We have also observed spinodal decomposition for alloys with x=0.43 and 0.51 along the $<1\bar{1}0>$ as well as the $<111>$ directions. Figure 2 shows electron diffraction patterns of a $Pb_{0.57}Eu_{0.43}Te$ alloy showing periodicities of 14.62Å (Fig. 2a)) and 18.72Å (Fig. 2b)) along the $<111>$ and $<1\bar{1}0>$ directions, respectively.

The observed directions for spinodal decomposition can be understood in terms of the elastic contribution to the free energy. The total free energy of an isotropic solid solution is [5]

$$F=\int [f'(c) + \frac{\eta^2 E}{1-\nu}(c-c_o)^2 + \kappa(\nabla c)^2]dV \quad (5)$$

where $f'(c)$ is the free energy of a homogeneous solid solution
with composition c, $\kappa(\nabla c)^2$ represents the increase in free energy
due to a gradient in the composition ∇c and E and v are Young's
modulus and Poison's ratio, respectively. For an isotropic cubic
lattice with a composition modulation the strain associated
with the composition modulation is given by [12]

$$\varepsilon = A\eta\cos(\beta.r) \tag{6}$$

and the strain energy is given by

$$F_\sigma = Y_{lmn}\varepsilon^2 \tag{7}$$

where

$$Y_{lmn} = \frac{1}{2}(C_{11}+2C_{12})\left[3 - \frac{(C_{11}+2C_{12})}{(C_{11}+2(2C_{44}-C_{11}+C_{12})(l^2m^2+m^2n^2+n^2l^2))}\right] \tag{8}$$

for a modulation along the z-axis. C_{ij} are the elastic constants
and l, m and n are the direction cosines between β and the cube
axes.[12] Thus, the minimum value of the elastic energy depends
on the sign of the anisotropy term $2C_{44}-C_{11}+C_{12}$.[12] If $2C_{44}-C_{11}+C_{12}>0$ the elastic energy is minimum for β along <100> and
maximum for β along <111>. This implies that the system becomes
unstable first to fluctuations along <100>. The opposite is true
for both PbTe and EuTe where $2C_{44}-C_{11}+C_{12}<0$. Therefore a solid
solution of $Pb_{1-x}Eu_xTe$ becomes unstable first to fluctuations in
the composition along the <111> directions, then along the <110>
directions and last along <100> in agreement with our TEM
observation. For PbTe and EuTe $Y_{111}=0.56\times10^{12}$ d/cm^2 and 0.61×10^{12}
d/cm^2, $Y_{110}=0.78\times10^{12}$ d/cm^2 and 0.74×10^{12} d/cm^2 and $Y_{100}=1.15\times10^{12}$
d/cm^2 and 0.99×10^{12} d/cm^2, respectively. As a first approximation
the Y_{lmn} for the $Pb_{1-x}Eu_xTe$ alloys are obtained by taking the
weighted averages between the Y_{lmn} of PbTe and EuTe. In this case
Y_{111} and Y_{110} are comparable and give rise to modulations along
both the <111> and <110> directions.

Transmission electron microscopy results obtained from
samples with compositions in the range $0<x<0.35$ and $x=0.75$ did
not show evidence for spinodal decomposition. For these values of
x our TEM results show that the $Pb_{1-x}Eu_xTe$ alloys are stable and
form disordered solutions. In addition to spinodal decomposition
and disordered solutions we have observed ordered solutions in
small areas of the samples with $x=0.35$ and $x\approx0.5$. The ordered
solutions give rise to reflections that are not allowed in the
NaCl structure.[2] The observation of ordered and disordered
solutions, as well as spinodal decomposition for $Pb_{1-x}Eu_xTe$ alloys
with $0.35<x<0.75$ indicates that the $Pb_{1-x}Eu_xTe$ alloys are
marginally stable in this range of compositions.

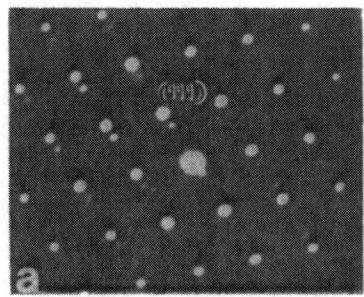

Figure 2. a) (1$\bar{1}$0) and b) (11$\bar{2}$) electron diffraction patterns of a $Pb_{0.57}Eu_{0.43}Te$ alloy showing periodicities of 14.62Å and 18.72Å along the <111> and <1$\bar{1}$0> directions, respectively.

CONCLUSIONS

The observation of spinodal decomposition indicates that the $Pb_{1-x}Eu_xTe$ system is unstable to fluctuations in the composition. The observation of ordered solutions on the other hand, indicates that the $Pb_{1-x}Eu_xTe$ alloys are strain stabilized. The observation of both, spinodal decomposition and ordered solutions indicates that the $Pb_{1-x}Eu_xTe$ alloys are only partially stabilized.

ACKNOWLEDGMENTS

We acknowledge Sahn Nahm and Helaleh Maghsoudlou for preparing the samples for TEM.

REFERENCES

1. D.L. Partin, J. Electronic Mat. 13, 493 (1984).
2. L. Salamanca-Young, D.L. Partin and J. Heremans, J. Appl. Phys.(in press).
3. G.B. Stringfellow, J. Appl. Phys. 54, 404 (1983).
4. K. Onabe, Jpn. J. Appl. Phys. 22, 287 (1983).
5. J.W. Cahn, Acta Met. 9, 795 (1961).
6. P.B. Littlewood, Phys. Rev. 34, 1363 (1986).
7. C.P. Flynn, Phys. Rev. Lett. 57, 599 (1986).
8. A.A. Mbaye, A.Zunger and D.M. Wood, Appl. Phys. Lett. 49, 782 (1986).
9. A. Zunger, Appl. Phys. Lett. 50, 164 (1986).
10. L. Salamanca-Young and M. Wuttig, to be published.
11. See for example G.B. Stringfellow, J. Cryst. Growth 58, 194 (1982).
12. J.W. Cahn, Acta Met. 10, 179 (1962).
13. O. Ueda, S. Isozumi and S. Komiya, J. Appl. Phys. 23, L241 (1984)
14. J.R. Pressetto and G.B. Stringfellow, J. Cryst. Growth 62, 1 (1983).
15. B. de Cremoux, P. Hirtz and J. Ricciardi, Proc. of the 1980 Int. Symp. on GaAs and Related Compounds, Vienna 1980 (The Institute of Physics, Bristol and London, 1981), p. 115.
16 M. Quillec, H. Launois and M.C. Joncour, J. Vac. Sci. Technol. B1, 238 (1983).
17. D. de Fontaine in Local Atomic Arrangements Studied by X-Ray Diffraction, ed. by J.B. Cohen and J.E. Hilliard, Gordon Breach, N.Y. (1966), p 51.

PART III

Phase Stability

STABILITY AND METASTABILITY IN SEMICONDUCTOR STRAINED-LAYER STRUCTURES

Brian W. Dodson, I.J. Fritz, S. Thomas Picraux, and Jeffrey Y. Tsao, Sandia National Laboratories, Albuquerque, NM 87185.

ABSTRACT

The physics governing stability properties and relaxation of mismatch strain in semiconductor strained-layer structures is reviewed. Experimental data on stability and rates of strain relaxation are examined. We conclude that essentially all observations on structural relaxation of semiconductor strained-layer structures can be explained by standard models of plastic deformation adapted to the special conditions controlling dislocation dynamics in these structures.

EQUILIBRIUM STABILITY BEHAVIOR

When a material is grown on a chemically similar but lattice mismatched substrate, a coherently strained layer may be formed if the lattice mismatch is sufficiently small[1]. The stability of such a layer is determined by the competing strain energies of two alternate structures. The coherently strained layer has a homogeneous strain energy associated with the misfit strain. The competing structure is one in which interfacial coherence has been lost, and the lattice mismatch is accommodated by a combination of homogeneous strain and misfit dislocations at the interface. The strain energy densities of the two possible structures depend on film thickness in different manners. As a result, there is a critical layer thickness below which the coherently strained overlayer is stable, but above which the incoherent structure is preferred.

A convenient approach to the problem of strained-layer stability was introduced by Matthews and Blakeslee[2]. They considered the balance of forces on a threading dislocation in the strained layer. Because there is no stress field acting on the portion of the dislocation in the substrate, motion in the strained overlayer can take place only by elongation along the substrate/film interface, forming a misfit dislocation. This elongation is opposed by the line tension of the misfit dislocation, which contributes another term to the total thermodynamic force acting on the threading dislocations. If there is a net force acting to move the threading dislocation in the strained layer, thus producing misfit dislocations at the layer interface, the interfacial coherence has become unstable. This force balance criterion for stability of strained-layer coherence is equivalent to the earlier energy balance arguments, but emphasizes the role of forces on dislocations for driving relaxation in strained heterostructures.

We have recently extended the viewpoint of Matthews and Blakeslee by observing that the force balance argument can be used to define an effective stress acting to drive dislocation motion in a strained layer[3,4]. The physical stress in a strained overlayer is related to the layer strain ϵ by

$$\sigma_m = \frac{2(1+\nu)}{(1-\nu)} \mu \, \epsilon, \qquad (1)$$

where ν is Poisson's ratio and μ is the shear modulus. As discussed above, the force generated on a dislocation segment by the physical stress in the overlayer is reduced by a contribution resulting from the misfit dislocation tension. This counterforce can be expressed as a contribution to the total thermodynamic force acting on dislocations in the overlayer. This term is simply

$$\sigma_d = - \frac{\mu \ (1-\nu\cos^2\beta)}{2\pi(1-\nu)} \ \frac{\ln(4h/b)}{(h/b)} \ , \tag{2}$$

where h is the layer thickness, b is the magnitude of the Burger's vector of the dislocation, and β is an angle describing the orientation of the dislocation segment. We define the sum of these two terms to be the excess stress in the strained layer ($\sigma_{ex} = \sigma_m + \sigma_d$).

The excess stress is a measure of the driving force for dislocation dynamics (and hence strain relief). As an overlayer is grown, the excess stress is initially negative, but increases with increasing thickness. When the excess stress becomes greater than zero, dislocation motion begins, resulting in strain relief. As a result, the structure cannot be in equilibrium if the excess stress is greater than zero. This has two consequences. The limit of the stability regime for the coherently strained overlayer is defined by an excess stress of zero. This is equivalent to the Matthews-Blakeskee criterion. In addition, if a structure thicker than the equilibrium critical thickness is grown, the amount of strain relaxation at thermodynamic equilibrium will be that producing a state of zero excess stress. Later we will describe the application of the excess stress for relaxation of nonequilibrium structures.

The application of continuum models to structures having very thin layers (perhaps less than 100 Å) may seem problematical, in that the discrete atomic structure should alter the energetics at some point. To test the regime of validity of the continuum viewpoint, the thermodynamic stability of thin siliconlike overlayers has been studied by Dodson and Taylor using atomic-scale simulations[5]. The equilibrium critical thickness was determined by comparing energies of the relaxed coherent and incoherent structures to determine which of these competing states has lower energy. By varying the overlayer thickness and lattice mismatch over a wide range, the dependence of the critical thickness on lattice mismatch was obtained. The primary result is that the continuum and atomistic stability models agree for films thicker than 6 monolayers (about 20 Å). In thinner films there is a gradual decrease in stability relative to the continuum models. This has two causes. First, the nature of the misfit dislocations changes from the conventional three-dimensional structures to the solitonlike structures characteristic of mismatched monolayers. Second, the stability properties in thin films are determined as much by surface and interfacial energies as by the volumetric strain energies. However, such effects are not expected to be important in the structures considered in the remainder of this paper.

EXPERIMENTAL RESULTS ON STABILITY AND METASTABILITY

Relaxation of Mismatch Strain

One of the first studies of relaxation in strained semiconductor structures was carried out by Sugita, Tamura, and Sugawara, who investigated the formation of misfit dislocations in silicon overlayers deposited at 1410 °K on boron-doped silicon (111) substrates[6]. The boron concentration was chosen to produce a mismatch of 9×10^{-5} relative to pure silicon. The misfit dislocations in the resulting structures were detected using x-ray topography, which provides direct images of the dislocations, and is thus sensitive to dislocation densities as small as one per cm. There is some indication that boron diffusion into the overlayer during the deposition process results in a smaller effective mismatch, and thus in onset of relaxation at about 4.5μm thickness, rather than the value of about 2.5μm expected from the assumption of equilibrium behavior and no boron diffusion. However, even taking this into account, it is clear that the progress of relaxation in still thicker structures is much slower than predicted by equilibrium models. For example, when an overlayer 16.4μm thick is examined, the amount of relaxation observed is about 5×10^{-6}, whereas at equilibrium (adjusted for the effects of boron diffusion) the relaxation would be $4\text{-}6 \times 10^{-5}$. The actual relaxation is thus about an order of magnitude less than expected at equilibrium.

Similar behavior was observed by Matthews and coworkers in germanium overlayers grown on GaAs(110) substrates[7]. These overlayers, having a mismatch of 7×10^{-4}, were deposited at a temperature of 620 $^{\circ}$K. This low temperature was chosen to emphasize the role of kinetic factors in the processes leading to relaxation of misfit strain. As in the boron-doped silicon bilayers described earlier, these structures exhibit strongly nonequilibrium relaxation behavior. The amount of strain relaxation was determined using topographic techniques having a sensitivity of about 2×10^{-5}. The equilibrium critical thickness of these structures is about 0.25μm. However, no relaxation is observed (within the sensitivity of the topographic analysis) for overlayers with a thickness of 2μm, even though the equilibrium relaxation would have been about 6×10^{-4}. An annealing study was also carried out, in which a structure having a germanium overlayer 1.75μm in thickness was heated to 870°K for 30 minutes after growth. Similar overlayers not subjected to this annealing procedure showed no relaxation. However, the annealed film exhibited strain relaxation of 4×10^{-4}, which is about 2/3 of the equilibrium value. This study directly established the role of temperature in affecting the rate of strain relaxation in strained semiconductor overlayers.

A study of $Si_{1-x}Ge_x$ films ($x \geq 0.15$) deposited on Si(100) substrates was performed by Kasper et al[8]. These structures, which were grown at temperatures of 1020 $^{\circ}$K, had an order of magnitude more mismatch than the earlier studies (up to mismatch values of 6×10^{-3}). The amount of strain relaxation was determined by measuring the separation of misfit dislocations using transmission electron microscopy, resulting in an effective sensitivity of about 3×10^{-5}. The relaxation again appears to be strongly nonequilibrium in nature, being roughly an order of magnitude less than expected at equilibrium.

These initial SiGe/Si(100) studies were extended by Bean and coworkers, who developed the capability of growing high-quality films via layer-by-layer growth for all SiGe compositions[9-11]. These films were grown at 820 $^{\circ}$K at a deposition rate of 5 Å/sec. A variety of analytical techniques were used to study strain relaxation, including x-ray diffraction, ion channeling, electron microscopy, and Raman scattering. Although these techniques involve very different interactions with the structure of the strained overlayer, they all have similar sensitivities to strain relaxation (about 1×10^{-3}). Consistent with the earlier results, Bean found that these structures showed very sluggish relaxation compared to expectations based on equilibrium models. Annealing experiments were also carried out[11], providing a direct demonstration that strained layers thicker than the equilibrium critical thickness are metastable.

A useful summary of the data from these various experimental studies of relaxation in strained-layer semiconductor structures is presented in Figure 1, where the observed relaxation is plotted against the excess stress of the unrelaxed structure. Despite local discrepancies (probably related to differences in deposition temperature and timescale), there is clearly a trend which is consistent over three orders of magnitude in excess stress. Roughly speaking, the amount of strain relaxation is proportional to the square of the initial excess stress. We shall see later that this relation is predicted by a new model for relaxation of mismatch strain based on the physics of plastic deformation.

Experimental Measurements of 'Critical Thickness'

Based on these measurements, Bean and his coworkers determined a 'critical thickness' as a function of alloy composition for nearly the entire range of compositions. The criterion for this 'critical thickness' was development of a detectable amount of strain relaxation (thus, about 1×10^{-3}). Because of the sluggish relaxation, which seems ubiquitous in this class of materials, the 'critical thickness' determined experimentally based on this criterion is expected to be much larger than the equilibrium critical thickness. This turns out to be the case.

Much attention has recently been focused on determination of this low-resolution 'critical thickness' for a range of systems. Kamigaki and coworkers[12], and Orders and Usher[13], have both presented x-ray diffraction studies of strain relaxation in strained InGaAs/GaAs(100) heterostructures. In addition, Abstreiter[14] has examined the case of SiGe strained layers grown on GaAs(110) substrates at low temperature (620 °K) and small deposition rates (about 5 Å/min). His group has used both LEED and Raman techniques for structural analysis. In all cases, the resulting 'critical thickness' values are in reasonable agreement with the broader study of Bean et al (Figure 2). This suggests that the detailed nature of the strained layer, the substrate, or the growth process does not strongly influence the observed relaxation behavior.

Role of Experimental Resolution in Critical Thickness Measurements

Every experimental study of strain relaxation in strained-layer semiconductor structures reveals that the extent of relaxation is much less than expected from equilibrium models. In addition, annealing studies provide strong evidence that the structures are metastable, since further heating results in additional strain relaxation. The sluggish nature of relaxation in these structures suggests the possibility that measurements of critical thickness may depend strongly upon the sensitivity of the technique used to detect the initial stanges of relaxation.

Discrepancies between experimental critical thickness values measured using various analytical techniques were noted by Laidig and coworkers[15], who studied relaxation in InGaAs/GaAs strained-layer superlattices. The probes used to determine quality were x-ray diffraction, having a resolution of perhaps 2×10^{-3}, and efficiency of stimulated photoluminescence, which should be sensitive to relaxations on the order of 10^{-5}. They found that the critical thickness determined by x-ray diffraction was more than a factor of two larger than that determined using the photoluminescence efficiency, and that the latter value was in reasonable agreement with equilibrium models. This effort was expanded by Fritz, Laidig, and coworkers[16], who studied a wide range of InGaAs/GaAs superlattices using high-resolution analytical techniques (low-temperature Hall mobility and photoluminecsence efficiency). They found that these techniques agreed in determining the onset of relaxation, and that relaxation begins near the equilibrium critical thickness. This is in strong constrast with the low-resolution measurements described above, in which the 'critical thickness' values are much greater than the equilibrium values.

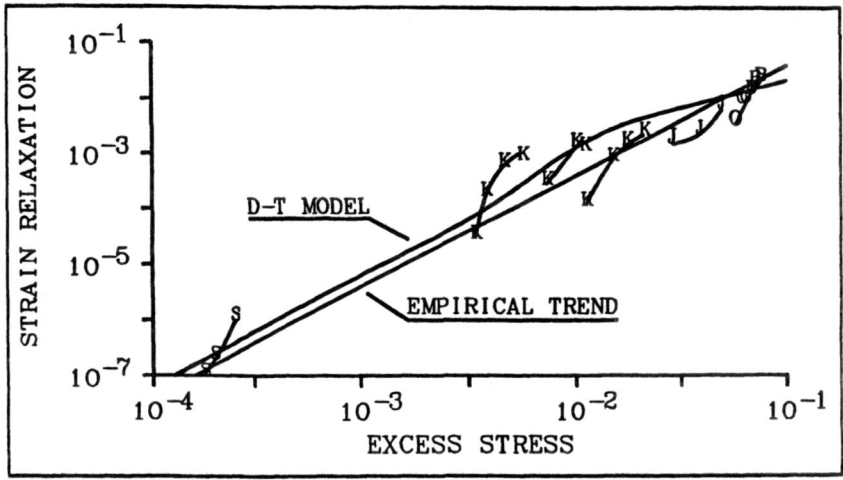

Figure 1. A summary of various experimental stucies of relaxation in strained-layer semiconductor sturctures, plotted in terms of strain relaxation versus initial excess stress (see text). The experimental points are S (Ref. 6), K (Ref. 8), J (Ref. 12), O (Ref. 13), and B (Ref. 10). The line marked 'Empirical Trend' sets relaxation proportional to the square of the excess stress. The curve marked 'D-T Model' is a result of the Dodson-Tsao kinetic relaxation model.

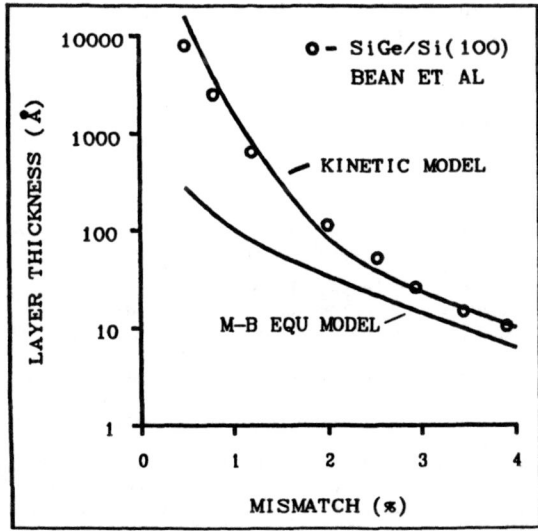

Figure 2. Low-resolution (10^{-3}) critical thickness measurements for the SiGe/Si(100) system taken by Bean and coworkers (Ref. 10). These data points differ strongly from the equilibrium critical thickness (M-B Equ. Model), but are in good agreement with the Dodson-Tsao kinetic relaxation model (Kinetic Model).

The experiments of Laidig and Fritz are certainly suggestive that the resolution of the analytical technique has a strong influence on measurements of the onset of relaxation in strained-layer structures. The results are not entirely satisfactory, however, since neither lattice relaxation nor misfit dislocation density were directly measured. This limitation has recently been addressed by Fritz, Gourley, and Dawson, in a series of papers[17-19] examining the relation between experimental resolution and critical thickness measurements. Gourley[17] has developed a technique, called photoluminescence microscopy, in which the misfit dislocations in a strained-layer structure are imaged as dark-line defects on a background of recombination radiation. When used in a scanning mode, this technique is capable of detecting relaxation on the order of 10^{-8}. The technique is easy to use, non-destructive, and is eminently suited to detection of the first stages of strain relaxation.

Photoluminescence microscopy has been used to examine relaxation in two series of InGaAs/GaAs structures, namely single strained quantum wells and strained-layer superlattices[18,19]. In both cases, misfit dislocations begin to appear near the equilibrium critical thickness. As the layer thickness continues to increase, the misfit dislocation density is much less than predicted by the equilibrium models. Thus, even though relaxation begins where equilibrium theory indicates it should, the extent of relaxation is much less than expected. In some sense, then, relaxation of misfit strain is sluggish in strained semiconductor structures. Most recently, the photoluminescence microscopy results have been compared with x-ray diffraction measurements carried out on the same samples. As expected from earlier studies (described above), the 'critical thickness' measured using the low-resolution x-ray technique was about three times larger than that measured using photoluminescence microscopy, which was in good agreement with the equilibrium value. It is now clear that low-resolution measurements of the onset of relaxation do not provide information on the critical thickness, but rather determine the conditions which produce sufficient relaxation to be detected. The extreme sluggishness of strain relaxation in these structures makes this distinction crucial.

MODELS FOR RELAXATION IN METASTABLE STRUCTURES

Nucleation Models

Models in which metastability in strained semiconductor structures is limited by onset of nucleation of misfit dislocations have, on occasion, been proposed. The best known of these models is probably that recently introduced by People and Bean[20]. However, the data on relaxation of metastable structures does not encourage the use of nucleation models. First, the evidence shows that relaxation begins as the appropriate equilibrium conditions are exceeded, which is inconsistent with a nucleation threshold occurring far from equilibrium. Second, the energy criterion suggested in Ref. 20 is based on providing sufficient strain energy to nucleate dislocations locally. However, this level of strain energy is enough to nucleate dislocations everywhere. The result would be abrupt and nearly complete relaxation of the misfit strain as the nucleation threshold is exceeded. In fact, the relaxation proceeds very slowly, and seems limited by kinetic factors rather than nucleation mechanisms. Dislocation nucleation may serve some role in strain relaxation. However, the experimental data allows us to rule out models based primarily on nucleation in favor of descriptions based primarily on the kinetics of relaxation.

Fritz's Sluggish Relaxation Model

Fritz[21] has developed a model which predicts experimental 'critical thickness' values based on the sensitivity of the analytical technique and assumptions concerning the sluggish nature of strain relaxation. The assumptions made are that the analytical technique has a resolution R, and that the strained structures exhibit an amount of strain relaxation which is smaller than the equilibrium value by a constant factor Q. The equilibrium strain relaxation is calculated as described earlier. The resulting model has a single adjustable parameter R/Q. When the model is fit to the low-resolution 'critical thickness' data of Laidig, Bean, and Orders and Usher, a reasonable fit is obtained by setting $R/Q \approx 7.5 \times 10^{-3}$. As the typical resolution for these measurements was $\sim 10^{-3}$, this fit suggests that the amount of relaxation displayed by these structures was nearly an order of magnitude smaller than the equilibrium value, which is in good agreement with the experimental studies of relaxation. It is therefore reasonable to conclude that the measurements of anomalously large critical thicknesses in strained-layer structures result from the combined effects of finite experimental resolution and initially sluggish strain relaxation.

Dislocation Dynamics and Plastic Deformation

Although the model developed by Fritz (described above) presents a good argument for the dual role of experimental resolution and sluggish relaxation in understanding experimental observations of metastable strained-layer structures, no mechanism for the slow kinetics of strain relaxation is included. In order to obtain a predictive model for strain relaxation, a description of the relaxation process is required. Relaxation of mismatch strain takes place through introduction of misfit dislocations. A kinetic model of such processes will thus depend on the physics of dislocations in a semiconductor strained-layer structure. We begin by outlining mechanisms for dislocation dynamics and plastic flow in bulk diamond-phase semiconductors developed by Haasen and coworkers[22].

Dislocation glide is mediated by thermally activated nucleation of double kinks in this class of materials. The dislocation velocity can be expressed as the linear product of a dislocation mobility and a driving stress σ, which is just the magnitude of the resolved shear stress driving dislocation motion

$$V = \alpha(\sigma,T) \; \sigma, \tag{3}$$

where V is the dislocation velocity, the mobility $\alpha(\sigma,T) = A \; \exp(-E(\sigma)/kT)$ depends on stress only through the activation energy, $E(\sigma)$ is the (stress dependent) activation energy for dislocation glide, and T is the temperature.

Another phenomenon required to describe plastic flow in semiconductors is dislocation multiplication. This describes the observed tendency for moving dislocations to produce new dislocations in their wake. The mechanisms resulting in this behavior are not presently understood. However, a useful phenomenological description of the rate of dislocation multiplication is

$$d\rho/dt = K \; \rho \; V \; \sigma_{ex} , \tag{4}$$

where ρ is the dislocation density, V is the dislocation velocity, and K is a phenomenological parameter. The combination of this description of dislocation multiplication with the thermally-activated dislocation mobility forms the basis of the model developed by Haasen's group to describe plastic deformation in semiconductors.

Matthews' Dislocation Dynamics Model

The first attempt to apply dislocation dynamics to the problem of relaxation of metastable structures was presented by Matthews et al[7]. This model describes the kinetics of strain relief in strained-layer structures driven by thermally-activated glide of preexisting threading dislocations replicated from the substrate. This model does not fully take into account the relevant dislocation dynamics, in that dislocation multiplication is not included and the relation between misfit dislocation density and strain relief is not appropriate for strained-layer structures. They obtained

$$\gamma(t) = \frac{(1-\nu)}{2(1+\nu)} \sigma_{ex} (1-e^{-\beta t}) , \tag{5}$$

where γ is the amount of structural relaxation and β is a characteristic temperature-dependent timescale. However, this model does not reproduce the late stages of relaxation kinetics (see Figure 3b).

Dodson-Tsao Kinetic Relaxation Model

In a recent paper[3], Dodson and Tsao have presented a model in which strain relaxation in metastable strained-layer structures is described with considerable accuracy using a phenomenological model based on the treatment of plastic flow outlined above. This strain relaxation model includes the combined effects of thermally-activated dislocation motion and dislocation multiplication under stress. The excess stress σ_{ex} in the strained layer drives relaxation via plastic flow in the strained-layer structure. Dislocation multiplication is required to explain the observation that the dislocation density increases by orders of magnitude in the early stages of plastic flow of a high-quality bulk semiconductor crystal. The resulting model describes the time dependence of the strain relaxation γ by a single non-linear ordinary differential equation

$$d\gamma/dt = \alpha(\sigma_{ex},T) [\sigma_{ex}(\gamma)/\mu]^2 \gamma . \tag{6}$$

Since we consider overlayers which are initially coherent and nearly dislocation-free, it is appropriate to introduce a term representing a source of misfit dislocations. Physically the presence of such a term is reasonable, as even dislocation-free materials are observed to have such sources, perhaps in the form of 'microdefects'[23]. The dislocation source term is included as a background 'dislocation source density' γ_o. This is an ad hoc procedure, but the results are quite insensitive to the value chosen for γ_o.

COMPARISON WITH EXPERIMENT

The description of strain relaxation in semiconductor strained-layer structures provided by the above model was first applied to a series of experimental measurements of strain relaxation in metastable SiGe/Si[100] structures studied by Bean and coworkers[10]. Data on the residual strain of a $Si_{0.5}Ge_{0.5}$ film grown past the critical thickness (about 40Å) to thicknesses of 100, 500, 1000, and 2500Å appears in Figure 3a, along with a relaxation curve calculated using the current model. The results of the Dodson-Tsao model agree quite well with the experimental data. This impression is reinforced by calculation of strain relief in SiGe films of constant thickness as a function of alloy composition (and thus lattice mismatch). The equilibrium critical mismatch for the 500Å films shown in Figure 3b is about 0.0027. Despite this, the initial relaxation is sufficiently sluggish that no observable deviation from perfect coherence is seen until a mismatch of nearly 0.01 is attained. Then both the kinetic model and the experimental results diverge abruptly from coherence, followed by more gradual relaxation of the films as the mismatch continues to increase. The present kinetic model clearly provides at least a semiquantitative description of relaxation in semiconductor strained-layer structures.

The low-resolution 'critical thickness' can also be calculated using this kinetic model for relaxation. This is accomplished by determining the overlayer thickness which produces a detectable amount of strain relief. Figure 2 shows the predictions of the kinetic model for the 'critical thickness' in the SiGe/Si(100) system given an experimental sensitivity to strain relief of 10^{-3} (a representative value for conventional diagnostics). The agreement between our model and the experimental results of Bean et al is excellent.

When the excess stress in the strained layer is comparable to the zero-temperature flow stress, the activation energy for dislocation glide becomes smaller, making thermally-activated plastic flow occur faster than expected based on the small-stress activation energies. This effect is well-known in metals[24], where the dependence is roughly linear on the shear stress σ

$$E(\sigma) \approx E(0) \ [1 - (\sigma/\tau_o)] \ , \tag{7}$$

where $E(\sigma)$ is the stress dependent glide activation energy and τ_o is the zero-temperature flow stress. Because of difficulties with fracture at large stresses, this relation has not been evaluated for siliconlike materials. Dodson and Tsao[25] have recently extended the kinetic model for strain relaxation described above to include stress-dependent dislocation dynamics. This extension is important for strained-layer structures because of the large values of mismatch strain, and hence excess stress, which can be accommodated. They find that the existing data is consistent with Eq. 7 when excess stress is substituted for the shear stress and $\tau_o \sim 0.1\mu$. The resulting model is compared with data taken over a wide range of deposition temperature in the stability diagram[4] appearing in Figure 4. A stability diagram is a collection of iso-relaxation contours on an excess stress-temperature plot. Figure 4 demonstrates that low-temperature deposition offers great promise for production of high-quality structures very far from equilibrium.

SUMMARY

We have briefly reviewed the current experimental and theoretical understanding of stability and metastability in semiconductor strained-layer structures. Observations indicate that the onset of strain relaxation coincides with the equilibrium critical thickness. The experimental data on strain relaxation in metastable structures reveals that the degree of relaxation observed under typical growth conditions is always much less than that expected at equilibrium. Direct comparison of low-resolution (10^{-3}) with high-resolution (10^{-8}) analytical techniques demonstrates that the very large "metastable critical thickness" are not fundamental in nature, but are simply the thicknesses for which relaxation has proceeded far enough to be detected using low-resolution techniques. Various models which have been proposed to describe the stability properties of strained-layer structures are compared with the experimental data. The recent model of Dodson and Tsao, which is

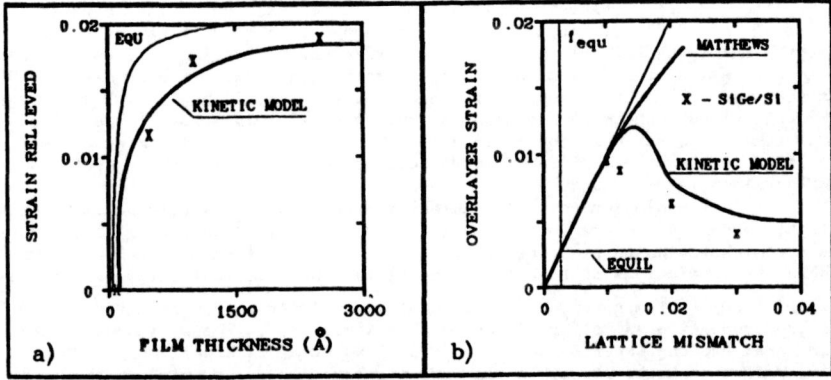

a) FILM THICKNESS (Å)

b) LATTICE MISMATCH

Figure 3. Experimental measurements of relaxation in SiGe/Si(100) strained-layer structures (Bean and coworkers, ref. 10). In 3a) the predictions of the Dodson-Tsao kinetic relaxation model are compared with the observed relaxation of $Si_{0.5}Ge_{0.5}$ strained layers of various thickness. In 3b) a similar comparison is made with experimental data on relaxation of a series of 500 Å strained overlayers as a function of lattice mismatch. In both cases, the kinetic relaxation model is in semiquantitative agreement with the data. Neither equilibrium relaxation nor the earlier kinetic relaxation model of Matthews agrees with the observed behavior.

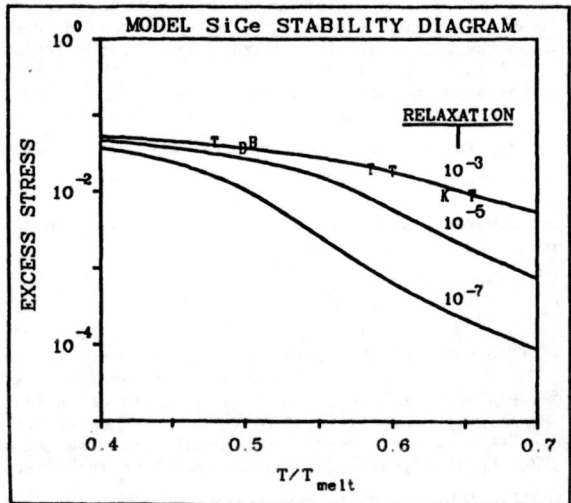

Figure 4. Stability diagram for the SiGe/Si(100) system. Isorelaxation curves are plotted on a graph of excess stress versus deposition temperature. The isorelaxation curves are calculated using the Dodson-Tsao kinetic relaxation model modified to include stress-dependent glide activation energies. The data points (T (Ref. 4), B (Ref. 10), and K (Ref. 8)) are all from experiments sensitive to relaxation of about 10^{-3}.

based on conventional mechanisms of plastic deformation, can semiquantitatively reproduce essentially all experimental data.

ACKNOWLEDGEMENTS

This work was performed at Sandia National Laboratories and was supported by U.S.D.O.E. contract DE-AC04-76DP00789.

REFERENCES

1. C.A.B. Ball and J.H. van der Merwe, in Dislocations in Solids, Chapter 27, F.R.N. Nabarro, ed. (North-Holland, New York, 1983).
2. J.W. Matthews and A.E. Blakeslee, J. Crystal Growth 27, 118 (1974).
3. B.W. Dodson and J.Y. Tsao, Appl. Phys. Lett. 51, 1325 (1987).
4. J.Y. Tsao, B.W. Dodson, S.T. Picraux, and D.M. Cornelison, Phys. Rev. Lett. (in press).
5. B.W. Dodson and P.A. Taylor, Appl. Phys. Lett. 49, 642 (1986).
6. Y. Sugita, M. Tamura, and K. Sugawara, J. Vac. Sci. Tech. 6, 585 (1969).
7. J.W. Matthews, S. Mader, and T.B. Light, J. Appl. Phys. 41, 3800 (1970).
8. E. Kasper, H.J. Herzog, and H. Kibbel, Appl. Phys. 8, 199 (1975).
9. J.C. Bean, T.T. Sheng, L.C. Feldman, A.T. Fiory, and R.T. Lynch, Appl. Phys. Lett. 44, 102 (1984).
10. J.C. Bean, L.C. Feldman, A.T. Fiory, S. Nakahara, and I.K. Robinson, J. Vac. Sci. Tech. A2, 436 (1984).
11. A.T. Fiory, J.C. Bean, R. Hull, and S. Nakahara, Phys. Rev. B31, 4063 (1985).
12. K. Kamigaki, H. Sakashita, H. Kato, M. Nakayama, N. Sano, and H. Terauchi, Appl. Phys. Lett. 49, 1071 (1986).
13. P.J. Orders and B.F. Usher, Appl. Phys. Lett. 50, 980 (1987).
14. K. Eberl, G. Krotz, T. Wolf, F. Schaffler, and G. Abstreiter, Semicond. Sci. Tech. 2, 561 (1987).
15. W.D. Laidig, C.K. Peng, and Y.F. Lin, J. Vac. Sci. Tech. B2, 181 (1984); N.G. Anderson, W.D. Laidig, R.M. Kolbas, and Y.C. Lo, J. Appl. Phys. 60, 2361 (1986).
16. I.J. Fritz, S.T. Picraux, L.R. Dawson, T.J. Drummond, W.D. Laidig, and N.G. Anderson, Appl. Phys. Lett. 46, 967 (1985).
17. P.L. Gourley, R.M. Biefeld, and L.R. Dawson, Appl. Phys. Lett. 47, 482 (1985).
18. I.J. Fritz, P.L. Gourley, and L.R. Dawson, Appl. Phys. Lett. 51, 1004 (1987).
19. P.L. Gourley, I.J. Fritz, and L.R. Dawson (Preprint).
20. R. People and J.C. Bean, Appl. Phys. Lett. 47, 322 (1985).
21. I.J. Fritz, Appl. Phys. Lett. 51, 1080 (1987).
22. H. Alexander and P. Haasen, in Solid State Physics (Academic, New York, 1968), Vol 22.
23. J.R. Patel, Disc. Faraday Soc. 38, 201 (1964).
24. H.J. Frost and M.F. Ashby, Deformation-Mechanism Maps (Pergamon, Oxford, 1982).
25. B.W. Dodson and J.Y. Tsao (Preprint).

STABILITY DIAGRAMS FOR EPITAXIAL STRAINED LAYERS

J.Y. TSAO, B.W. DODSON AND S.T. PICRAUX
Sandia National Laboratories, Albuquerque, NM 87185

ABSTRACT

For bulk materials, plastic deformation mechanisms and rates depend in a complex way on resolved shear stress and temperature, but can be succinctly described using the deformation mechanism maps pioneered by Ashby and Frost [1]. In this paper, we describe the use of such maps to demarcate the various stability and metastability regimes of single and multilayered strained epitaxial structures, and to interpret experimental work in the SiGe system.

INTRODUCTION

The conditions under which a thin film of one lattice constant can be grown coherently on a substrate with a different lattice constant is a simple yet basic question in materials science. The thermodynamic part of the question -- for what thickness/misfit combinations is the equilibrium configuration fully strained rather than partially relieved -- has been solved for some time. That question was first treated in the 1940's by Frank and van der Merwe [2], more recently by Matthews and Blakeslee [3], and even more recently by Dodson [4], Grabow [5] and co-workers using molecular dynamics simulation.

Essentially, these workers established that coherent films will be stable provided they are thin enough. In simple terms, their reasoning was the following. In order for the film to grow coherently, its in-plane lattice constant must be the same as that of the substrate, and so it must grow strained. This costs some strain energy, but that energy is small if the film is thin. The alternative, which is to relieve some strain through the introduction of misfit dislocations at the interface, also costs energy, but that energy isn't small even if the film is thin, because of the core energy of the dislocation. Hence, for thin films, the energy of the system is minimized by avoiding the introduction of misfit dislocations.

Recently, however, it has been found that in some semiconducting materials, most notably the SiGe system studied by Kasper [6], Bean [7] and co-workers, it is possible to grow strained films much thicker than expected on the basis of equilibrium theory. Moreover, the amount of relief, even when relaxation has begun, is not in accord with equilibrium theory.

We believe that the principal explanation for these observations lies in the kinetic part of the question -- how fast are the rates of relaxation towards equilibrium? This view is based on the idea that there are two independent halves to the problem. The first half is growth of fully strained layers by epitaxy; the second half is relaxation from the fully strained to the strain relieved structure by plastic deformation [8]. Although these two halves are independent, they may occur simultaneously, and, indeed, will be in competition with each other. If relaxation is fast compared to growth, then as soon as the critical layer thickness is exceeded, the film will become defective; if relaxation is slow compared to growth, then thicker, metastable films will be grown.

DEFORMATION MECHANISM MAPS

The ideas outlined above are deceptively simple; it is not obvious how to apply them in a simple way to the problem of strained layer metastability. The difficulty with invoking sluggish relaxation kinetics is that plastic deformation is so complex. Even in bulk materials, there are many deformation mechanisms, and it is not a priori clear that the same mechanisms should apply to thin films. However, there is one greatly simplifying feature: no matter what the mechanism, there are only two parameters important in plastic deformation [9]. These are stress, which is the driving force for plastic flow, and temperature, which determines the mobility of the dislocations which mediate plastic flow.

Indeed, a standard way of describing plastic deformation in the bulk is using stress/temperature diagrams, called deformation mechanism maps, illustrated here in Fig. 1 for the case of silicon. For our purpose, the most important information on such a map is summarized by the iso-strain rate contours. Two features are important to point out. First, there is a general trend that at higher temperatures less stress is needed to produce a given strain rate. This is reasonable, since dislocations become more mobile at higher temperatures. Second, there are distinct regimes of stress and temperature in which different flow mechanisms are dominant. For example, there is a low-temperature plasticity regime in which relatively high stresses are required to drive flow, as well as a power-law creep regime in which relatively low stresses can cause flow.

It would appear straightforward to apply such deformation mechanism maps to the question of strained layer metastability. Every experimental growth sequence defines an experimental time scale. For strain relief to be observable on that time scale, there is a minimum required strain rate. To produce such a minimum strain rate at a fixed growth temperature, we require, by analogy to Fig. 1, a critical stress. The critical stress described above does not, however, refer to the actual stress in the film, but rather to an "excess" stress [10]. In the following, we derive these excess stresses first for the case of single strained layers and then for the case of strained multilayers.

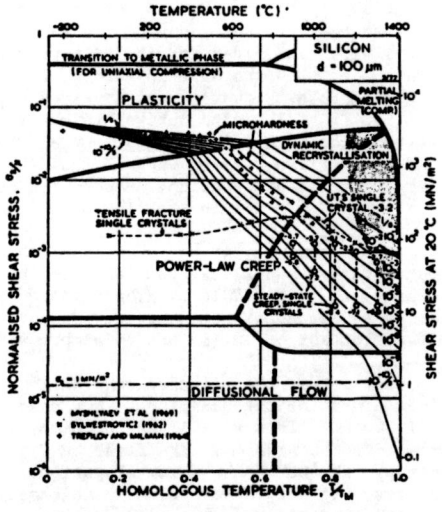

Fig. 1. Stress/temperature map for silicon of grain size 100 μm. Data are labelled with $\log_{10}(\dot{\gamma})$. After Ref. 1, Fig. 9.1.

FORCE BALANCE IN SINGLE STRAINED LAYERS

Consider the action of the stress field due to biaxial film strain on a 60°
dislocation threading through a single diamond-cubic strained layer. The
glide component of the Peach-Kohler "misfit-strain" force parallel to the
(001) film plane, integrated over the length of such a dislocation, is

$$\bar{F}_M = \frac{hb\sigma_M}{2} \; [1/\sqrt{2} \; -1/\sqrt{2} \; 0], \qquad (1)$$

where h is the film thickness, $\bar{b} = (a_o/\sqrt{2}) \; [1/\sqrt{2} \; 0 \; 1/\sqrt{2}]$ is the Burger's
vector, a_o is the alloy lattice constant, the dislocation lies along the
unit vector $\xi = [0 \; 1/\sqrt{2} \; 1/\sqrt{2}]$, $\sigma_M \approx 2e\mu(1+\nu)/(1-\nu)$ and e are the principal
in-plane stresses and strains, μ is the shear modulus and $\nu \approx 0.3$ is
Poisson's ratio.

Because there are no forces acting on the dislocation in the (nearly)
unstrained substrate, the dislocation must elongate (approximately) along
the substrate/film boundary in order to move laterally in the film above.
As originally noted by Matthews, this elongation generates an opposing force
due to the "image" (or self) energy [11] of the dislocation,

$$F_D = \frac{hb\sigma_D(h)}{2}, \qquad (2)$$

where we have introduced an "effective" stress (with $\beta = 60°$)

$$\sigma_D(h) \approx \frac{\mu}{2\pi} \; (\frac{1 - \nu\cos^2\beta}{1 - \nu}) \; \frac{\ln(4h/b)}{h/b}. \qquad (3)$$

The difference between these two forces is the excess force on the
dislocation, $F_{exc} = hb\sigma_{exc}/2$, where the "excess" stress, the driving force
for strain relief [12], is:

$$\sigma_{exc} = \sigma_M - \sigma_D(h) = 2e\mu\frac{1+\nu}{1-\nu} - \frac{\mu}{2\pi}(\frac{1 - \nu\cos^2\beta}{1 - \nu})\frac{\ln(4h/b)}{h/b}. \qquad (4)$$

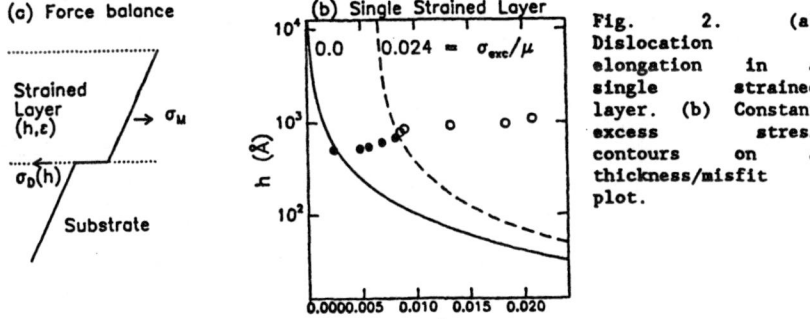

(a) Force balance

(b) Single Strained Layer

Fig. 2. (a)
Dislocation
elongation in a
single strained
layer. (b) Constant
excess stress
contours on a
thickness/misfit
plot.

To make this concept of excess stress more concrete, Fig. 2(b) shows a
thickness/misfit diagram, on which contours of constant excess stress are
plotted. Thickness/misfit combinations that lie along a contour represent
films with the same degree of metastability. Films to the left of the zero-
stress isobar are absolutely stable. Films to the right, if they have not
broken down, are metastable.

FORCE BALANCE IN STRAINED MULTILAYERS

The same reasoning may be applied towards a definition of excess stress in strained multilayers [13]. In this case, however, it is convenient conceptually to define two kinds of stability, illustrated in Fig. 3(a): (1) stability of the superlattice as a whole, and (2) relative stability of the individual layers that make up the superlattice.

First, the excess stress that governs stability of the superlattice as a whole can be deduced in exactly the same manner as that described above for a single strained layer, except that we must sum the forces on the individual layers to find the total Peach-Kohler misfit-strain force:

$$\bar{F}_M^{\Sigma i} = \frac{hb\sigma_M^{\Sigma i}}{2} \; [1/\sqrt{2} \; -1/\sqrt{2} \; 0], \tag{5}$$

where

$$\sigma_M^{\Sigma i} = \sum_i \frac{h^i \sigma_M^i}{h} \tag{6}$$

is the average stress in the multilayer and $h = \Sigma_i h^i$ is the total height of the multilayer. Within each layer i, h^i is the thickness, $\sigma_M^i \approx 2e^i \mu^i(1+\nu)/(1-\nu)$ and e^i are the principal in-plane stresses and strains, and μ^i is the shear modulus. The superlattice excess stress is then

$$\sigma_{exc}^{\Sigma i} = \sigma_M^{\Sigma i} - \sigma_D(h), \tag{7}$$

where $\sigma_D(h)$ has the same definition as in Eq. (3) and Fig. 2(b) applies. Eq. (7) generalizes the result of Hull and co-workers [14].

Second, the excess stress that governs stability of the individual layers can be deduced by subtracting the average misfit stress from that of each of the layers to get the "residual" misfit stress in the layer:

$$\Delta\sigma^i = \sigma_M^i - \sigma_M^{\Sigma i}. \tag{8}$$

Since, as can be seen in Fig. 3(a), dislocation motion within each of the superlattice layers is now opposed by the elongation of _two_ misfit dislocation segments, the excess stress in each layer is

$$\Delta\sigma_{exc}^i = \Delta\sigma^i - 2\sigma_D(h^i). \tag{9}$$

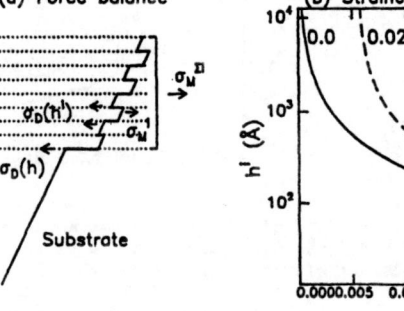

(a) Force balance

(b) Strained Multilayer

Fig. 3. (a) Dislocation elongation in a strained layer superlattice. (b) Constant excess stress contours on a thickness/misfit plot.

Fig. 3(b) shows contours of constant individual layer excess stress on a thickness/misfit diagram. Again, thickness/misfit combinations that lie along a contour represent layers with the same degree of metastability. Note that Eq. (9) differs from Eq. (4) by a factor two in the σ_D term; therefore, for equivalent thickness/misfit combinations, individual layers in a multilayer will be stabler than single strained layers [3].

STABILITY DIAGRAMS

With the above definitions of excess stress, we can now summarize, on a stress-temperature diagram, regimes of stability and metastability. In Fig. 4 we show on such a diagram measurements from a range of sources of the onset of strain relief in SiGe films. Although there is certainly some scatter, probably at least partly due to the wide range of experimental growth conditions in the various measurements, there is a remarkable self-consistency associated with the measurements, provided they are looked at in terms of excess stress and temperature rather than film thickness and misfit. As with bulk deformation mechanism maps, there is a general trend for the critical stresses to decrease as temperature increases. Furthermore, there also appears to be a kink in the curve at a scaled temperature of 0.6, indicating that the relief mechanisms may be different at high and low temperatures, again consistent with what is known about bulk deformation.

We have called plots like these "stability diagrams", because they can be used to demarcate the various regimes of stability and metastability. The boundary between absolute stability and metastability we call the van der Merwe/Matthews boundary. Films below this boundary are absolutely stable; films above this boundary have a tendency to relieve. Note, however, that the boundary dividing unrelieved and partially relieved structures is fuzzy, and will depend on the definition of the onset of strain relief. If the definition is 0.1% relief, then the Kasper/Bean/Tsao boundary applies. If the definition is one misfit dislocation per cm^2, then, as noted by Fritz [15], a more stringent boundary, which we call the Fritz/Gourley boundary, applies. Note that for SiGe films this boundary has been calculated [13], not measured, and is drawn here only schematically by the dotted line.

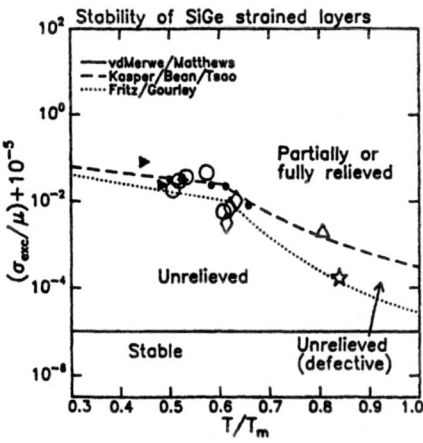

Fig. 4. Stability diagram for SiGe single strained layers. The data are taken from various sources, as discussed in Ref. 10.

CONCLUSION

In summary, we have introduced the concept of excess stress, which is the driving force for plastic flow of coherent strained layers. This is a crucial concept, even beyond its use in describing strained layer stability, in that it provides a connection between dislocation dynamics in the bulk with that in thin films. Much is known about bulk deformation, but little about thin film deformation, so that connections between the two are important. The quantitative definition of excess stress, however, depends on the detailed mechanism for flow. Here, we have considered only glide of 60° dislocations threading through diamond-cubic strained layers.

We have also extended our original treatment of relief of single strained layers to relief of strained multilayers. There are two kinds of relief: that of the multilayer as a whole, and that of the individual strained layers relative to each other. Each kind of relief has a different excess stress associated with it.

Finally, based on this concept of excess stress, and an analogy to deformation mechanism maps in bulk materials, we have introduced stress/temperature stability diagrams. Such diagrams embody a unified view of strained layer stability and metastability which is consistent with available experimental data in the SiGe system.

ACKNOWLEDGEMENTS

The portion of this work performed at Sandia National Laboratories was supported by the U.S. Department of Energy under contract DE-AC04-76DP00789.

REFERENCES

1 H.J. Frost and M.F. Ashby, Deformation-Mechanism Maps (Pergamon Press, Oxford, 1982).
2 C.A.B. Ball and J.H. van der Merwe, in Dislocations in Solids, Chap. 27, F.R.N. Nabarro, ed. (North-Holland, 1983).
3 J.W. Matthews and A.E. Blakeslee, J. Crystal Growth 27 (1974) 118.
4 B.W. Dodson and P.A. Taylor, Appl. Phys. Lett. 49, 642 (1986).
5 M.H. Grabow and G.H. Gilmer, Mat. Res. Soc. Symp. Proc. 56, 13 (1986).
6 E. Kasper, Surf. Science 174 (1986) 630.
7 J.C. Bean, Science 230 (11 October 1985) 127.
8 B.W. Dodson and J.Y. Tsao, Appl. Phys. Lett. 51, 1325 (1987).
9 For a given microstructure.
10 J.Y. Tsao, B.W. Dodson, S.T. Picraux and D.M. Cornelison, Phys. Rev. Lett. 23 November, 1987.
11 J.P. Hirth and J. Lothe, Theory of Dislocations, 2nd Ed. (Wiley-Interscience, New York, 1982), Eq. 3-87.
12 This excess stress is the driving force for motion of threading dislocations. However, since such motion generates misfit dislocations either directly through elongation at the film/substrate interface, or indirectly through the creation of new threading dislocations in its wake, this stress can also be considered the driving force for strain relief. It is not a measure of the propensity for misfit dislocations to nucleate spontaneously at the interface without threading dislocation intermediaries.
13 B.W. Dodson and J.Y. Tsao, manuscript in preparation.
14 R. Hull, J.C. Bean, F. Cerdeira, A.T. Fiory and J.M. Gibson, Appl. Phys. Lett. 48, 56 (1986).
15 I.J. Fritz, P.L. Gourley and L.R. Dawson, Appl. Phys. Lett. 51, 1004 (1987).

Interface Structure and Solid State Reactions of Fe/Zr Multilayers

Bruce M.Clemens* and D.L.Williamson**

* Physics Department, General Motors Research Laboratories Warren, Michigan 48090-9055

** Department of Physics, Colorado School of Mines, Golden Colorado 80401

Abstract

Iron zirconium multilayer films were prepared by electron beam evaporation and by sputter deposition. Layer thicknesses were varied from 50 monolayers of each constituent down to 2 monolayers. Conversion electron Mössbauer spectroscopy, x-ray diffraction, and Auger spectroscopy have been used to characterize multilayers in the as-deposited and annealed state. Amorphous phase formation occurs during thermal anneals, and samples with layer thicknesses of 5 monolayers or less of each constituent are amorphous as deposited. Amorphous interfaces are observed in all samples, with this interface region being larger in the electron beam evaporated samples than in the sputter deposited samples.

Introduction

The first iron-based amorphous alloy formed by the process of solid state reaction has been reported[1]. Elemental layers of crystalline Fe and Zr were partially transformed to an amorphous phase by a solid state anneal. More recently, complete transformation of Fe/Zr multilayers with an average composition of $Fe_{50}Zr_{50}$ was observed, with evidence that this process produced a phase separated amorphous alloy[2]. Solid state formation of amorphous alloys has been observed in several early transition metal-late transition metal pairs. Metastable free energy diagrams have been used to demonstrate that, due to the large negative heat of mixing typical of these alloy systems, the amorphous phase has a lower free energy than the two phase starting material[1,3,4]. It has also been observed that one element is the dominant moving species in the reaction[5,6].

Amorphous phases have also been observed in metal multilayers when the layer thickness is less than about 10 atomic planes of each constituent per layer. This has been observed in multilayer pairs where the large atomic structural mismatch precludes epitaxy between the layers, including Ti/Ni[4], Zr/Ni[4], Hf/Cu[7,8], and Nb/Co[9]. Absence of long range crystalline order is even observed at small layer thicknesses in Mo/Ni multilayers, where at larger layer thicknesses the interfaces are crystalline[10].

In this paper we investigate the structure and properties of Fe/Zr multilayers in both the annealed and as deposited state. We have prepared several samples with different layer thicknesses by both electron-beam evaporation and dc magnetron sputter deposition. A Mössbauer technique known as conversion electron Mössbauer spectroscopy, CEMS[11], is used to examine the state of the Fe. X-ray diffraction is used to determine the crystal structure of the layers and to examine the composition modulation. Auger electron electron spectroscopy is used to characterize the composition modulation and possible contamination by oxygen and carbon. We compare the amorphous phase produced by solid state reaction during thermal anneals with the amorphous phase in the small layer thickness as-deposited samples, as well as amorphous phases formed at the interface between crystalline Fe and Zr.

Experiment

Sample Preparation

Samples were prepared by either dc magnetron sputter deposition or electron beam evaporation. The electron beam evaporation was performed in a manner described previously[1] in a system with a base pressure below 4×10^{-8} Torr, and with two sources which the substrates (oxidised Si (100) and Kapton) were rotated over to produce compositionally modulated films. A vertical plate extending from below the sources to within about 1 cm of the substrate table shielded the sources from each other and prevented extensive co-deposition. Sputter deposition of multilayers was performed by rotation of the substrates under two sources which were shielded so that co-deposition was completely avoided. The sputter deposition chamber was evacuated to a base pressure less than 1×10^{-7} Torr by a liquid nitrogen trapped diffusion pump, and the depositions were performed at a pressure of 2.3 mTorr of Ti getter cleaned Ar.

In both electron beam evaporation and sputter deposition the deposition rate from each source was monitored with an oscillating quartz crystal with feedback control to the sources. The fluctuations in rate for sputter deposition were about 0.01 nm/s and about 10 times that for electron beam evaporation. The deposition rates were between 0.25 and 0.5 nm/s. For both deposition methods, the rate and substrate rotation were set to produce an equal number of close packed planes (Fe (110) and Zr (002)) of each constituent. This resulted in an overall composition of $Fe_{61}Zr_{39}$. The Electron beam evaporated samples in this study will be designated as E–n where E refers to electron beam evaporated and n is the number of close packed planes of each constituent per layer. Two series of sputtered samples were prepared and these are designated S_i–n, where i refers to the series. A third sample with $n = 10$ was prepared and is referred to as S_3–10.

X-Ray Diffraction

X-ray diffraction was performed in symmetric reflection geometry, where the scattering vector is perpendicular to the plane of the sample, and symmetric transmission geometry, where the scattering vector lies in the plane of the sample. The later experiments were performed on samples deposited on kapton, whereas the former were performed on both the oxidised Si (100) and kapton substrate samples with no difference observed between the samples on the two different substrates. In symmetric reflection geometry, the structure of the sample is probed in the direction of growth, and at low angles, superlattice peaks are observed corresponding the the artificially induced composition modulation.

Peak positions were used to determine strain in the direction appropriate to the diffraction experiment. Peak widths were obtained by dividing the intensity into the integrated peak area. In the cases where two or more orders of a given peak were observable, peak broadening was used to determine the degree of inhomogeneous strain and finite size broadening via the equation[12]:

$$\Delta k = [<\epsilon^2> +(\frac{2\pi 0.9}{t})^2]^{1/2},$$

where Δk is the observed peak width, $<\epsilon^2>^{1/2}$ is the rms value for the inhomogeneous strain, and t is the size of the diffracting crystal. This treatment is only approximate and is used here since a complete Fourier analysis is not feasable due to overlapping peaks.

Auger electron spectroscopy was combined with ion milling to provide information on the composition versus depth into the sample. Composition modulation was investigated by alternating ion sputter with Auger spectroscopy. Impurity concentrations of oxygen and carbon were monitored during continuous milling to avoid contamination of the highly reactive zirconium by the imperfect vacuum of the analysis chamber.

Mössbauer Measurements

A conventional constant acceleration spectrometer with a 25 mCi Rh:^{57}Co source was used to acquire the conversion electron Mössbauer spectra (CEMS). The dection of the electrons was made by ionization of He + 6% CH_4 in a gas-flow detector in which the specimen is mounted[13]. Calculations appropriate to integral CEMS[11,14,15] from a pure Fe absorber show that about 80% of the conversion electrons come from the layer extending from the surface to a depth of about 100 nm. This depth is essentially determined by the mass density of the absorber[11]. The average density of the Fe/Zr multilayer specimens will be less than Fe and we estimate the 80% fraction to come from about 110 nm of total layer thickness. Since the compositional modulation wavelength for the largest n is about 22 nm the CEMS data is coming from at least 5 layers of Fe.

Each spectrum was least-square fit with a sum of Lorentzian lines to yield the usual spectral parameters: δ = isomer shift (relative to α-Fe), Δ = quadrupole splitting, Γ = full line-width at half maximum, H = magnetic hyperfine field, F = fractional resonance area (integrated intensity).

Results

X-Ray Diffraction

Figure 1 shows the high angle symmetric reflection x-ray diffraction for samples S_1-2, S_1-5, S_1-10, S_1-15, and S_1-20. Samples S_1-2 and S_1-5 have diffraction patterns with one broad maxima characteristic of an amorphous Fe/Zr alloy. The S_1-10 diffraction maxima has split into two resolvable peaks which correspond to iron rich and zirconium rich regions. It is difficult to say from these results whether this sample is crystalline or amorphous, but it is clear that there is considerable disorder. The diffraction results for S_1-15 and S_1-20 clearly show the Zr(002) and Fe(110) peaks. Analysis of the peak widths for the Fe(110) and (220) peaks shows that the peak broadening is mainly due to particle size. The layer thickness extracted from this analysis is 19-20 monolayers of Fe for S_1-20 and 13-15 monolayers of Fe for S_1-15. There is some inhomogeneous strain in the Fe layers which is found to be about 0.7% for S_1-20 and 1.5% for S_1-15. The Zr (004) peak is extremely weak, so similar analysis is not possible for the Zr layers. However, the Zr (002) peak is substantially broader than expected for layers of the intended thickness in both the S_1-15 and S_1-20 samples. This indicates disorder in the Zr layers from either interdiffusion of Fe (inhomogeneous strain) or reaction to form an amorphous interface region, or both. From the positions of the peaks we find that the Fe lattice constant in the growth direction is expanded by about 0.7% in S_1-20 and by 1% in S_1-15. The Zr lattice parameter has a smaller expansion of 0.1% in S_1-20 and 0.3% in S_1-15.

The small angle diffraction data for this series of samples is shown in Figure 2. Clearly evident in each case are superlattice satelites due to scattering from the artifically induced composition modulation. Particularly remarkable is clear indication of composition modulation in S_1-2. This sample has a modulation wavelength less than 1 nm, and like sample S_1-5 can be considered a compositionally modulated amorphous alloy. The modulation wavelength found from the positions of these peaks agrees to within less than 5% with the intended modulation wavelength.

The diffraction patterns of all samples prepared with layer thickness greater than 10 atomic planes of each constituent were characteristic of diffraction from individual layers of crystalline elements. Even though the small angle x-ray scattering demonstrates that there exists strong composition modualtion we see no high angle superlattice lines. Recently it has been demonstrated that this type of diffraction result is due to amorphous or disordered interfaces [16].

The transmission diffraction results for S_1-15 and S_1-20 show rather broad bcc Fe and hcp Zr crystalline peaks. Their positions indicate a 0.8% expansion for Fe and a 1% contraction for Zr in S_1-15 and a 0.6% expansion for Fe and a 0.4% expansion for Zr in S_1-20. Peaks which correspond to lattice planes perpendicular to the bcc [110] and hcp [001] are observed, indicating the orientation preference for close packed planes to lie in the plane of the sample. This is confirmed by rocking curve measurements which show a dropoff of intensity of the Fe (110) and Zr(002) lines with increasing deviation from symmetric reflecting angles.

The reflection diffraction results for a series of electron beam samples are similar to those of the sputtered samples. The sample E-5 looks like an amorphous alloy. E-10 shows considerable disorder with evidence for an amorphous phase, but also shows a Fe (110) peak which is more prominant than that in S_1-10. The Zr (002) and Fe (110) peaks are broader in E-20 than in S_1-20. These differences reflect the greater rate fluctuations and poorer control in the electron beam deposited samples relative to the sputtered samples.

Auger Profiling

Composition modulation was observed for multilayer samples with wavelengths as small as 5 monolayers of each constituent. For layer thicknesses of 10 monolayers or greater the layers could be seen as rings on the etch pit crater when imaging with total absorbed sample current. Oxygen and carbon were detected in these films with concentrations less than 5 at%.

Conversion Electron Mössbauer Spectroscopy

Figure 3 compares CEMS data from two series of as-deposited multilayer films. Both series show a transition from the magnetic, crystalline bcc(α)-Fe phase to a paramagnetic phase. The Mössbauer spectra of the paramagnetic phase can be described as a doublet with unequal intensities. The relative intensities of the six-line magnetic resonance (M) from the α-Fe are close to the 3:4:1:1:4:3 which are expected for magnetic moments lying parallel to the film

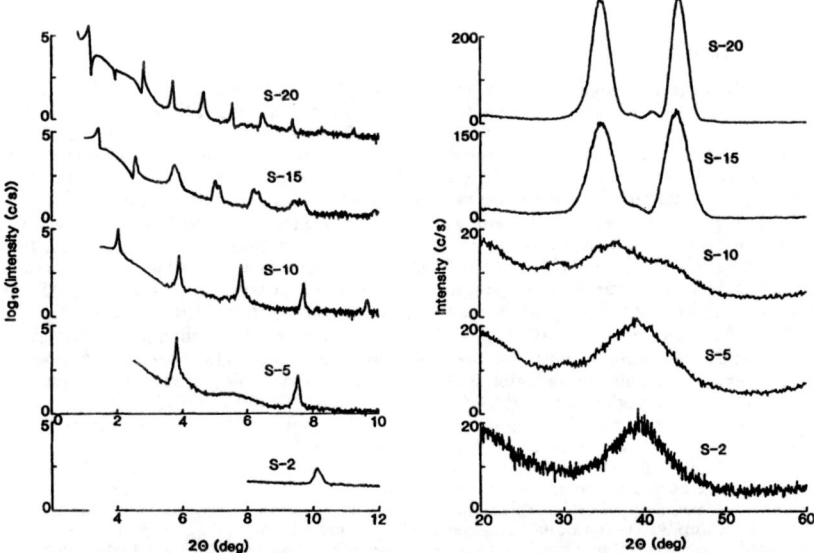

Figure 1: High angle reflection x-ray diffraction for series of sputtered samples.

Figure 2: Small angle reflection x-ray diffraction for a series of sputtered samples.

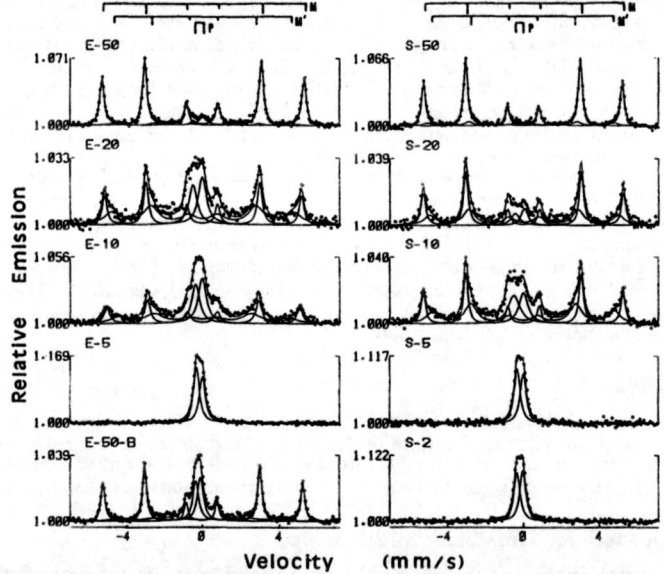

Figure 3: Conversion electron Mössbauer spectra from (a) sputter-deposited multilayers and (b) electron-beam deposited multilayers. The solid line passing through the data points is the computer fit of a superposition of the subspectra indicated. Also shown is the result for sample E-50 after an anneal of 1 hour at 350° C (E-50-B).

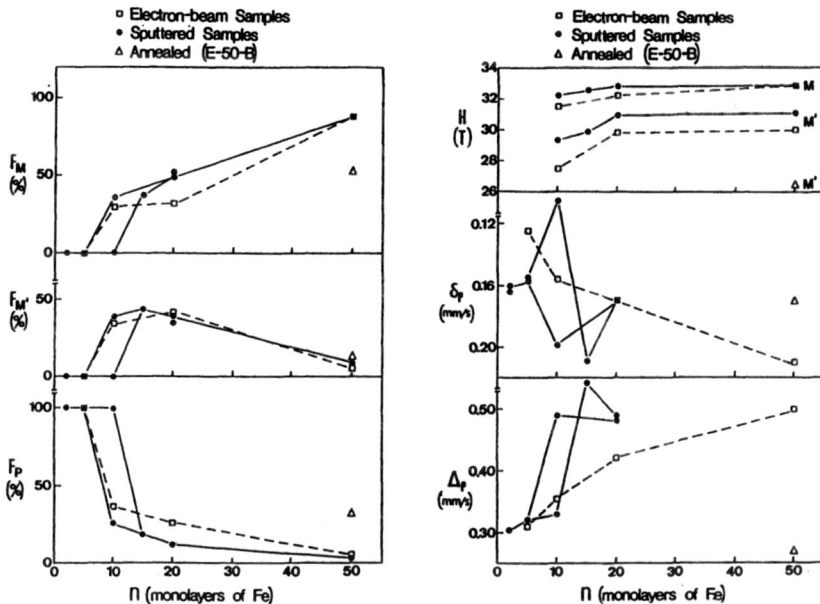

Figure 4: Mössbauer resonance fractions versus number of monolayers. F_P is the fractional resonance area of the paramagnetic phase, F_M and $F_{M'}$ are the fractional resonance area of the magnetic components. The lines are drawn to suggest systematic behavior.

Sample	$c(\delta)$	$c(\Delta)$	$c(I_-/I_+)$	$\bar{c}$	x_p
S_2-2	0.5–0.8	0.6–0.7	0.6	0.6	4[a]
S_2-5	0.5–0.8	0.5–0.8	0.6	0.6	10[a]
S_2-10	0.3–0.4	0.3–0.4	< 0.6	0.4	3.3
S_3-10	0.9	0.5 or 0.9	> 0.6	0.9	20[a]
S_2-20	0.4–0.5	0.3	< 0.6	0.4	3.0
S_2-50	≲0.3	b	b	≲0.3	2.5
E-5	0.85–0.9	0.5–0.9	> 0.6	0.9	10[a]
E-10	0.5–0.8	0.4–0.3	< 0.6	0.5	3.7
E-20	0.3–0.5	0.3–0.4	< 0.6	0.4	6.5
E-50	0.3–0.5	0.2–0.4	< 0.6	0.3	5.0
E-50-A[d]	0.5–0.8	c	≲0.6	0.5	15
E-50-B[e]	0.5–0.8	c	≲0.6	0.5	17

Table 1: Composition estimates of paramagnetic phase based on comparison with a-$Fe_c Zr_{1-c}$ (Ref. [18]). Values of $\bar{c}$ and x_p for the $S_1 - n$ series were reported in ref. [17].

a. For these samples the entire Fe and Zr multilayer film is paramagnetic (amorphous).

b. Spectral parameters not obtained here due to weak resonance.

c. No match possible here.

d. Annealed 10 min at 350°C.

e. Annealed 1 hour at 350°C.

plane (perpendicular to the γ-ray direction). An additional magnetic resonance (M') with a reduced hyperfine field compared to α-Fe occurs for all samples showing magnetic resonance. The major qualitative difference in the two series is that the sputtered multilayers show less of the paramagnetic phase for the same n. Note that both series show the transition to 100% paramagnetic phase between $n = 10$ and $n = 5$. The other series of sputtered samples S_1 shows the transformation to 100% paramagnetic phase between $n = 10$ and $n = 10$[17]. A third sputtered sample $(S_3\text{-}10)$ is also completely paramagnetic.

Figure 3 also shows the results of vacuum annealing sample E-50 for 1 h at 350°C. Formation of a paramagnetic phase occurs and the M' resonance is significantly smaller than those seen in the as deposited films.

In order to examine the paramagnetic phases in more detail, spectra were acquired on an expanded velocity scale. Differences in peak positions and relative intensities were observed. The paramagnetic phases are well fit with a superposition of two Lorentzian lines constrained to have the same linewidths.

Figure 4 displays systematics of the spectral parameters and relative phase fractions obtained from the fits displayed in Figure 3 thow obtained on the expanded velocity scale data. Also included are data from sample E-50 which was first annealed for 10 min at 350°C. Note that little change occurred as a result of the longer 1 h anneal. Spectral parameters of the paramagnetic phase of the other sputtered series (S_1) were reported in ref. [17] and are in good agreement with these values for the same n. This demonstrates the reproducibility of the process (except for $n = 10$). The following quantitative differences in the sputter-deposited multilayers compared to the electron-beam deposited multilayers can be summarized:

(i) The paramagnetic fraction is lower for $n = 50, 20$, and 10.

(ii) The M' fraction is very similar, however it is a sharper resonance.

(iii) The magnetic hyperfine fields are larger for a given n (except for M at $n = 20$ where they are the same).

(iv) Significant differences in isomer shift and quadrupole splitting occur for the paramagnetic phase, the annealed sample yielding the smallest value of Δ.

Interpretation and Discussion

The Mössbauer spectral parameters of the paramagnetic phase can be compared to those reported by Unruh and Chien[18] for $a\text{-}Fe_cZr_{1-c}$ alloys for $0.9 > c > 0.2$. The validity of the comparison is improved by having both studies use the same procedure for fitting the paramagnetic phase data (two Lorentzian lines). The combined use of δ, Δ, and the relative intensity of the quadrupole doublet, I_-/I_+, to estimate c yields the values shown in Table 1. Using the average concentration $\bar{c}$ as estimated in Table 1 we calculate the thickness x_p of the interfacial amorphous phase (in monolayers) from $2x_p\bar{c} = nF_P$, where F_p is the Mössbauer resonance phase fraction of the paramagnetic phase. This assumes no density difference in the α-Fe and the $a - Fe_cZr_{1-c}$ phases and no difference in recoilless fractions at the Fe sites in these two phases. The results are included in Table 1. Note that x_p is about a factor of two smaller in the sputtered multilayers (for $n = 50, 20$, and 10) compared to the electron beam deposited multilayers. The thickness of the amorphous phase increased about a factor of 3 upon anealing the E-50 sample for either 10 min. or 1 hr, and its composition became more enriched in Fe.

The larger linewidths obtained for Γ_p from the $S_2\text{-}20$, $S_2\text{-}10$, E-20, and E-10 specimens suggest that a larger range of c exists for the paramagnetic phase in these samples compared to the $n = 5$ and $n = 2$ samples. The quadrupole splitting obtained from the annealed E-50 sample is significantly lower than any value reported. However, this discrepancy may be due to errors in Δ that occur when using the transmission geometry and thick absorbers[19,20]. The CEMS technique does not suffer from this problem. Absorber thicknesses used in ref.[18] were not given so we cannot estimate possible errors in their reported quadrupole splittings.

The spectral parameters of the crystalline phases $FeZr_3$, $FeZr_2$ and Fe_2Zr[21,22] clearly do not match those observed here.

The origin of the M' resonance is attributed to an interface region where Fe atoms in the α-Fe experience a reduced hyperfine field due to the presence of non-magnetic Zr neighbors. It is well known[25] that substitutional non-magnetic impurities in α-Fe will reduce the hyperfine field at nearest and next-nearest neighbor Fe sites by about 2 or 3 Tesla, similar to the reductions seen here for $H(M')$. However, an atomically sharp interface between the α-Fe and the amorphous Fe-Zr paramagnetic phase would not produce the observed fractions of M and M' subspectra (Fig. 4). This conclusion is based on the model described by Jaggi et al.[23] who show that for x_α monolayers of α-Fe with (110) in-plane orientation in sharp registry with V (also with (110) in-plane orientation) the fraction $(x_\alpha - 2)/x_\alpha$ would have no V nearest or next nearest neighbors. This predicts 90% and 80% fraction of M for $n = 20$ and $n = 10$ respectively. Clearly, the low values of $F(M)$ (or the correspondingly high values of $F(M')$) for the $n = 20$ and $n = 10$ must be due to an interdiffused interface. The degree of interdiffusion required to enhance $F(M')$ relative to $F(M)$ can be estimated from Jaggi et al's model[23] to be only 1 or 2 monolayers.

In addition to the observation of less amorphous phase formation at the interface between crystalline Fe and Zr in the sputtered samples, two experimental results show that the sputtered multilayers have less interdifussion than the electron-beam deposited ones: (i) the ratios $F(M')/F(M)$ are lower, (ii) the values of $H(M')$ are larger therby implying fewer Zr neighbors on the average. Sputter deposition below 10 mTorr leads to higher kinetic energies in the arriving species. This can lead to higher density, more defect free films than evaporation onto room temperature substrates. Molecular dynamics simulations show that, even with kinetic energy as large as twice the cohesive energy of the solid, the arriving atom does not embed into the substrate[26]. Recent experiments show that increasing the sputter deposition pressure, which decreases the energy of the arriving species, leads to poorer layering in Ti/Ni multialyers[27]. This would suggest that evaporation, which has lower energy arriving species, would produce poorer multilayers than sputtering. In the present case we certainly see evidence for greater diffusion and poorer layering in the electron beam evaporated samples, although these effects could be at least partially caused by the poorer rate control and shielding in our evaporation apparatus.

The consistance between the different series of sputtered samples is very good except for the $n = 10$ samples, where S_1-10 and S_3-10 are completely paramagnetic while S_2-10 has a paramagnetic fraction of 0.26. X-ray diffraction also shows a bcc Fe peak for the S_2-10 sample, while the other two sputtered $n = 10$ samples show only broad peaks characteristic of highly disordered material. This suggests that 10 monolayers of iron is a critical value for stabilization of bcc Fe. This is also evidenced by the sudden increase in paramagnetic fraction which occurs with decreasing n near this value (Table 4). Slight variations in sputter deposition conditions, most likely the tooling factor which relates the rate monitor readings to the deposited thickness, produce samples with either slightly more or less than this critical value resulting in either mostly bcc Fe or disordered paramagnetic phase material. Further work utilizing in-situ structural characterization is needed to fully understand the mechanism for this sudden onset of crystalline order with increasing n.

Conclusions

The interface between crystalline iron and zirconium is amorphous, and about twice as large in electron beam evaporated samples as in sputter deposited samples, where it is about 3–4 atomic planes thick. The crystalline structure in sputter deposited iron layers is abruptly destroyed when the layer thickness decreases below 10 atomic planes, with samples near this layer thickness showing inconsistant results. An amorphous Fe-Zr phase grows during solid state reaction between crystalline iron and zirconium. This phase appears to be about 50 at% zirconium.

Acknowledgments

The Auger profiling was performed by A. Dow and S. Simko of the Analytical Chemistry Department of General Motors Research. We would also like to thank M. Devour for his part in sample preparation.

References

1. B. M. Clemens and M. J. Suchoski, Appl. Phys. Lett. **47**, 943 (1985).
2. H. U. Krebs, D. J. Webb and A. F. Marshall, Phys. Rev. B **35**, 3592 (1987).
3. R. B. Schwarz and W. L. Johnson, Phys. Rev. Lett. **51**, 415 (1983).
4. B. M. Clemens, Phys. Rev. B **33**(11), 7615 (1986).
5. H. Schröder and K. Samwer, Phys. Rev. Lett. **54**, 197 (1985).
6. Y.-T. Cheng, W. L. Johnson, and M.-A. Nicolet, Appl. Phys. Lett. **47**, 800 (1985).
7. S. M. Heald, J. M. Tranquada, B. M. Clemens, and J. P. Stec, *Proceedings of The Fourth International Conference on EXAFS, Abbaye Royale de Sontevraud, France,* (July 1986).
8. B. M. Clemens, J. P. Stec, S. M. Heald, and J. M. Tranquada, *Interfaces, Super-lattices, and Thin Films, Materials Research Society Proceedings,* (1987).
9. C.-J. Lin, F. Spaepen, and D. Turnbull, J. Non. Cryst. Solids **61&62**, 767 (1984).
10. M. R. Kahn, C. S. L. Chun, G. P. Felcher, M. Grimsditch, A. Kueny, C. M. Falco, and I. K. Schuller, Phys. Rev. B **27**, 7186 (1983).
11. D. Liljequist, T. Ekdahl, and U. Båverstam, Nucl. Instrum. Meth. **166**, 49 (1978).
12. S. F. Bartram, Crystallite size determination from line broadening and spotty patterns In *Handbook of X-Rays For Diffraction, Emission, Absorption, and Microscopy* Emmett F. Kaelble, editor, (McGraw Hill, New York, 1967) P. 17-13.
13. A. Sette-Camora and W. Keune, Corros. Sci. **15**, 441 (1975).
14. F. L. Deeny and P. J. Mcarthy, Nucl. Instrum. Meth. **166**, 49 (1979).
15. D. L. Williamson, F. M. Kustas, D. F. Fobare, and M. S. Mishra, J. of Appl. Phys. **60**, 1493 (1986).
16. B. M. Clemens and J. G. Gay, Phys. Rev. B Rapid Comm. **35**, 9337 (1987).
17. D. L. Williamson and B. M. Clemens, Hyperfine Interactions (In press), (1987).
18. K. M. Unruh and C. L. Chien, Phys. Rev. B **30**, 4968 (1984).
19. R. E. Meads, B. M. Place, F. W. D. Woodhams, and R. C. Clark, Nucl. Instrum. Meth. **98**, 29 (1972).
20. S. A. Wender and N. Hershkowitz, Nucl. Instrum. Meth. **98**, 105 (1972).
21. F. Aubertin, U. Gonser, S. J. Campbell, and H-G. Wagner, Z. Metallk. **76**, 237 (1985).
22. M. Maurer, J. M. Friedt, and J. P. Sanchez, J. Phys. F **15**, 1449 (1985).
23. N. K. Jaggi, L. H. Schwartz, H. K. Wong, and J. B. Ketterson, J. Magn. and Magn. Matrls. **49**, 1 (1985).
24. K. Kawaguchi, R. Yamamoto, N. Hosoito, T. Shinjo, and T. Takada, J. Phys. Soc. Jpn. **55**, 2375 (1986).
25. G. K. Wertheim, Y. Jaccarino, J. H. Wernick, and R. C. Clark, Phys. Rev. Lett. **12**, 24 (1964).
26. R. P. Reimer and J. G. Gay, (private communication) 1987.
27. B. M. Clemens, J. of Appl. Phys. **61**, 4525 (1987).

IN-SITU AND HIGH-RESOLUTION TEM OBSERVATION OF INTERFACIAL REACTIONS IN METAL-SILICON MULTILAYERS

KAREN HOLLOWAY, KHIEM BA DO AND ROBERT SINCLAIR
Department of Materials Science and Engineering, Stanford University, Stanford, California 94305

ABSTRACT

Titanium-silicon and molybdenum-silicon multilayers have been investigated in cross-section using a high resolution TEM with a point-to-point resolution of .22 nm. This resolution has allowed us to confirm the amorphous nature both of the intermixed layers which form at the metal-silicon interfaces on sputter deposition, and of the Ti-Si reaction product which forms at temperatures below which crystalline silicides nucleate.

The bulk studies are complemented by *in-situ* annealing experiments in the high-resolution microscope. Phase reactions in the multilayers can be followed in real time, allowing direct observation of the formation of the amorphous alloy, the appearance of rows of Kirkendall voids, and the crystallization and growth of silicide phases.

INTRODUCTION

Extensive investigation into reactions of refractory metals and silicon has been motivated by the technological applications of silicides in integrated circuit integrated schemes.[1] In particular, the disilicides of molybdenum and titanium have high melting points and low resistivities which make them primary candidates for these purposes. The intensive study of these systems has created a body of information through which fundamental questions of the kinetics and thermodynamics of phase transitions in thin films can be addressed. For example, the formation and growth of amorphous M-Si alloys has recently been demonstrated by transmission electron microscope (TEM) studies of thin film layers of amorphous silicon with rhodium,[2] nickel,[3] and titanium.[4] These reactions are similar to solid state amorphization reactions which have been reported in metal-metal systems such as Au-La [5] and Ni-Zr. [6]

Cross-section high resolution TEM is particularly powerful in revealing the microstructures and extent of reactions at interfaces. HRTEM can determine whether a small volume of material is truly amorphous or rather microcrystalline, since lattice fringes arising from the presence of long-range order can be observed in microcrystals.

The reaction mechanism can be dynamically observed in real time while heating the TEM sample *in-situ* in the microscope. There are some advantages over individual sequential micrographs taken from a series of different samples, so long as the results are shown to be representative of the bulk behavior. Transient events such as the nucleation of phases and voids can be studied, and the sequence of reactions demonstrated, while such occurrences would possibly be missed in a bulk sequence. In this article we describe the results of both *in-situ* and bulk annealing studies of Ti-Si and Mo-Si multilayers.

EXPERIMENTAL PROCEDURE

Titanium-silicon and molybdenum-silicon multilayers were sputter deposited in a cryo-pumped system equipped with a rotating table (described elsewhere [7]). Alternating layers of metal and silicon were deposited at room temperature onto 3 inch <100> silicon wafers. The relative thicknesses of metal and Si were controlled by the power applied to the targets; the bilayer spacings of the multilayers were controlled by the rotation speed of the table. Base vacuum pressure was 6×10^{-7} torr; the argon sputter gas pressure was 2.2 microns. The substrate temperature does not rise above 50-100°C during deposition, since the samples spend only a fraction of the time under the targets. The overall composition of the Ti-Si film is very close to 50 at.% Ti, 50 at.% Si, as measured by Rutherford backscattering spectrometry, and the bilayer spacing is 25.0 nm. Sputter deposition rates were chosen so that the Mo-Si multilayer composition would also be close to equiatomic; the bilayer spacing is about 13.0 nm.

Coupons scribed from each substrate were rapid-thermal annealed in an AG Associates Heatpulse 210 in 99.998% pure flowing argon for 30 seconds at temperatures ranging from 300°C to 550°C.

Cross-section samples of annealed and as-deposited multilayers were prepared for TEM analysis by mechanical thinning and ion-beam milling using the technique developed by Bravman and Sinclair. [8] A specimen of each as-deposited film was also annealed *in-situ* in the microscope using a Philips PW6592 heating holder equipped with a resistively-heated stage. The temperature was measured with a thermocouple built into the holder and located next to the specimen. The temperatures reported herein are nominal; measured temperature can differ from the true local value.[9] Any difficulty arising from specimen drift during heating is circumvented by recording the images (one every 1/30 second) on video tape. All experiments were performed in a medium voltage TEM (a Philips EM430ST operating at 300kV) with a resolution of 0.22 nm. The cross-sections were aligned perfectly with respect to the beam using the low-angle electron diffraction pattern which arises from the bilayer periodicity of the film.

RESULTS

Titanium-silicon interfacial reaction

The results of the bulk and *in-situ* annealing of the Ti-Si multilayers are summarized in Figure 1. A high degree of texture allows at least one set of fringes of the hexagonal α-Ti lattice to be imaged in most grains. The lighter silicon layers have the typical amorphous HREM appearance. At each α-Ti - a Si interface in the as-deposited multilayer (Figure 1a), a largely planar 2.9 nm thick amorphous interlayer has formed on sputter deposition of the layers. After rapid thermal annealing at 400°C for 30 seconds, 6.0 nm thick planar amorphous Ti-Si alloy layers have formed (Figure 1b). At 450°C, a 30-second anneal produces the microstructure shown in Figure 1c. The alloy layer has grown to 7.5 nm, completely consuming the amorphous silicon. A nearly continuous 2.0-2.5 nm band of crystalline α-Ti remains, as is evident from the lattice image in the center of each layer. Rows of Kirkendall voids with an average diameter of 4.4 nm, formed by the interdiffusion process, are present in the place of the silicon layers. The presence and location of the voids confirms that silicon is the more mobile element in this process, as is the case for the growth of the crystalline titanium silicide.[10]

The amorphization reaction was observed *in-situ* in a cross-section TEM sample heated in the microscope. The sequence and morphology of the reaction observed *in-situ* correspond exactly to that observed in cross-section samples fabricated from bulk-annealed material. This experiment allowed us to closely follow the various stages of the amorphization reaction. Figures 1d, e, and f show the Ti-Si multilayers at room temperature, and at nominal temperatures of about 510°C, and 590°C, respectively. Each of these figures was produced by photographing the videotape on playback using a 1/2 second exposure, averaging 15 frames. The crystalline α-Ti region is distinguished in the darker layers by the various diffracting conditions of the titanium grains. The amorphous alloy is seen to grow in a planar fashion in the nominal temperature range of 380 to 490°C. After 18 minutes at 510°C (Figure 1e), the interface between the amorphous alloy layers and the remaining amorphous silicon had developed a roughness of about 1 nm, giving the interfaces a 'wavy' appearance. When the alloy layers had grown until the silicon is completely consumed (after 15 minutes at 590°C), the interfaces from neighboring layers meet,

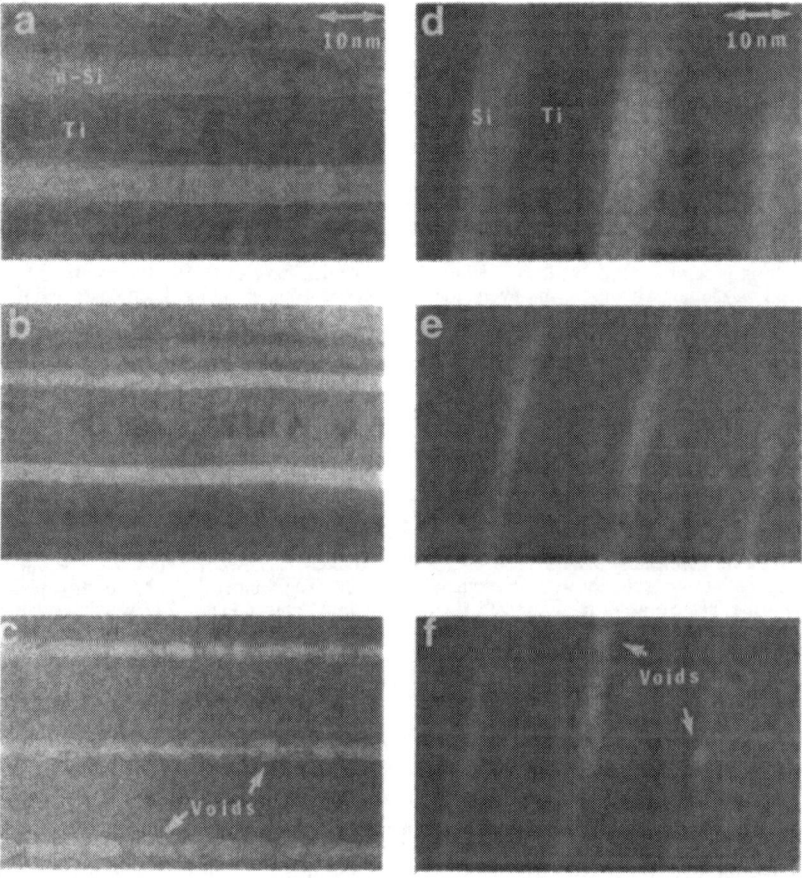

Figure 1. Cross-section TEM micrographs of Ti-Si multilayers with a bilayer spacing of 25.0 nm: (a) high resolution micrograph of unannealed structure showing 3.0 nm thick amorphous interfacial layers, after rapid-thermal annealing in bulk at 400°C for 30 s (b), the amorphous layers are 6.0 nm thick; and (c) after annealing in bulk at 450°C for 30 s, they are 7.5 nm thick. Kirkendall voids have formed between the layers. Video photographs show the multilayers at various stages of *in-situ* annealing: (d) before annealing, (e) at a nominal temperature of 510°C, showing that the a-Si - a-TiSi interfaces are roughening, and at 590°C (c), Kirkendall voids are forming between the layers.

and voids formed in the troughs of the 'waves' (Figure 1f). A thin, continuous band of crystalline titanium remains in the middle of each layer, as was observed in the bulk study. Further details of this work are given elsewhere.[11]

<u>Molybdenum-silicon interfacial reaction</u>

The as-deposited structure of the molybdenum-silicon multilayers is similar in many respects to that of the Ti-Si structure described above. On sputter deposition, amorphous Mo-Si intermixed layers form at each interface (Figure 2a). However, these layers are consistently thicker on the 'bottom' (toward the substrate) side of the Mo layers than on the 'top'. The average thicknesses are 1.9 nm and 1.1 nm, respectively. This asymmetry arises from the sputtering process. The molybdenum atoms, being heavier, have a greater momentum as they deposit on the underlying silicon layers, than vice-versa. This effect has been described in a previous HREM study.[12]

The reaction at molybdenum-silicon interfaces on rapid thermal annealing proceeds quite differently from that observed in Ti-Si multilayers. The solid state amorphization reaction does not occur, or at least amorphous alloy growth is very limited, in spite of the presence of an already nucleated amorphous phase at each interface. A rapid thermal anneal at 400°C for 30 s (not shown) does not change the microstructure of the multilayers. Instead, both interfaces crystallize, then the growth of the crystalline silicide proceeds. After annealing at 450°C for 30 s (Figure 2b), the 'top' and 'bottom' interfacial layers have grown, to 1.4 and 2.7 nm respectively. Fringes arising from long-range order are present in both layers. At 500°C the new phase has grown enough to be identified as hexagonal $MoSi_2$.

The reaction morphology, observed on *in-situ* heating in the TEM, is the same as that observed after annealing in bulk. The as-deposited structure is shown in Figure 3a. The *in-situ* experiment allows us to observe the sequence of events. The thinner ('top') Mo-Si amorphous layer is the first to crystallize. Figure 3b, taken at a nominal temperature of 275°C, shows lattice fringes of the new phase. The thicker ('bottom') amorphous interlayer does not crystallize until 375°C (Figure 3c). These results are preliminary; further work will determine the specifics of this reaction.

DISCUSSION

In an *in-situ* experiment, care must be taken to consider the proximity of the surfaces of the TEM foil to any point in the material. The kinetics and perhaps the morphology of the reaction may change due to surface diffusion and capillarity effects. However, bulk and *in-situ* experiments lead to the same observations in the metal-silicon multilayers studied here. In these cases, the *in-situ* results can be interpreted with confidence that conditions specific to the TEM thin foil does not affect the reaction. Indeed, as interfacial reactions involve interdiffusion of only a few atomic distances, the presence of a surface a few tens of nanometers away would not be expected to have a significant influence in many cases. As an example, the formation of Kirkendall voids in the Ti-Si multilayer TEM sample confirms that bulk diffusion is taking place during the course of the *in-situ* experiment. If the proximity of the foil surfaces allowed surface diffusion to dominate, vacancies, or their analogue in amorphous materials, created by the unequal diffusion flux, would migrate to the surfaces of the foil rather than nucleate voids. Nevertheless, bulk experiments should be undertaken as well to rule out the possibility of thin foil effects.

The great strength of the *in-situ* approach lies in the opportunities one has to observe the sequence of reactions in detail. For example, our rapid thermal annealing study of Mo-Si multilayers revealed that the initial amorphous interlayers crystallize before growth, but the intriguing fact that the thinner layers are first to do so could only be revealed by an *in-situ* experiment, unless the experimenter is very fortunate. In this way, such an experiment

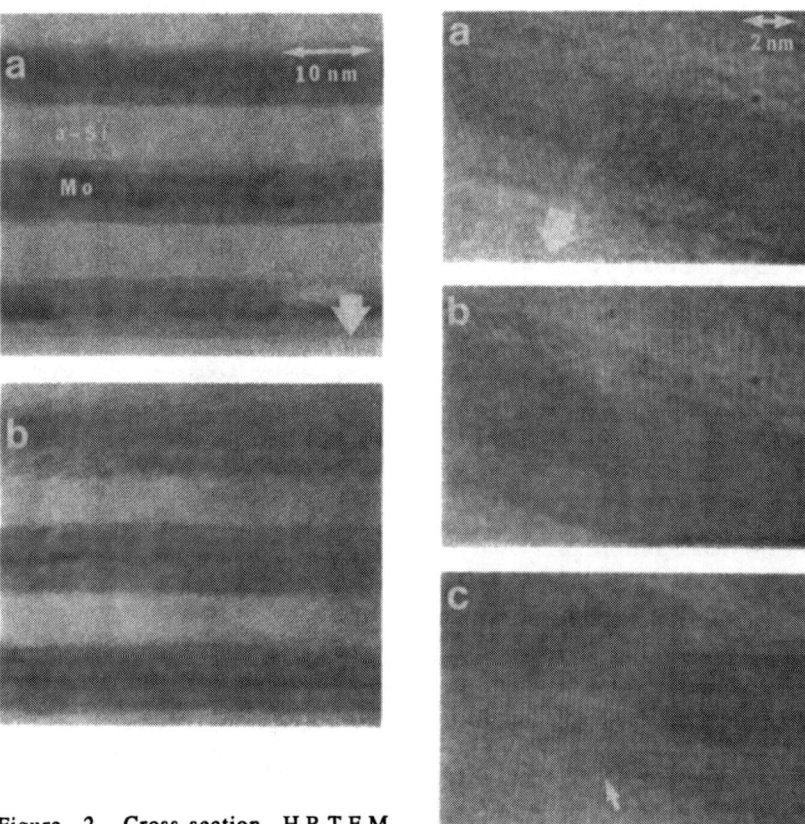

Figure 2. Cross-section HRTEM micrographs of Mo-Si multilayers with a bilayer spacing of 13.0 nm: the as-deposited structure (a) shows amorphous layers at each interface. These layers are thicker on the substrate side of the Mo layers. The substrate direction is indicated with a thick arrow. After rapid-thermal annealing at 450°C for 30 s (b), fringes indicating crystallinity are present in both interlayers.

Figure 3. Cross-section TEM video photographs of the Mo-Si multilayers at various stages of in-situ annealing: (a) before annealing, there are amorphous layers at each interface. The interlayers on the substrate side of the Mo layers are thicker. The substrate direction is indicated with a thick arrow. At a nominal temperature of 275°C, lattice fringes reveal a crystalline phase in the thinner interlayers. The thicker interfacial layers do not crystallize until 375°C (c). The small arrow points to a set of fringes in a crystal grain.

complements more conventional approaches. It is also significant to note that we were able to follow this reaction with high resolution in this mode.

CONCLUSIONS

1. The formation of an amorphous Ti-Si alloy by solid-state reaction in Ti-amorphous Si multilayers has been observed by high-resolution TEM after rapid thermal annealing in bulk, and by observing the reaction while heating *in-situ* in the microscope.

2. The structure of as-deposited Mo-amorphous Si multilayers show amorphous intermixed layers at each interface. These layers are thicker on the 'bottom' (toward the substrate) side of the Mo layers than on the 'top'. This asymmetry arises from the sputtering process. The solid state amorphization reaction does not occur in Mo-Si. Instead, the amorphous interfacial layers crystallize, and growth of the crystalline silicide proceeds. *In-situ* TEM shows that the thinner amorphous layers crystallize first.

3. The sequence and morphology of both reactions observed *in-situ* correspond exactly to those observed in cross-section from samples which were rapid thermal annealed in bulk. The two types of annealing experiments complement each other: bulk experiments assure bulk behavior and allow detailed study of an annealed microstructure; *in-situ* experiments allow the reactions to followed in real time, and reveal the sequence of events that make up the reaction.

ACKNOWLEDGEMENTS

The authors would like to acknowledge the use of the facilities of the Materials Science and Engineering Department and the Center for Materials Research at Stanford University. Funding has been provided by the NSF - MRL program through the Center for Materials Research. We thank Kenneth P. MacWilliams and Bruce Fishbien (Stanford University) for assistance with the Heatpulse 210 RTA system in the Center for Integrated Systems at Stanford University. Funding for the Philips EM430 electron microscope was kindly provided by the NSF-MRL program, the Schools of Engineering, Humanities and Sciences, and Earth Sciences, and the Departments of Electrical Engineering and Materials Science and Engineering at Stanford University.

REFERENCES

1. S.P. Murarka in *Silicide for VLSI Applications*, Academic Press, New York (1983).
2. S. Herd, K.N. Tu, and K.Y. Ahn, Appl. Phys. Lett. **42**, 599 (1983).
3. M. Natan, J. Vac. Sci. Technol. B **4**, 1404 (1986)
4. K. Holloway and R. Sinclair, J. Appl. Phys. **61**, 1359 (1987).
5. R.B. Schwarz and W.L. Johnson, Phys. Rev. Lett. **51**,415 (1983).
6. B.M. Clemens, J. Non Cryst. Solids **61&62**,817 (1984).
7. T. Barbee and D.L. Keith, in *Synthesis and Properties of Metastable Phases*, edited by E.S. Machlin and T.J. Rowland (AIME, New York, 1980), p.93.
8. J.C. Bravman and R. Sinclair, J. Electron Microsc. Tech. **1**, 53 (1984).
9. R. Sinclair, M.A. Parker, and K.B. Kim, Ultramicroscopy, in press.
10. W.K. Chu, S.S. Lau, J.W. Mayer, J. Muller, and K.M. Tu, Thin Solid Films **25**, 393 (1975).
11. K. Holloway and R. Sinclair, J. Less Common Metals, in press.
12. A. Petford-Long, M.B. Stearns, C.-H. Chang, D.G. Stearns, N.M. Ceglio, and A.M. Hawryluk, J. Appl. Phys. **61**, 1422 (1987).

THE EFFECT OF ION IMPLANTATION ON THE INTERDIFFUSION IN Si/Ge AMORPHOUS ARTIFICIAL MULTILAYERS

B. PARK[1], F. SPAEPEN[1], J.M. POATE[2], F. PRIOLO[2], D.C. JACOBSON[2], C.S. PAI[2], A.E. WHITE[2] AND K.T. SHORT[2]

[1]Division of Applied Sciences, Harvard University, Cambridge, MA 02138
[2]AT&T Bell Laboratories, 600 Mountain Road, Murray Hill, NJ 07974

ABSTRACT

Amorphous Si/Ge artificial multilayers have been implanted with Si and B at liquid nitrogen temperature, and partially ion mixed with Ar at different temperatures. In all cases, the square of the mixing length was found to be proportional to the dose. Annealing of Si-implanted samples showed that after relaxation the diffusivity appeared unaffected by the implantation process. Annealing of the B-implanted samples showed an enhancement of the diffusivity at the higher dose. The diffusive component of the square of the mixing length in the Ar-ion mixed samples has an Arrhenius-type temperature dependence, with an activation enthalpy of 0.22 eV.

INTRODUCTION

Diffusion in amorphous covalent network formers, such as Si and Ge, is governed by structural defects (broken bonds, vacancies or interstitials) that are present as a result of growth or implantation processes. The presence of dopants may also affect the diffusivity of the host atoms. It is therefore of interest to study the effects on the diffusivity of processes that change these defect concentrations: self-implantation, implantation of dopants, and noble gas ion mixing.

Since these amorphous systems must be kept below the temperature at which they crystallize rapidly, the values of the diffusivities to be measured are necessarily small. Techniques based on determining the broadening of a single composition profile, such as Rutherford backscattering, have a resolution limit of about 20Å, which allows measurements of diffusivities as low as 10^{-19} cm^2/s. An artificial multilayer, on the other hand, contains several hundred identical diffusion couples and changes in their composition profile can be detected by x-ray diffraction with a resolution of 0.1Å; diffusivities down to 10^{-23}cm^2/s can therefore be measured by this technique [1].

Last year we presented a first report on a study of self-implantation on the interdiffusion of amorphous Si/Ge multilayers of equiatomic average composition [2]. We demonstrated there that the artificial multilayer technique can be used for very accurate determinations of ion beam mixing lengths, and that the interdiffusivity of implanted and unimplanted samples was the same. In this paper we present further data on the effects of self-implantation, as well as on the effects of implantation of boron and of ion mixing with high energy argon.

EXPERIMENTAL TECHNIQUE

The artificial multilayers were prepared by ion beam sputtering from alternating elemental targets onto a (100) polished and oxidized Si substrate [3]. The layer repeat lengths were around 60 Å, and were kept constant within 4%. The total film thickness was usually around 2000Å, and the films were capped with a 150Å Si film. The average composition of the films was equiatomic, and they contained about 1 at. % Ar. The sputtering gas was cleaned in a Ti gettering furnace immediately before use. Annealing of the samples was done in a quartz lamp furnace, after evacuating to 10^{-7} torr and backfilling with helium. The cross-sectional transmission electron micrograph of Figure 1 clearly shows that the layer structure is very well defined; TEM investigation of irradiated or annealed samples confirmed that the planar layer structure was preserved in these processes, and that no crystallization had occurred.

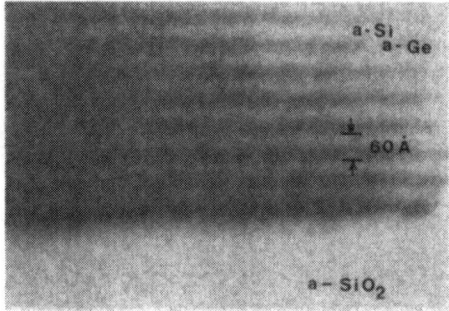

Fig. 1. Cross-sectional transmission electron micrograph of an artificial multilayer of amorphous Si and Ge with a repeat length of 60Å.

The samples were investigated by x-ray diffraction in θ–2θ scans on a GE horizontal diffractometer, using Cr K_α radiation. A series of peaks is observed at

$2\theta=2\cos^{-1}[n\cos(\sin^{-1}(Nn\lambda/2d))]$, where λ is the x-ray wavelength, N is the order of the peak, n is the average index of refraction of the material, and d is the repeat length in the multilayer [1,4]. In the kinematic approximation, the integrated intensity of each reflection is proportional to the square of the corresponding coefficient in the Fourier expansion of the composition profile. If, as a result of annealing or ion mixing, the intensity of the first peak decreases from I_1 to I_2, diffusive broadening with a characteristic length $x=[\ln(I_1/I_2)d^2/8\pi^2]^{1/2}$ has occurred; in thermal diffusion $x=(Dt)^{1/2}$, where t is the annealing time.

SELF-IMPLANTATION

Figure 2 shows that the square of the diffusive length, x^2, resulting from implantation of ^{29}Si (at 70 and 160 keV at liquid nitrogen temperature), is proportional to the dose, Φ, for doses up to $6 \times 10^{14}/\text{cm}^2$. The implantation energies were chosen to have a reasonably uniform defect distribution throughout the layers [2]. This shows that the displacements induced by the ion bombardment can be described by a random walk process, as in thermal diffusion.

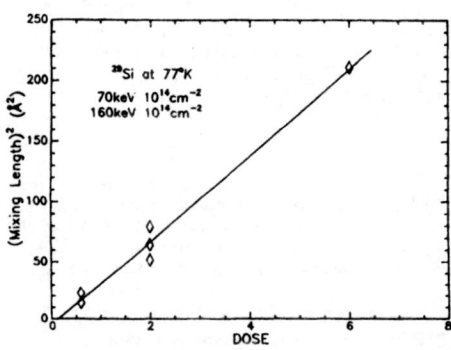

Fig. 2. Dose dependence of the square of the mixing length measured after self-implantation.

A sample, implanted at $6 \times 10^{13}/\text{cm}^2$ (corresponding to x=4.7Å), was annealed, together with a simultaneously deposited unimplanted control sample. Figure 3 shows the results; the slope of lnI(t) (I: the intensity of the first diffraction peak; t: annealing time) is proportional to the interdiffusivity. During the first anneal, at 650K, the implanted sample exhibits somewhat stronger relaxation (i.e., decrease of the diffusivity with time) than the control sample. During the second anneal, at 670K, the interdiffusivity in the two samples is the same within experimental error. These observations are qualitatively similar to those reported in an earlier paper [2]. The relaxation in the earlier samples was less strong, possibly because they were produced without the use of a Ti gettering system. TEM on the earlier samples confirmed that no crystallization had occurred even at the longest annealing times.

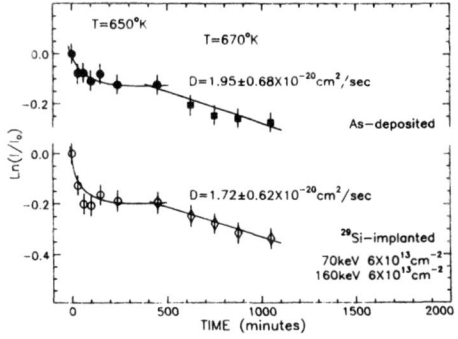

Fig. 3. Intensity of the first diffraction peak resulting from the composition modulation as a function of annealing time at two temperatures. Open symbols: self-implanted sample. Closed symbols: control sample.

This similarity in diffusion can be explained in several ways. For example, the implantation-induced defects could annihilate rather quickly, during implantation and during the earliest stage of relaxation, so that their concentration falls below that of growth-related ones.

BORON IMPLANTATION

Figure 4 shows that the square of the diffusive length, x^2, resulting from implantation of ^{11}B (one dose Φ at 20keV, and 2Φ at 50 keV at liquid nitrogen temperature), is also proportional to Φ, up to $2 \times 10^{15}/cm^2$. Again, the energies were chosen to have a reasonably uniform dopant

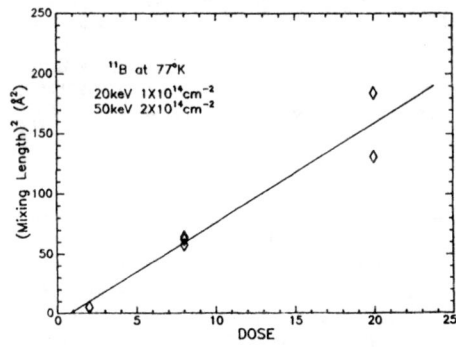

Fig. 4. Dose dependence of the square of the mixing length measured after B-implantation.

distribution throughout the layers. Figures 5 and 6 show the interdiffusivity during annealing of two implanted samples(maximum 0.05 and 0.2 at % B concentration, respectively) and their unimplanted controls. The respective mixing lengths, x, are 2.3 and 7.6Å. In both cases considerable relaxation observed during the first anneal at 650K is similar to what is observed for self implantation: slightly more relaxation in the implanted samples than in their controls. For the

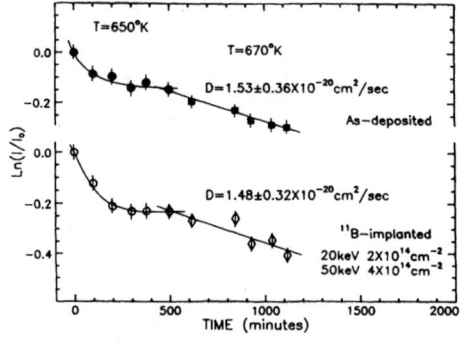

Fig. 5. Intensity of the first diffraction peak resulting from the composition modulation as a function of annealing time at two temperatures. Open symbols: B-implanted sample (0.05 at.%). Closed symbols: control sample.

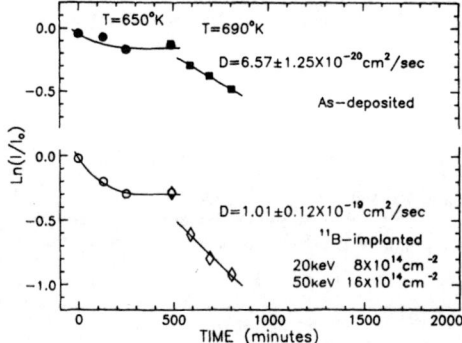

Fig. 6. Intensity of the first diffraction peak resulting from the composition modulation as a function of annealing time at two temperatures. Open symbols: B-implanted sample (0.2 at.%). Closed symbols: control sample.

second anneal, at higher temperature, no difference was observed between the 0.05%B sample and its control; in the 0.2%B sample, on the other hand, the diffusivity was about a factor 1.5 higher than in its control.

It is known that incorporation of dopant elements in amorphous semiconductors causes enhancement of the solid phase epitaxial (SPE) regrowth rate. For 0.2 at. % B, for example, the rate is enhanced by a factor of 5 in Si [5] and by a factor of 2.5 in Ge [6]. The minimum B concentration for SPE enhancement is 0.04 at. % in Si. A comparison of these, admittedly preliminary, diffusion results and the SPE data suggests that a fraction of the dopant-induced defects contributing to SPE growth is contributing to the diffusion process as well.

ION BEAM-ENHANCED DIFFUSION

Some of the artificial multilayers were partially mixed by bombarding them with 1.5 MeV Ar ions. At this energy, the Ar ions pass completely through the films. Figure 7 shows that the square of the mixing length, x^2, in this process is proportional to the dose, as in the other diffusive processes discussed above. Figure 8 shows the dependence of x^2 on the temperature of the substrate during implantation. Between liquid nitrogen temperature and room temperature, the mixing process is entirely ballistic in nature (temperature independent). The diffusive component of x^2, obtained by subtracting out the ballistic part, has an Arrhenius-type temperature dependence (see Figure 9), with an activation enthalpy of 0.22 eV. Dt for thermal diffusion over the duration of the implantation process is still two orders of magnitude smaller than the diffusive component of x^2 at the highest temperatures investigated. These results are

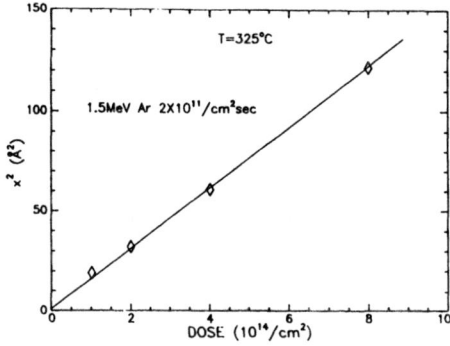

Fig. 7. Dose dependence of the square of the mixing length measured after ion mixing with 1.5 MeV Ar at 325 °C.

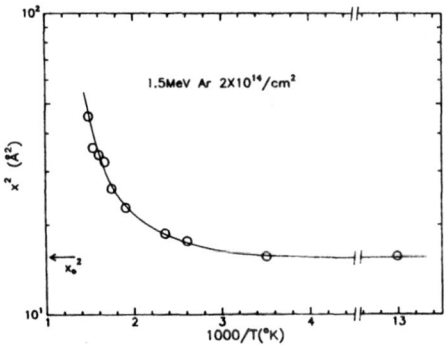

Fig. 8. Temperature dependence of the square of the mixing length measured after ion implantation with 2 x 10^{14} cm^{-2} 1.5 MeV Ar at a dose rate of 2 x 10^{11} cm^{-2}s^{-1}.

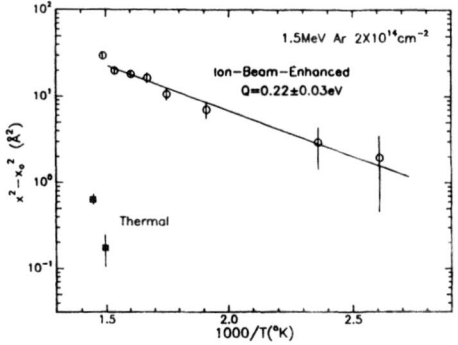

Fig. 9. Diffusive component of the data of Fig. 8. The thermal data are from the control samples after relaxation in Figs. 3, 5, and 6.

similar to the recent measurements of Priolo et al. [7] for the ion beam-enhanced diffusion of Au in amorphous Si.

The existence of a well-defined activation enthalpy for the diffusive component of x^2 supports the idea that atomic transport in amorphous semiconductors is indeed governed by well-defined atomic scale defects similar to those present in crystals.

ACKNOWLEDGEMENTS

We thank Eric Nygren and Y.Z. Lu for their help with the cross-sectional TEM. The work at Harvard has been supported by the National Science Foundation through the Materials Research Laboratory under contract number DMR-86-14003.

REFERENCES

1. A.L. Greer and F. Spaepen, in Synthetic Modulated Structures, edited by L.L. Chang and B.C. Giessen (Academic Press, New York, 1985) p.419.

2. B. Park, F. Spaepen, J.M. Poate and D.C. Jacobson, Beam-Solid Interactions and Transient Processes, edited by S.T. Picraux, M.O. Thompson, J.S. Williams (Mater. Res. Soc. Proc. 74, Pittsburgh, PA, 1987), pp. 493-497.

3. F. Spaepen, A.L. Greer, K.F. Kelton and J.L. Bell, Rev. Sci. Instr. 56, 1340 (1985).

4. A.M. Kadin and J.E. Keem, Scripta Metall. 20, 440 (1986).

5. G.L. Olson, Energy Beam-Solid Interactions and Transient Thermal Processing/1984, edited by D.K. Biegelsen, G.A. Rozgonyi, C.V. Shank (Mater. Res. Soc. Proc. 35, Pittsburgh, PA, 1985), p. 25.

6. I. Suni, G. Göltz, M.-A. Nicolet and S.S. Lau, Thin Solid Films 93, 171 (1982).

7. F. Priolo, J.M. Poate, D.C. Jacobson, J. Linnros and S.U. Campisano, to be published.

STABILITY AND CRYSTALLIZATION
OF AMORPHOUS SEMICONDUCTOR MULTILAYERS

P.D.Persans*, A.F.Ruppert# and B.Abeles#

* Physics Department and Center for Integrated Electronics
Rensselaer Polytechnic Institute, Troy, NY 12180-3590

Exxon Research and Engineering Company
Annandale, NJ 08801

ABSTRACT

We discuss recent measurements of relaxation, interdiffusion and crystallization in amorphous hydrogenated semiconductors and insulators prepared as periodic multilayers. The stability of multilayer structures depends upon temperature, repeat distance and the nature of the materials. Crystallization of amorphous silicon in amorphous silicon/amorphous silicon dioxide layers is inhibited when the silicon thickness is reduced below 20 nm.

INTRODUCTION

Alternating deposition of hydrogenated amorphous silicon (a-Si:H) with other amorphous semiconductors and insulators such as germanium (a-Ge:H), silicon nitride (a-SiN$_x$:H) and silicon oxide (a-SiO$_x$) under the proper conditions yields compositionally modulated solids (amorphous superlattices) with a high density of reproducible layers and interfaces [1-4]. Sub-layer thicknesses down to 1 nm with compositionally abrupt (1 monolayer transitional composition) interfaces have been demonstrated [2,5,6]. These structures hold promise for applications in photovoltaic, photoreceptor and electrophotographic technologies and therefore their electronic properties are of intense current interest. Structural and electronic stability under thermal and optical stress must also be understood in order to exploit amorphous thin films for these applications.

The high density of interfaces has two important ramifications. Interface and layer thickness become increasingly important in determining the structure and stability of each layer. For example composition and strain gradients near the interface could lead to enhanced atomic diffusion. Additionally the interfaces make up a significant and experimentally controllable fraction of the deposited structure and therefore probes which are normally used for thick film and bulk measurements (i.e.- optical absorption, Raman scattering, IR absorption) can be applied to study the nature of the interface region[5].In this paper we focus on the stability of ultrathin layers against crystallization. We report observations of crystallization of thin amorphous silicon sub-layers in hydrogenated amorphous silicon (a-Si:H)/hydrogenated amorphous silicon oxide (a-Si:H/a-SiO$_x$) periodic multilayer structures.

EXPERIMENTAL DETAILS

Samples for the present study were prepared by plasma assisted chemical vapor deposition from pure SiH$_4$ and SiH$_4$:N$_2$O in the ratio 1:50 for a-Si:H and a-SiO$_x$ respectively [1-3]. Periodic multilayers were deposited onto Corning 7059 glass substrates by periodically changing the composition of the plasma gases. Typical growth conditions were: substrate temperature, 525 K, total gas pressure, 30 mT; plasma rf power, 5 W at 13.56

MHz; growth rate, 0.1 nms^{-1}. The repeat distance d_r for the periodic structure was varied from 1 to 100 nm and the total sample thicknesses were about 1000 nm. Samples were annealed in a vacuum tube furnace with a pressure of 10^{-5} T at temperatures up to 1375K. Techniques for Raman and optical measurements are described in detail elsewhere[5,7,8].

RESULTS AND DISCUSSION

Stability of Ultrathin Amorphous Silicon Layers

In Fig.1 we show Raman spectra for an a-Si:H/SiO$_x$ periodic multilayer with silicon sub-layer thickness d_s=5.0 nm and oxide sublayer thickness d_o=2.5 nm after a series of 30 minute thermal anneals. As annealing temperature is increased from 975 K to 1125 K the development of a sharp peak at 520 cm^{-1} is observed which indicates the formation of Si crystallites. The broad peak at 470 cm^{-1} is due to optic-like modes of the amorphous solid. Normally, thick films (d_s > 50 nm) of a-Si:H are fully crystallized after annealing for 30 min. at only 975 K. In Fig.1 we observe that this sample is stable against crystallization up to 1075 K. This data suggests that thin films of a-Si:H clad by a-SiO$_x$ are more stable against crystallization than thick films. This observation is born out by more systematic investigation of the thickness dependence of the crystallization temperature for isochronal 30 min. annealing[9]. For thickness less than 5 nm the crystallization temperature increases drastically by more than 300 K to 1300 K.

The fact that ultrathin a-Si:H layers do not easily crystallize tells us about the relative interface energies for plasma deposited a-Si:H/SiO$_2$, a-Si:H/x-Si and x-Si/SiO$_2$ (where x-indicates the crystal form.) The free energy for the formation of a disk shaped crystallite of thickness l and diameter 2r, bounded on top and bottom by oxide is given by:

$$\Delta H = H_{ax}\pi r^2 l + \sigma_{ax}2\pi rl + \Delta\sigma 2\pi r^2 \tag{1}$$

H_{ax} (-10^3 J cm^{-3} [9]) is the free energy of the amorphous Si to crystalline phase transition, σ_{ax} (3x10^{-5} J cm^{-2} [10]) is the surface tension of the crystal/ amorphous Si interface, and $\Delta\sigma = (\sigma_{ox} - \sigma_{oa})$ is the difference in surface tension between the x-Si/oxide and a-Si/oxide interfaces. If $\Delta\sigma$ were negative then the crystallization would be enhanced in thinner films relative to thicker films. Since crystallization is inhibited in thin films we conclude that $\Delta\sigma$ is positive. That is, the amorphous/ oxide interface energy is lower than the crystal/ oxide interface energy in the form produced by plasma deposition and annealing.

For a spherical crystallite embedded in amorphous silicon matrix we can estimate the size of the smallest thermodynamically stable crystallite by balancing free energy gain from volume conversion against free energy loss due to interface energy[9]: $d_{min}=(6\sigma_{ax}/\Delta H)=2\pm.5$nm. For Si sub-layers more than a few nanometers thicker than d_{min} we do not expect crystal stability to be affected by layer thickness. Equation 1 applies when $\Delta\sigma < \sigma_{ax}$. If $\Delta\sigma > \sigma_{ax}$ then the crystal interface will form inside of the Si layer and $\Delta\sigma$ will be replaced by σ_{ax} in eq. 1 and l will be reduced accordingly.

We have found by transmission electron microscopy of transverse sections that crystallites which form at Ta=1300 K are always larger than 3 nm, even when the silicon sub-layer is thinner than this. At the temperatures necessary to form crystallites in thin layers in reasonable time the diffusion coefficient of Si in SiO$_2$ is sufficiently great that crystallites form by the aggregation of islands of Si . Due to these types of kinetic effects it has not yet been possible for us to determine the minimum stable crystallite size in this system.

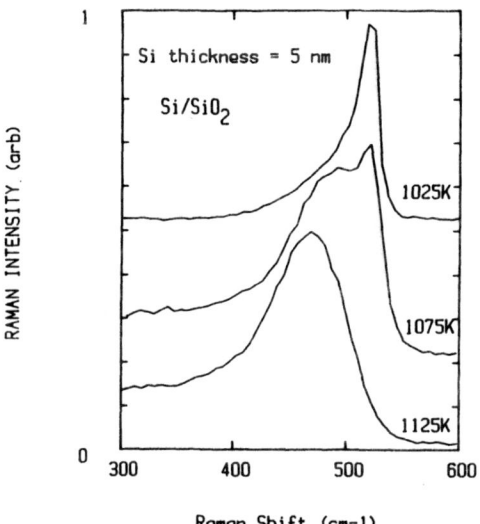

Figure 1. Raman spectra of an a-Si/a-SiO$_x$ multilayer sample with d$_S$= 5 nm and d$_{oxide}$=2.5 nm. The sample has been annealed for 30 minutes in 50 K steps. Top- T$_a$=1125 K, Middle- T$_a$= 1075 K, Bottom- T$_a$= 1025 K.

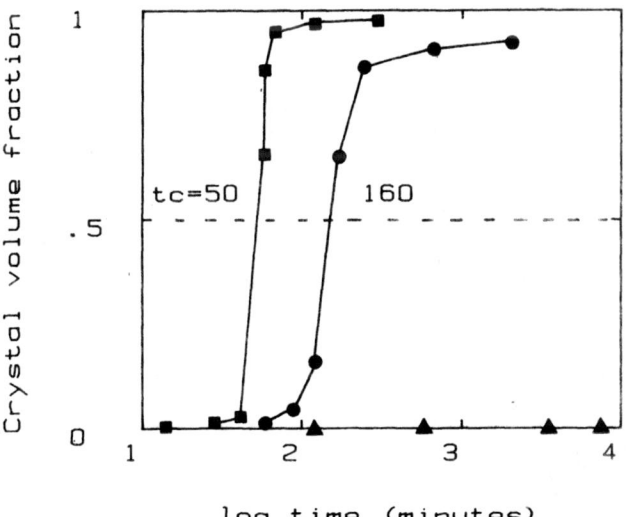

Figure 2. Crystallized volume fraction of the Si Sublayers plotted against annealing time at 975K; d$_S$=5 nm, d$_S$= 10 nm, d$_S$= 50 nm.

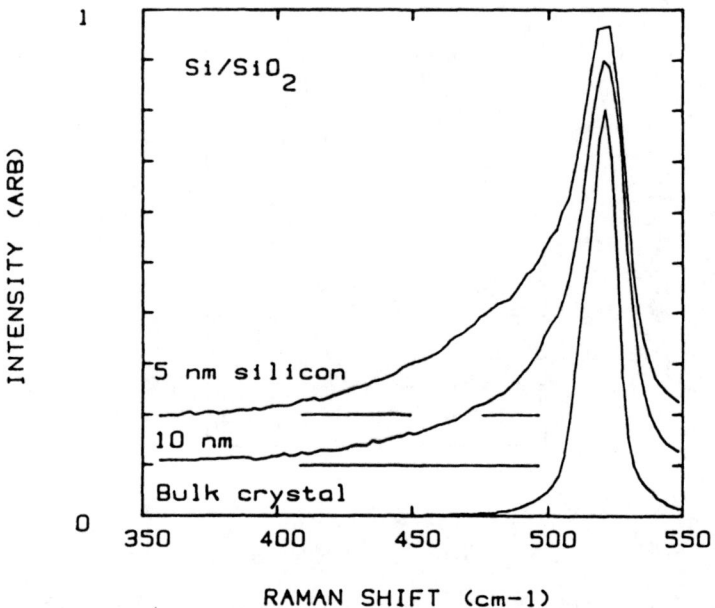

Figure 3. Raman spectra of three fully annealed samples with silicon sublayer thicknesses as marked.

Crystallization Kinetics

Striking differences in relative stability are observed in the crystallization time at fixed temperature. By integrating the area under the broad amorphous peak at 470 cm^{-1} and the sharp crystal peak at 520 cm^{-1} it is possible to analyse the volume fraction of Si which is crystallized [10]. In Fig.2 we show the crystallized volume fraction deduced from Raman analysis as a function of annealing time at 975 K for three films with silicon sublayer thickness of 5, 10 and 50 nm respectively. The $d_s = 5$ nm film did not crystallize during the course of more than 200 hours annealing at 975 K. In this thickness regime the crystallization time is extremely dependent upon sub-layer thickness. (A film with slightly thicker sub-layers (5.5 nm) crystallized in about 6 hours at the same temperature.) Thinner sub-layer films (<3 nm) were stable for several hundred hours at temperatures as high as 1125 K.

We might expect large changes in crystallization time even for sub-layer thicknesses much greater than the minimum stable crystallite size. For example, even if the crystal nucleation rate per unit a-Si volume N_v and the crystal growth rate v_g are independent of thickness the crystallization rate will change due to geometric considerations. If the layer thickness is less than $N_v^{-1/3}$ then the average distance between nucleation centers will increase as $l^{-1/2}$ and crystallization time could increase by the same factor if crystallization were limited by the growth rate.

Transmission microscopy shows that the lateral crystallite size is comparable to layer thickness l for 50 nm>l>5 nm. From this observation we conclude that growth of crystallites along the interfaces is slowed relative to bulk materials, probably due to the additional energy necessary to form the interface.

Structure of Microcrystal Surfaces

In Fig.3 we show Raman spectra for three samples with different Si sub-layer thicknesses (and thus crystallite sizes) of 5 nm, 10 nm and 50 nm respectively. All of these samples have been annealed at 1125 K until the specrum no longer changed. In the layered samples we observe a tail in the spectrum on the low energy side of the bulk crystal transverse optic mode at 520 cm^{-1}. We find that the spectra can be fit by a broad peak and a sharp peak centered at 490 cm^{-1} and 520 cm^{-1} respectively. It has been suggested that the broad low energy peak is due to a disordered surface phase[13]. Integration of the area under the sharp and broad portions of the spectrum respectively suggests that about one half of the volume of the $d_s = 5$ nm sample and one- quarter of the volume of the $d_s = 10$ nm sample is disordered . If we consider the disordered volume to in fact lie on the surface of crystalline spheres then we can easily compute the effective thickness of this layer. For both samples we deduce a disordered surface layer thickness of .6 nm. This is roughly equivalent to two to three monolayers on the surface of each crystallite. Thus the present results are quantitatively consistent with the hypothesis that the 490 cm^{-1} peak is indeed due to a disordered surface phase with structure intermediate between bulk amorphous and crystalline materials.

Supported in part by the National Science Foundation under grant number EET-8714842.

REFERENCES

1. B.Abeles and T.Tiedje, Phys. Rev. Lett., 51,2003,(1983).

2. B.Abeles, T.Tiedje, K.S.Liang, H.W.Deckman, H.Stasiewski, J.Scanlon, J.Non-Cryst.Sol. 66, 351, (1984).

3. P.D.Persans, B.Abeles, H.Stasiewski, and J.Scanlon, Proc. of the 17th Int. Conf. on the Phys. of Semiconductors, ed. J. Chadi and W.Harrison,(Springer, New York, 1985), p.449.

4. J.Kakalios, H.Fritzsche, N.Ibaraki, and S.Ovshinsky, J. Non-Cryst. Sol. 66, 339,(1984).

5. P.D.Persans, A.F.Ruppert, B.Abeles, and T.Tiedje, Phys. Rev. B32, 5558, (1985), and J. de Phys., Coll.C8, 46, 597 (1985).

6. D.D.Allred, J.Gonzalez- Hernandez, O.V.Nguyen, D.Martin, and D.Pawlik, J. Mat. Res., 1, 468, (1986).

7. P.D.Persans and A.F.Ruppert in Phase Transitions in Condensed Systems- Experiments and Theory, ed. G.Slade Cargill III, F.Spaepen, K.-N. Tu,(Materials Research Society, Pittsburgh, 1987), 329.

8. P.D.Persans and A.F.Ruppert, in Semiconductor- Based Heterostructures: Interfacial Structure and Stability, ed. M.L.Green, (Materials Research Society, Pittsburgh, 1987).

9. P. Persans, A.F. Ruppert, and B. Abeles, J. Non-Cryst. Sol., (in press)

10. R.Tsu, J.Gonzalez- Hernandez, S.S.Chao, S.C.Lee, and K.Tanaka, Appl. Phys. Lett., 40, 534, (1982).

11. E.Donovan, F.Spaepen, and D.Turnbull, J. Appl. Phys.,57, 1795, (1985).

12. F. Spaepen, Acta Met. 26, 1167 (1978).

13. T.D.Moustakas, D.A.Weitz, E.B.Prestridge, and R.Friedman, in Plasma Synthesis and Etching of Electronic Materials, (Materials Research Society, Pittsburgh,1985)

THE MECHANISMS OF RELAXATION IN STRAINED LAYER GeSi/Si SUPERLATTICES: DIFFUSION VS. DISLOCATION FORMATION.

F.K. LeGoues, S.S. Iyer, K.N. Tu, and S.L. Delage
IBM T.J. Watson Research Center, Yorktown Heights, New York 10598, USA

ABSTRACT

Si_xGe_{1-x} strained layer superlattices are known to be metastable in that they can be grown fully commensurate with layer thickness higher than the equilibrium, calculated T_c at which dislocation formation becomes energetically favorable. In this paper, we describe the mechanism of relaxation in such multilayers. Both plane-view and cross-sectional transmission electron microscopy (TEM) were used to examine the formation of dislocation at the different interfaces. RBS was used to follow interdiffusion. We found two competing mechanisms for relaxation: The preferred mode for relaxation is the creation of dislocation networks at each of the interfaces. This process can be stopped or considerably inhibited by the difficulty of forming new dislocations in samples which are perfectly commensurate after growth; Some dislocations appear necessary in order to generate more dislocations during annealing. When this is not the case, the only possible way to attain relaxation is through diffusion. In such a case, stress-enhanced diffusion is observed, with a diffusion coefficient 200 times higher than expected.

I. INTRODUCTION

Recently, a considerable amount of work has been published on strained-layer semiconductor superlattices. The interest in these structures comes from the possibility of modifying the band structure and enhancing mobility for specific multilayers compositions and thickness. This has indeed been proven in the Si_xGe_{1-x} /Si case[1]. More recently, the SiGe superlattices have received renewed interest due to the demonstration of an heterojunction bipolar transistor[2]. Van de Merve proposed[3], based on equilibrium calculations and using the idea that, at a critical thickness T_c, it will be energetically more favorable to create dislocations than to strain the layer as a whole. In semiconductors, the theories generally predict value of T_c much lower than experimentally observed, because of the difficulty of forming dislocations at the low growth temperatures that can now be attained by MBE and CVD[4,5,6]. These metastable strained layer superlattices may however relax during thermal annealing[7]. Two mechanisms compete during this relaxation process: The first one is the introduction of dislocations at the interfaces. The second one is interdiffusion between the layers. As was pointed out by Matthews and Blakeslee[8,9,10], the interface between the substrate as a whole is inherently different from the individual interfaces in the superlattice because two types of misfits need to be considered: one is between the substrate and the superlattice as a whole, and the other one is between individual layers in the superlattice. It was indeed shown experimentally that there can be dislocation formation at the substrate/superlattice interface even in cases when the individual layers are completely coherent[7,9]. Fiory et al[7] studied the stability and thermal relaxation of such layers predominantly via the introduction of new dislocations during annealing.

Mat. Res. Soc. Symp. Proc. Vol. 103. ©1988 Materials Research Society

In this study, we show that the difficulty in introducing dislocations is a major factor in determining the mechanism of relaxation during thermal annealing. Indeed, when the interfaces are perfect to start with, we show that no dislocations are introduced during annealing. The only way to relieve the misfit is thus through interdiffusion. At the temperature used in the study, the diffusion coefficient is about 200 times higher than expected. Thus, we propose that stress-enhanced diffusion occurs with a much larger diffusion coefficient than normal. On the other hand, when the interfaces contain a network of dislocations to start with, the introduction of new dislocations is facilitated, and this is the relaxation mode chosen. In this case, no enhanced diffusion is observed.

II. EXPERIMENT

Si_xGe_{1-x} multilayers were grown by MBE on Si(100). Two sets of samples were studied. The first one (which will be referred to as sample 1) had a total of 4 layers, 500Å thick, with about 15% Ge in the SiGe layer and was grown at 607°C. The second one (referred to as sample 2) was grown at 615°C, had a total of eleven layers 1900Å thick and contained about 12% Ge. Both sets of samples were annealed for times varying between 10 min and 24 hours at 900°C in a quartz tube furnace, in flowing purified He. Cross-sectional TEM was used to study the perfection of the different interfaces and to locate the misfit dislocations. Plane-view TEM was used to obtain the total density of dislocations. RBS was used to study interdiffusion.

III. RESULTS

Figure 1 shows cross-sectional TEM of both samples, at different annealing times. It is clear from the micrographs of the as deposited samples that sample 1 is very nearly perfect while sample 2 contains dislocations at all the interfaces. It is also worthwhile to note that the contrast between the different layers is stronger for sample 1 than for sample 2 even though, in both case, the Ge content is similar. This may be due to the more stressed state of sample 1 (since they are no dislocations at the individual interfaces, this sample is more highly stressed). Figure 2 shows planar views of sample 1 and 2, as deposited. Two observations can be made from these micrographs. First, the total number of dislocations is lower for sample 1 than for sample 2: $6.5X10^3$ /cm versus $7.0X10^4$/cm. Second, sample 2 shows paired dislocations, similar to those described by Matthews and Blakeslee[7,8], indicating that the dislocations are present on more than one interface (paired dislocations were described as being portions of the same dislocation line lying on different interfaces and thus separated , in plane view, by a distance of nXhXcot55°where n is an integer and h is the layer thickness - see ref. 8). The fact that this is not observed for sample 1, combined with evidence obtained by an extensive cross-sectional TEM search for the presence of any dislocation at the interfaces other than the substrate/multilayer one, shows that all of the dislocations in this sample are located at the interface between the substrate and the multilayer. Upon annealing, the appearance of the multilayer changes drastically for sample 1: The contrast between the different layers disappears gradually, is barely discernible after 2 hours, and is not visible at all after about 12 hours. At the same time, more dislocations appear at the substrate/multilayer interface, but no dislocations are detected in the multilayer itself. Plane-view samples showed that, after two hours, the density of dislocations at the substrate/multilayer interface is $2.7X10^4$ /cm and increases to $5.1X10^4$/cm after 12

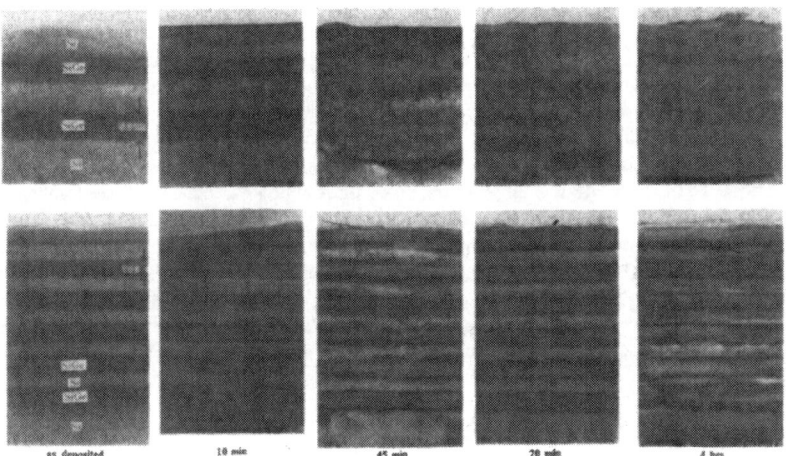

Figure 1:

Cross sectional TEM micrographs of the multilayers, as a function of annealing time, at 900°C.

a) Sample 1 b) Sample 2

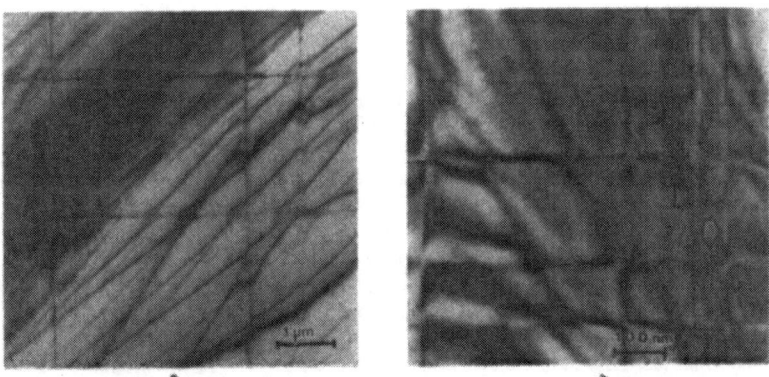

Figure 2:

Flat-on micrographs of the multilayers, as deposited.

a) Sample 1

b) Sample 2: the arrows indicate a pair of bunched dislocations.

hours. For sample 2, an increased number of dislocations is observed, at all the interfaces as a function of time. The contrast between the different layers remains about constant. It is nearly impossible to obtain the dislocation density at one specific interface after annealing because one has to neglect paired and bunched dislocations, and this becomes increasingly difficult as the density increases. The loss of contrast for sample 1 can be explained by the intermixing of the Ge equally in all layers and by concurrent stress relaxation. RBS shows that, after 2 hours, some intermixing has occurred (see figure 3a), but the layers still have distinct compositions. Still, very little contrast remains, which shows that the contrast in the as-deposited sample was due, at least in part to stress. RBS results, shown in figure 3a for sample 1 show that, after about 12 hours, the multilayer has been replaced by a uniform layer of about half the initial concentration of the SiGe layer. Fig 3b shows similar RBS results for sample 2, which shows that little or no intermixing is detectable for this sample, even at the longest time.

IV. DISCUSSION AND CONCLUSIONS

It has been proposed by Hagen and Strunk[11], and Vdovin et al[12], and found to happen in SiGe multilayers by Rajan et al[13], that, when an orthogonal network of dislocations is present, the intersection between dislocations with identical Burger vectors can be a source for dislocation multiplication. Thus, when one starts with an imperfect multilayer containing some dislocations at the interfaces, it is easy to imagine the relaxation process as being facilitated by the presence of the original dislocation network, and the introduction of new dislocation as being the easiest way to relax the multilayer. On the other hand, when no dislocations are present at the start, the creation of one new dislocation is very difficult, and another way has to be found to relax the multilayer. This is what happens in sample 1: because no dislocations are present between the different layers, diffusion is the only way to relieve the stress. The diffusion process is then enhanced by the stress itself so that intermixing occurs much faster than the diffusion coefficient at 900°C would suggest. Figure 4 shows simple diffusion calculations using the superposition and reflection method, simulating the diffusion experiment for sample 1, using different diffusion coefficients. Figure 4a shows that, even after 1000 hours, the multilayer is not completely homogenized when the diffusion coefficient of Ge in Si is used (following Dorner et al[14] D_o is taken as $103000.cm^{-2}$ and the activation energy as 5.34eV). Furthermore, in order to remove the "valley" between the two Ge rich layers, significant diffusion into the substrate has to occur (see curve for 2000 hours). Figure 4b show that, when the diffusion coefficient is increased by a factor of 200, a homogenized layer is obtained in about 12 hours. Again though, this is accompanied by significant diffusion into the substrate. The RBS data presented in figure 3 shows that the multilayer itself is practically homogeneous after about 12 hours (within detection limits), but that not much diffusion into the substrate has occurred. This leads to the conclusion that two different diffusion coefficients are acting in this system: one is effective in the high stress region, in the multilayer itself, and the other is active between the multilayer and the substrate, where the stress is relieved mainly by a network of dislocations. This explains why the multilayer itself can be homogenized by diffusion, without significant diffusion into the substrate. The highest density of dislocations at the substrate, measured after 12 hours, can only account for about half the misfit between the substrate and the multilayer. Thus, "normal" diffusion must relieve at least some of this misfit. Figure 3b showed that very little diffusion was present for sample 2, even after 4 hours at 900°C. This is explained by the fact that, in this case, all the stress can be relieved by the intro-

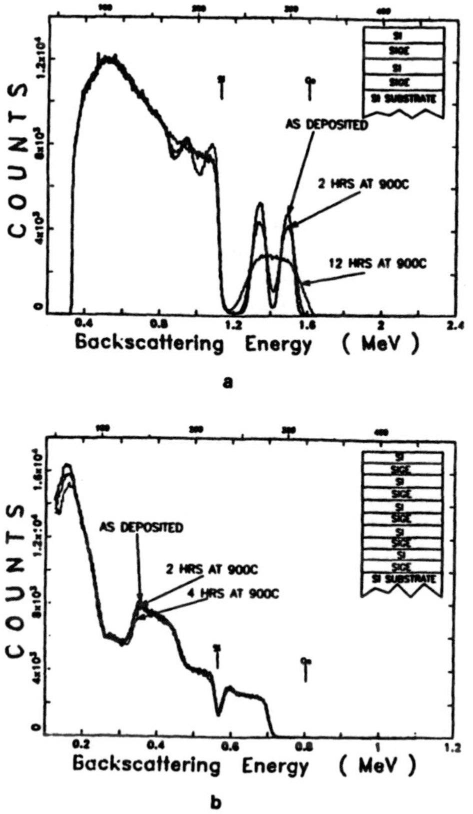

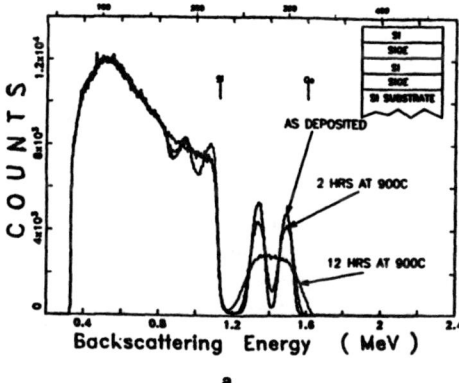

Figure 3:
RBS of the multilayers, as a function of annealing time at 900°C.

a) Sample 1 b) Sample 2

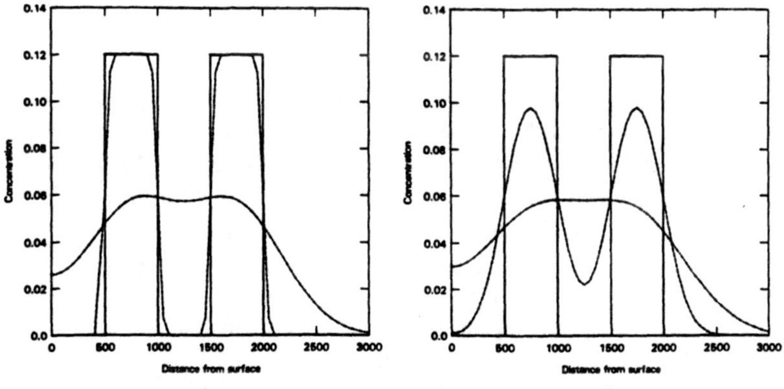

Figure 4:
Simulated diffusion experiment using different diffusion coefficients.
a) Diffusion coefficient for Ge into Si at 900°C: $Q(eV)= 5.33, Do(/cm^2)= 103000$. The curves correspond to annealing times of 0, 12 and 2000 hours. b) Diffusion coefficient about 200 time larger than for plot a.
The curves correspond to annealing times of 0, 2 and 12 hours.

duction of new dislocations before significant enhanced diffusion can occur, and by the fact that, as illustrated in figure 4a, not enough "normal" diffusion would occur in 4 hours to be detectable by RBS.

In conclusion, we have shown that two mechanisms compete during thermal relaxation of SiGe multilayer. When dislocations can be generated, this is the preferred mode of relaxation. This only occurs when proper sources of dislocations are present. Our evidence indicates that the main source of new dislocations is the intersection between dislocations formed during growth so that, if no dislocation are formed during the growth process, it is impossible to generate new dislocations during annealing. In this case, relaxation occurs by diffusion of Ge in the pure Si layer. Diffusion is enhanced by a factor of about 200 by the stress. These two mechanisms can be observed in one single sample if dislocations are present at the substrate/multilayer interface, but not between the different layers. This was the case for sample 1 where we observed two active diffusion coefficients.

Acknowledgement.

The authors wish to thank Idajean Fisher, S. Bruce Ek and J. Ott for technical support.

REFERENCES

1. L. Esaki anf R. Tsu, IBM J. Res. Dev.,**14**, 61 (1970)

2. S.S. Iyer, G.L. Patton, S.L. Delage, S. Tiwari, and J.M.C. Stork, 2ⁿᵈ Int. Si MBE Conf. Proc., Electrochem. Soc., Hawaii, Oct.1987.

3. J.H. van der Merwe, Surf. Sci, **32**, 198 (1972)

4. E. Kasper and H.J. Herzog, Thin Solid Films, **44**, 357 (1977)

5. J.C. Bean, T.T. Sheng, L.C. Feldman, H.T. Fiory and R.T. Lynch, Appl. Phys. Lett. **44**, 102 (1984)

6. K.J. Uram and B.S. Meyerson, Proceedings of the 1987 Fall Materials Research Society Meeting, Boston, MA Nov. 30- Dec. 4 1987, in press.

7. A.T. Fiory, J.C. Bean, R. Hull and S. Nakahara, Phys. Rev. B, **31**, 4063 (1985)

8. J.W. Matthews and A.E. Blakeslee, J. of Cryst. Growth, **27**, 118 (1974)

9. J.W. Matthews and A.E. Blakeslee, J. of Cryst. Growth, **29**, 273 (1975)

10. J.W. Matthews and A.E. Blakeslee, J. of Cryst. Growth, **32**, 265 (1976)

11. W. Hagen and H. Strunk, Appl. Physics, **17**, 85 (1986)

12. V.I. Vdovin, L.A. Matveeva, G.N. Semenova, N.Y. Skorohod, Y.A. Tkorik and L.S. Khozon, Phys. Stat. Sol. (a), **92**, 379 (1985)

13. K. Rajan and M. Denhor, J. Appl. Phys. Sept. (1987)

14. P. Dorner, W. Gust, B. Predel, U. Roll, A. Lodding, H. Odelius, Philosophical Magazine, A, vol. 49 (1984) 557.

THE EFFECT OF LAYER THICKNESS ON THE REACTION KINETICS OF NICKEL/SILICON
MULTILAYER FILMS

L.A. Clevenger, C.V. Thompson and R.C. Cammarata
Department of Materials Science and Engineering, Massachusetts Institute of
Technology, Cambridge, Massachusetts 02139

K.N. Tu
IBM Thomas J. Watson Research Center, Yorktown Heights, New York 10598

ABSTRACT

Differential Scanning Calorimetry (DSC) and Debye-Scherrer X-ray
diffraction have been used to characterize silicide formation in
nickel/amorphous-silicon multilayer films. Two different Ni:Si layer thickness
ratios were investigated, 3:11 and 1:1. Films with layer thickness ratio of
3:11 first formed Ni_2Si followed by NiSi at a temperature $25^\circ C$ higher.
Multilayer films with 1:1 thickness ratios formed only Ni_2Si. Activation
energies for these reaction were determined and found to be in agreement with
previous results on bilayer films. The temperature at which Ni_2Si formation
was complete in the 1:1 films was found to decrease with decreasing layer
thickness. Analysis of this phenomenon allowed determination of the
interdiffusivity during silicide formation, also in agreement with previous
results. Films with 1:1 layer thickness ratios and layer thickness of 125 Å
or less were found to sometimes undergo explosive silicidation. This
presumably occurs because the rate of heat generation at the reacting
interfaces exceeds the rate of heat dissipation.

INTRODUCTION

The kinetics of single layer nickel/crystalline-silicon reactions have
been investigated by many workers[1,2] usually using Rutherford Backscattering
and Seemann-Bohlin X-ray diffraction. In this study, freestanding
nickel/amorphous-silicon multilayer films were prepared and silicidation was
studied using Differential Scanning Calorimetry coupled with Debye-Scherrer X-
ray diffraction for phase identification. Use of multilayer films allows
observation of the heat released from many reacting interfaces at once. It
also allows easy variation of the overall atomic concentration ratio as well
as the individual layer thicknesses so that the effects of these parameters on
reaction kinetics can be studied.

EXPERIMENTAL

Freestanding multilayer thin films were prepared using alternating
electron-beam evaporation of nickel and silicon at room temperature onto
photoresist coated glass slides. Each film was composed of a total of ten
layers (5 Ni and 5 Si). The base pressure before the evaporation was never
higher than $1x10^{-7}$ torr. After deposition, the glass slides were soaked in
acetone to dissolve away the photoresist and remove the multilayer films.
Most of the films were then heated in the DSC for thermal analysis and X-ray
sample preparation.

Atomic concentration ratios of $NiSi_2$ and Ni_2Si (thickness ratios of 3:11 and 1:1 respectively) were chosen in order to determine the effects of the overall atomic concentration on the silicide formation sequence as well as the kinetics of the reactions. Films which formed $NiSi_2$ had nickel layer thicknesses of 204 Å and amorphous-silicon layer thicknesses of 746 Å for a total thickness of 4750 Å. Films which formed Ni_2Si (1:1 layer thickness ratios) had layer thickness ranging from 70 Å to 500 Å and total film thicknesses varying from 700 Å to 5000 Å.

RESULTS AND DISCUSSION

DSC traces for the 204 Å-nickel/746 Å-amorphous-silicon multilayer films and 500 Å-nickel/500 Å-amorphous-silicon films heated from 1.25 to 20°C/min are shown in figure 1. The 20°C/min traces for both multilayer films have six peaks, labeled 1-6. Samples of the 204 Å-nickel/746 Å-amorphous-silicon films were heated at 20°C/min in the DSC to 400 K, 516 K, 537 K, 710 K and cooled at 320°C/min to room temperature for X-ray analysis. X-ray data for the sample heated to 400 K indicated the presence of amorphous-silicon and nickel. X-ray data for the sample heated to 516 K indicated the presence of Ni_2Si, indicating that peak number 2 in the DSC output corresponds to the formation of Ni_2Si. X-ray data for the samples heated to 537 K and 710 K indicated that peak number 3 was due to formation of NiSi and peak number 6 was due to crystallization of unreacted amorphous-silicon. The other peaks in figure 1a can not yet be explained.

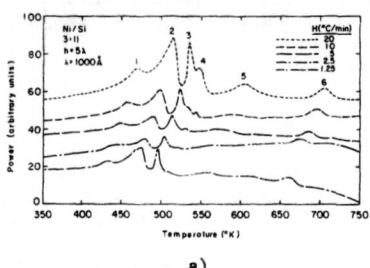 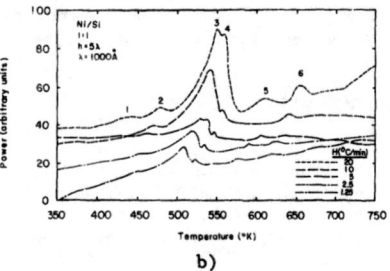

a)   b)

Figure 1: DSC traces for a) 204 Å Ni/746 Å Si and b) 500 Å Ni/500 Å Si multilayer films heated from 1.25 to 20°C/min.

Samples of the 500 Å-nickel/500 Å-amorphous silicon films were heated at 20°C/min to 425 K, 545 K, 700 K and cooled at 320°C/min to room temperature for X-ray analysis. X-ray data of the sample heated to 425 K confirmed that the initial multilayer film was made up of nickel and amorphous-silicon. X-ray data for the 545 K sample indicated that peak number 3 was due to the formation of Ni_2Si. X-ray analysis did not allow correlation of the other peaks in Fig. 1b with specific reactions. As will be discussed later, the activation energy corresponding to peak number 6 (2.5 eV) allows its tentative correlation with crystallization of unreacted amorphous silicon.[3] However we were not able to confirm this using X-ray analysis due to the small amount of unreacted silicon and overlap between the Ni_2Si and crystalline-silicon diffraction peaks.

Figure 1 shows that, for both multilayer films, as the heating rate is decreased, the peaks shift to lower temperatures. This peak temperature lowering with decreasing scan rate can be related to the activation energy of the reaction using an analysis due to Kissinger.[4] His analysis allows for the activation energy of a reaction to be determined from the slope of a plot of $\ln(H/T_p^2)$ vs $1/T_p$ where H is the heating rate and T_p is the peak temperature observed at heating rate H. Using this method for analysis of the 204 Å-nickel/746 Å-amorphous-silicon multilayer film gives an activation energy for Ni_2Si formation of 1.5 eV (peak 2 in Fig. 1a), NiSi formation of 1.4 eV (peak 3 in Fig. 1a) and an activation energy for crystallization of amorphous-silicon of 2.5 eV (peak 6 in Fig. 1a). For the 500 Å-nickel/500 Å-amorphous-silicon multilayers, the activation energy of Ni_2Si formation was 1.5 eV (peak 3 in Fig. 1b) and analysis of peak 6 indicates an activation energy of 2.5 eV, presumably corresponding to crystallization of amorphous-silicon. These values are in excellent agreement with previously reported data.[1,2,3]

For multilayer films with 1:1 layer thicknesses, decreasing the modulation period lowered the observed peak temperature for Ni_2Si formation. Figure 2 shows DSC data for 500 Å-nickel/500 Å-amorphous-silicon, 250 Å-nickel/250 Å-amorphous-silicon, 125 Å-nickel/125 Å-amorphous-silicon and 70 Å-nickel/70 Å-amorphous-silicon multilayer films heated at 20°C/min. Again each film had 5 layers of nickel and 5 layers of silicon. X-ray diffraction experiments on samples heated at 20°C/min and quenched in the DSC confirmed in all cases that peak 3 in Fig. 2 corresponds to Ni_2Si formation. The peak maximum for the 500 Å-nickel/500 amorphous silicon multilayer is at 545 K while the 70 Å-nickel/70 Å-amorphous-silicon peak maximum is at 430 K. Kissinger analyses for the 250 Å-nickel/250 Å-amorphous-silicon and 70 Å-nickel/70 Å-amorphous silicon films indicated that the activation energy for Ni_2Si formation was not a function of the layer thickness.

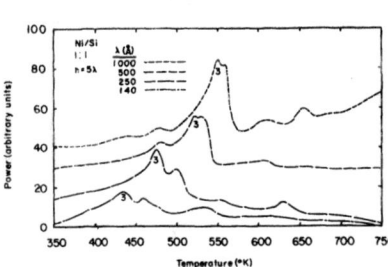 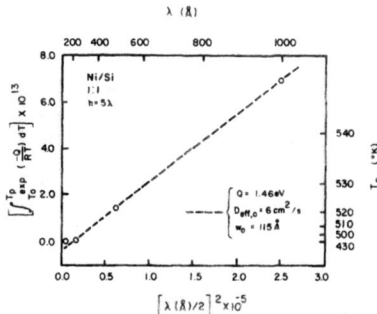

Figure 2: DSC traces for 500 Å Ni/500 Å Si, 250 Å Ni/250 Å Si, 125 Å Ni/125 Å Si and 70 Å Ni/70 Å Si multilayer films heated at 20°C/min.

Figure 3: Plot of equation 7 using the results shown in figure 2 with $T_o = 350$ K and Q=1.46 eV.

The layer thickness dependence of the Ni_2Si peak temperature described above is similar to results reported for amorphous phase formation in multilayer metal films[5]. Following an analysis similar to that given by Highmore et al.[6] the rate of increases in width, w, of the silicide layer is given by,

$$w \frac{dw}{dt} = D_{eff} = D_{eff,o} \; exp^{(-Q/kT)} \tag{1}$$

where D_{eff} is the effective interdiffusivity for Ni_2Si formation and is related to the actual interdiffusivity, the silicide concentration range and the mole fraction of nickel and silicon in Ni_2Si.[6,7] $D_{eff,o}$ is a temperature independent coefficient and Q is the activation energy for interdiffusion. The area under the Ni_2Si DSC peak, A_p, is given by

$$A_p = \gamma \; \alpha \; \Delta H_t \; A_f \; w \tag{2}$$

where γ is a proportionality constant, ΔH_t is the heat of transition, A_f is the area of the film and α is the number of interfaces at which the reaction is occurring ($\alpha = 9$ for the films studied here).

By differentiating equation 2 with respect to time and combining the result with equation 1, we get

$$A_p \frac{dA_p}{dt} = \gamma^2 D_{eff,o} \; \alpha^2 (\Delta H_t)^2 \; A_f^2 exp^{(-Q/kT)}. \tag{3}$$

This equation can be integrated from $A_p = A_o$ at $t=0$ to $A_p = A_t$ at $t=t_t$, where A_o corresponds to the heat released, if any, due to formation of Ni_2Si during deposition. By assuming that the heat of transition is not a function of temperature or time, equation 3 becomes,

$$A_t^2 - A_o^2 = \frac{2\gamma^2}{H} D_{eff,o} (\Delta H_t)^2 A_f^2 \alpha^2 \int_{T_o}^{T_p} exp^{(-Q/kT)} dT \tag{4}$$

given

$$A_t = \gamma\alpha \; A_f \; \Delta H_t \tag{5}$$

and

$$A_o = \gamma\alpha \; A_f \; w_o \; \Delta H_t \tag{6}$$

where w_o is the initial silicide thickness. We can combine equations (4), (5) and (6) to give

$$\int_{T_o}^{T_p} exp^{(-Q/kT)} dT = \frac{H}{2D_{eff,o}} [(\lambda/2)^2 - (w_o)^2] \tag{7}$$

where λ is the modulation period of the multilayer films (twice the layer thickness for 1:1 films) and T_o is the temperature at which Ni_2Si formation starts in the DSC.

Figure 3 shows calculations using equation 7 and the results shown in figure 2 where $T_o=350$ K. It can be seen that given $Q=1.46$ eV the values for $\lambda=1000$ Å, 500 Å and 250 Å fall on a straight line. From the slope of this line $D_{eff,o}$ was determined to be $6 cm^2/sec$, and from the y-intercept of the line, w_o was determined to be 115 Å. Our $D_{eff,o}$ value is in very good agreement with a value of 4 cm^2/sec derived from the results of Tu et al.[1] Also, the value of Q which provides the best fit to the data using this analysis, 1.46eV, is in exact agreement with the value obtained using the Kissinger method.

Films with $\lambda=140$ Å did not behave as expected from equation (4) and the results for films with thicker layers. This may be related to the relatively high value of w_o. For thicker films, the agreement between theory and experiment suggests that the observed decrease in the silicidation temperature with decreasing layer thickness is due to the decreased reaction time required for complete silicidation of the multilayer film.

During the course of the experiments described above, we made the surprising observation that freestanding 1:1 multilayer films with λ <250 Å can undergo an explosive reaction to form Ni_2Si. This reaction can initiate at room temperature due to mechanical impact or local heating using pulsed laser irradiation. The transformation is accompanied by a brief emission of light, suggesting substantial heating during silicidation. Explosive silicidation has also been reported for Rh/Si multilayer films.[8] Figure 4 shows a plane-view transmission electron micrograph of a 70 Å-nickel/70 Å-amorphous-silicon multilayer film before (left) and after (right) explosive silicidation. The starting material is a fine grained nickel/amorphous-silicon multilayer structure and the transformed film is Ni_2Si (as identified

Figure 4: Planar-view transmission electron micrographs of an unreacted 70 Å Ni/70 Å Si multilayer film (left) and an explosively reacted 70 Å Ni/70 Å Si multilayer film (right).

using X-ray diffraction) with a columnar grain structure and an average grain size of about 1500 Å. Explosive silicidation was observed only when the layer thickness was 125 Å or less and when the films were free standing. We suspect that when the number of reacting interfaces per unit volume is high, the rate of heat generation exceeds the rate of heat dissipation and the reaction becomes self-accelerating. These structures therefore provide systems for study of explosive reactions in which the heat released per unit of transformed volume can be controlled.

SUMMARY

We have shown that Differential Scanning Calorimetry can be used to give complete kinetic information about silicide formation in multilayer films. By observing peak shifts caused by changes in the heating rate, the activation energies for Ni_2Si and $NiSi$ formation were determined. Observation of peak shifts caused by changes in the layer thicknesses allowed determination of *both* the activation energy *and* the effective interdiffusivity for Ni_2Si formation.

In addition to demonstrating the usefulness of calorimetry for kinetic analysis of silicide formation, we also found that films composed of sufficiently thin layers underwent explosive silicidation.

This work is part of an MIT/IBM joint research program, supported by IBM. The authors would like to thank A.L. Greer for pointing out references 5 and 6, J. Carter, M. Porter and D. Roan for sample preparation and J.A. Floro and F. Spaepen for helpful discussions.

REFERENCES

1.   K.N. Tu, W.K. Chu and J.W. Mayer, Thin Solid Films 25, 408 (1975).
2.   J.O. Olowolofe, M.-A. Nicolet and J.W. Mayer, Thin Solid Films 38, 143 (1976).
3.   E.P. Donovan, F. Spaepen, D. Turnbull, J.M. Poate and D.C. Jacobson, Appl. Phys. Lett. 50, 566 (1987).
4.   H.E. Kissinger, Analyt. Chem. 29, 1702 (1957).
5.   R.J. Highmore, J.E. Evetts, A.L. Greer and R.E. Somekh, Appl. Phys. Lett. 50, 566 (1987).
6.   R.J. Highmore, R.E. Somekh, J.E. Evetts and A.L. Greer, to be published in J. Less Comm. Met..
7.   U. Gosele and K.N Tu, J. Appl. Phys. 53, 3252 (1982).
8.   J.A. Floro, J. Vac. Sci. Technol. A, 4, 631 (1986).

PART IV

Synthesis, Characterization, Properties and Application

RELAXATION OF METASTABLE SEMICONDUCTOR STRAINED-LAYER STRUCTURES BY PLASTIC FLOW

BRIAN W. DODSON and JEFFREY Y. TSAO
Sandia National Laboratories, Albuquerque, NM 87185-5800

ABSTRACT

The relaxation of misfit strain in metastable structures by plastic flow is described using a continuum model based on Haasen's picture of plastic flow in bulk diamond-phase semiconductors and the concept of excess stress. This model provides a unified explanation of the equilibrium critical thickness, the relaxation behavior of metastable strained-layer structures, and the "metastable" critical thicknesses reported in many semiconductor strained-layer geometries.

INTRODUCTION

When a material is grown on a chemically similar but lattice mismatched substrate, a coherently strained layer may be formed if the lattice mismatch is sufficiently small[1]. The stability of such a layer is determined by the relative energies of two interfacial structures, coherent and incoherent. The coherent structure has a homogeneous strain energy associated with the misfit strain, whereas the strain energy of the incoherent structure is a combination of homogeneous strain and misfit dislocations at the interface. Despite a considerable degree of success in reproducing experimental data, the inadequacy of this simple energy balance picture is shown by firm experimental evidence for the growth and long-term stability of metastable strained-layer SiGe and III-V semiconductor structures[2,3]. These experimental data have driven recent efforts to refine and expand models for stability and metastability in strained-layer structures.

DRIVING FORCE FOR STRAIN RELAXATION

Consider the Matthews-Blakeslee mechanism for misfit dislocation formation[4]. The portion of a threading dislocation in a strained overlayer is subject to motion parallel to the surface if the net stress acting on the threading dislocation is large enough. However, because there is no stress field acting on the dislocation in the substrate, motion in the strained overlayer can occur only through elongation along the substrate/film interface, forming a misfit dislocation. This elongation is opposed by the line tension of the misfit dislocation, which contributes to the forces which act on mobile dislocations. The excess stress σ_{ex} acting to move dislocations is related to the local sum of all forces acting on a dislocation. Upon collecting the contributions to the excess stress, we obtain

$$\sigma_{ex} = 2\mu \frac{(1+\nu)}{(1-\nu)} |\epsilon| - \frac{b\mu(1-0.25\nu)}{2\pi h (1-\nu)} \ln(4h/b), \qquad (1)$$

where the first term comes from the layer strain and the second term comes from the misfit dislocation tension (μ is the shear modulus, ν is the Poisson ratio, ϵ is the layer strain, h is the film thickness, b is the magnitude of the Burgers vector, and angular factors have been calculated for the case of 60° dislocations).

The excess stress described above is the driving force for generation of misfit dislocations in strained layers. If the excess stress is less than zero, the misfit dislocation tension is greater than the force provided by relaxation of the layer strain. In this case misfit dislocations will not form, and the coherent overlayer is thermodynamically stable. Given enough time, misfit dislocations will form if the excess stress is greater than zero. This condition is precisely equivalent to the Matthews-Blakeslee relation for critical thickness of strained overlayers[4].

PLASTIC FLOW AND STRAIN RELAXATION

As described above, the factors determining the equilibrium critical thickness are reasonably well understood. However, there are experimental examples in which interfacial coherence apparently persists to film thicknesses much greater than the equilibrium critical thickness, and in which coherence appears to be lost abruptly at a second 'metastable' critical thickness[2,3]. The physics leading to this behavior has not been well understood. A convincing case has been made by Fritz and coworkers[5,6] that the 'metastable' critical thickness is that for which strain relief has proceeded, on the experimental timescale, just far enough to be observable using techniques having finite resolution. Films thinner than this 'metastable' critical thickness but thicker than the equilibrium critical thickness are indeed metastable, in that strain relaxation has not proceeded to the extent predicted by equilibrium theory, but are not quite coherent, because some strain relief has begun.

The experimental observation that sluggish strain relief in semiconductor strained-layer structures results in metastable structures makes development of a model of the kinetics of strain relief necessary. The present authors have developed such a model[7] based on a phenomenological model for dynamics of dislocation glide and plastic flow in bulk diamond-phase semiconductors[8]. This model depends on two main points. First, the glide velocity of a dislocation in a strain field is restricted by the Peierls drag, which is thermally activated, and results in dislocation velocities roughly linear in the local stress,

$$V = B \, \sigma \, \exp(-U/kT), \qquad (2)$$

where σ is the stress acting on the dislocation, U is an activation energy for dislocation motion, T is the system temperature, and B is a constant. Second, it is observed that the density of dislocations increases by orders of magnitude in the early stages of plastic flow of a high-quality crystal. This effect, called dislocation multiplication, is needed to explain the relaxation behavior of bulk semiconductors. A commonly used phenomenological description of this process is

$$d\rho/dt = K \, \rho \, V \, \sigma, \qquad (3)$$

where ρ is the dislocation density and K is a multiplication parameter. Equations 1-3 provide a complete phenomenological description of the dislocation dynamics controlling the relief of mismatch strain via plastic flow[7]. They can be combined into a single non-linear ordinary differential equation for the time dependence of the strain relaxation γ.

$$d\gamma/dt = C \, (\sigma_{ex}/\mu)^2 \, \gamma, \qquad (4)$$

where C is a combined (temperature-dependent) constant which is determined by fitting to experimental data.

Bean and coworkers[2] studied relaxation behavior in metastable SiGe/Si[100] structures grown at a substrate temperature of 550°C over a broad range of film thickness and alloy composition. Comparison of the predictions of Eq. 4 with the observed relaxation behavior has established the accuracy of the model.   The amount of strain relaxation is found by numerically integrating Eq. 4. Data on the amount of strain relief exhibited by a $Si_{0.5}Ge_{0.5}$ strained-layer structure as a function of thickness appears in Figure 1a, along with a relaxation curve calculated using the current model.   The overlayer strains in SiGe/Si[100] structures having constant thickness but varying alloy concentration are compared with the kinetic model in Figure 1b.   In both cases, the experimental behavior is well reproduced by our kinetic strain relaxation model.   We conclude that the current model for strain relaxation in metastable strained-layer geometries provides a good description of the relaxation kinetics.

The 'metastable' critical thicknesses which would be observed using an analytical technique with finite sensitivity can also be calculated using the kinetic relaxation model. One simply determines the film thickness for a particular lattice mismatch which produces strain relief equal to the experimental resolution, here taken to be 0.001 (the resulting critical thicknesses are only a weak function of this value).   A comparison of several studies of low-resolution 'critical thickness'[2,] and the predictions of the current model appear in Figure 2. Again, the agreement between the plastic flow model and the experiments is very good.

QUALITATIVE REGIMES OF STRAIN RELAXATION

A qualitative outline of the rich variety of stability behavior demarcated by differing values of excess strain is of interest. Overlayers with excess stress less than zero have no force driving the deformation mechanisms, and hence are thermodynamically stable.   These are called coherent equilibrium structures. A second regime is defined by structures with small enough excess stress that observable strain relaxation does not occur on the experimental time scale.   We shall call these structures 'metastable', as such films are thermodynamically metastable and nearly coherent. This regime is delimited by the 'metastable' critical thickness. As the excess stress continues to increase, the films undergo substantial relaxation, but are still metastable in that the amount of strain relief is much less than that expected in an equilibrium structure. This regime will be called 'relaxed'.   Finally, when the excess stress is large enough (roughly twice the critical excess stress at moderate temperatures), plastic flow will occur rapidly compared to the deposition timescales, so that the strain relaxation approaches the equilibrium value. In this regime, therefore, equilibrium models are applicable, and the kinetic description of strain relaxation is no longer of concern.   This regime will be called 'fully relaxed'.

Finally, we use the recently introduced strained-layer stability diagrams[9] to summarize the available experimental data on strained-layer relaxation and to display iso-relaxation contours calculated using the kinetic relaxation model. The excess stresses of interest at lower temperatures are a substantial fraction of the zero-temperature yield stress for the diamond-phase semiconductors.   Accordingly, it is necessary to extend the kinetic model described earlier by making the dislocation glide activation energy stress-dependent.   Furthermore, at higher temperatures it is necessary to include dislocation climb mechanisms for strain relief[10].   The

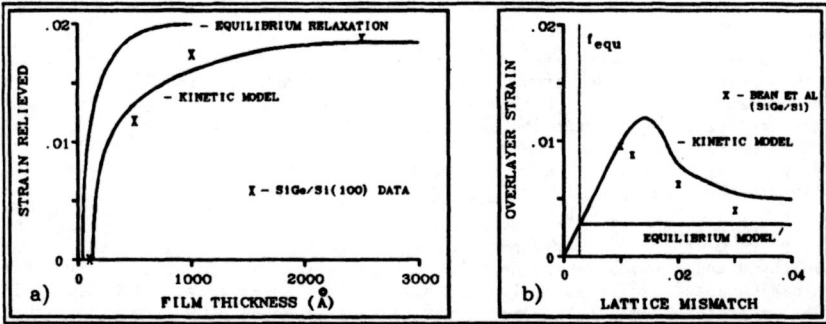

Figure 1. Equilibrium relaxation theory and the relaxation predicted by the current kinetic model (Eq. 4) are compared to data points due to Bean and coworkers (Ref. 2). a) Strain relaxation of a $Si_{0.5}Ge_{0.5}/Si[100]$ strained overlayer as a function of overlayer thickness. b) Overlayer strain versus lattice mismatch for a 500Å thick SiGe/Si[100] structure. The data agrees much better with the kinetic model developed here than with equilibrium theory.

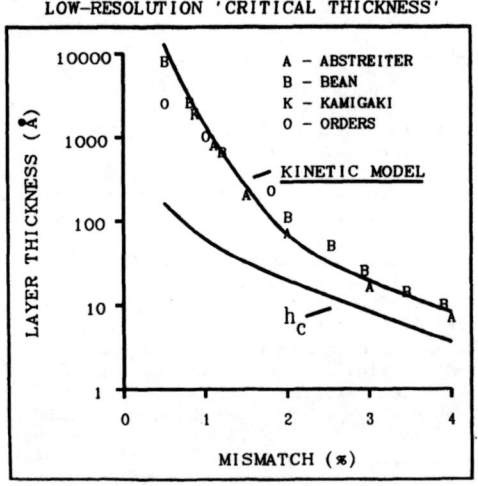

Figure 2. "Metastable" critical thickness as a function of lattice mismatch for SiGe/Si[100] strained-layer structures. The data is due to Abstreiter and coworkers (Ref. 12), Bean and coworkers (Ref. 2), Kamigaki et al (Ref. 13), and Orders and Usher (Ref. 14). The curve marked 'h_c' is the prediction of the equilibrium model, and the curve labeled 'kinetic model' is calculated by combining the slow relaxation kinetics of the present model with a finite experimental sensitivity for strain relaxation.

resulting iso-relaxation contours show two interesting features. The critical stress for a given amount of relaxation increases as the temperature
decreases, implying that, at low temperatures, structures further into the
metastable region can be grown. This effect is enhanced by the increased
spread between iso-relaxation contours with increasing temperature. Thus,
low temperatures makes growth of high-quality strongly metastable structures
possible.

A considerable amount of experimental data on 'metastable' critical
thickness exists for SiGe strained-layer structures[2,9,11]. These data are
displayed in Figure 3. Despite the scatter, which is presumably a consequence of a wide range of deposition rates and analytical techniques, the
data cluster near the 10^{-3} iso-relaxation contour, suggesting that the
kinetic ideas developed here have an appealingly wide range of applicability.

SUMMARY

The concept of excess stress in strained-layer structures has been
combined with a phenomenological model for plastic flow in bulk diamond-
phase semiconductors to obtain a model for the kinetics of strain relaxation
in metastable strained-layer structures. This model faithfully reproduces
the experimental data and offers a simple basis for understanding the
stability properties of strained-layer structures.

This research was performed at Sandia National Laboratories and was supported under U.S.D.O.E. contract DE-AC04-76DP00789.

REFERENCES

1. F.C. Frank and J.H. van der Merwe, Proceedings of the Royal Society
    A198 (1949) 205; A198 (1949) 216; and A200 (1949) 125.
2. J.C. Bean, L.C. Feldman, A.T. Fiory, S. Nakahara, and I.K. Robinson,
    Journal of Vacuum Science and Technology A2 (1984) 436.
3. P.J. Orders and B.F. Usher, Applied Physics Letters 50 (1987) 980.
4. J.W. Matthews and A.E. Blakeslee, Journal of Crystal Growth 27 (1974)
    118.
5. I.J. Fritz, P.L. Gourley, and L.R. Dawson, Applied Physics Letters 51,
    1004 (1987).
6. I.J. Fritz, Applied Physics Letters 51, 1080 (1987).
7. B.W. Dodson and J.Y. Tsao, Applied Physics Letters 51, 1325 (1987).
8. H. Alexander and P. Haasen, in Solid State Physics, vol 22 (1968).
9. J.Y. Tsao, B.W. Dodson, S.T. Picraux, and D.M. Cornelison, Physical
    Review Letters (in press).
10. B.W. Dodson and J.Y. Tsao, submitted to Phys Rev B.
11. E. Kasper and J.J. Herzog, Thin Solid Films 44 (1977) 357.
12. K. Eberl, G. Krotz, T. Wolf, F. Schaffler, and G. Abstreiter, Semicond.
    Sci. Tech. 2, 561 (1987).
13. K. Kamigaki, H. Sakashita, H. Kato, M. Nakayama, N. Sano, and H.
    Terauchi, Appl. Phys. Lett. 49, 1071 (1986).
14. P.J. Orders and B.F. Usher, Appl. Phys. Lett. 50, 980 (1987).

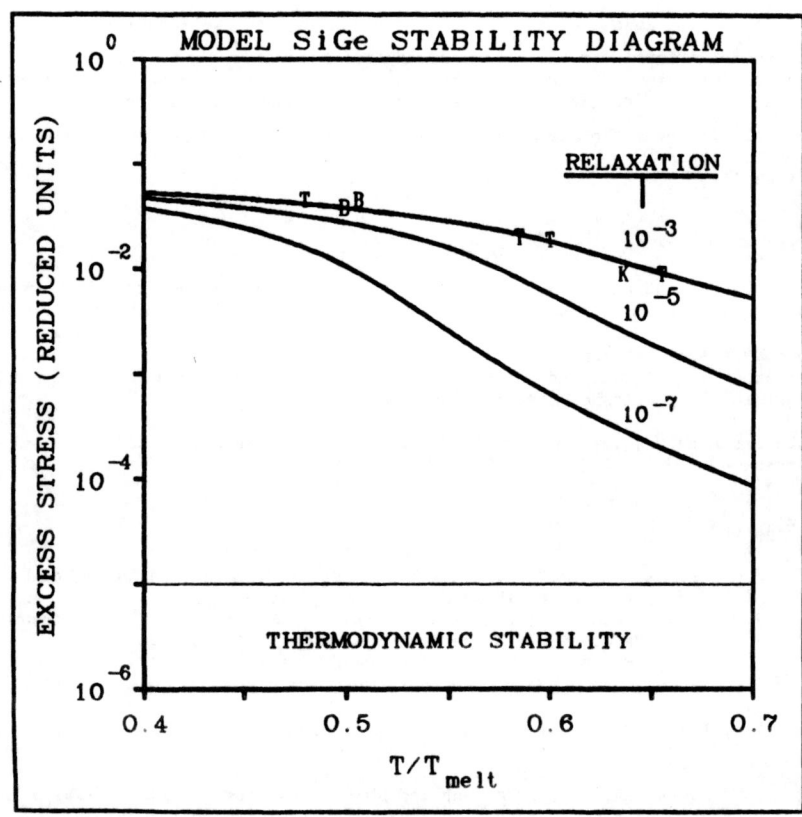

Figure 3. Excess stress versus reduced temperature diagram demarcating the stability behavior of SiGe epitaxial strained layers. The excess stress is divided by the shear modulus, and then 10^{-5} has been added so that the regime of thermodynamic stability can be indicated. Thus, the plotted excess stress is $(\tau_{ef}/\mu)+10^{-5}$. The data points are taken from Tsao et al[9] (T), Bean et al[2] (B), and Kasper and Herzog[11] (K). The iso-relaxation contours are calculated using stress-dependent dislocation glide activation energies.

EPITAXIAL GROWTH OF RARE EARTHS ON RARE EARTH FLUORIDES AND RARE EARTH
FLUORIDES ON RARE EARTHS : TWO NEW EPITAXIAL SYSTEMS ACCESSED BY MBE.

R.F.C. Farrow*, S.S.P. Parkin*, M. Lang* , K.P. Roche*

* IBM Almaden Research Center, 650 Harry Road, San Jose , Ca 95120-6099.

ABSTRACT

We report two new epitaxial systems, prepared by MBE: basal plane epitaxy of the rare
earth metal Dy onto LaF_3 films and vice versa. SQUID magnetometry studies indicate that
buried epitaxial Dy films , of ~300Å thick , order ferromagnetically at similar temperatures
to bulk Dy crystals.These epitaxial systems are one member of a new family of epitaxial
systems of basal plane epitaxy of rare earth metals on rare earth fluorides and vice versa.
Such systems may be used to probe the effects of strain and dimensionality on magnetic
ordering in rare earth metal films and multilayers.

INTRODUCTION

In this paper we report two new epitaxial systems, prepared by molecular beam epitaxy
: basal plane epitaxy of the rare earth metal Dy onto LaF_3 films and vice versa. The significance of these systems is that they are representative of a new family of epitaxial
metal-insulator systems which may be used to probe magnetic ordering in rare earth
metal films and multilayers under controlled strain.

The heavy rare earths (Gd through Lu) and several of their compounds with the transition metals have hexagonal close-packed crystal structures. On the other hand, the light
rare earth trifluorides have hexagonal (tysonite) structures[1,2,] with basal plane lattice
constants (a) close to a factor of two larger than the heavy rare earths.This suggests the
possibility of basal plane epitaxy between these two classes of materials. Apart from geometry, another factor favoring epitaxy is the exceptionally large free energies of formation of the light rare earth trifluorides. There is , in fact , a negligible thermodynamic
driving force for interfacial reactions between the heavy rare earths and the light rare
earth trifluorides.

Consider Figure 1. This illustrates the dependence of room temperature basal plane lattice
constant on the atomic number of the rare earths. What is plotted is 2a since it is this
quantity which has near coincidences with the basal plane lattice constant of appropriate
rare earth trifluorides.The latter values are indicated by horizontal lines. For example in
the case of Dy/LaF_3 the misfit defined as :

$$(2a_{Dy} - a_{LaF_3})/ a_{LaF_3} = -7.4 \times 10^{-4}$$

Other comparably matched combinations are Er/CeF_3 and Tm/PrF_3 for which the misfits
are -1.9×10^{-3} and -4.2×10^{-4} respectively. Furthermore, one can envisage , as in the case
of group II fluoride-semiconductor systems[3], exactly matching the lattice constant of the
metal to that of a solid solution of the fluorides. Inspection of Figure 1 shows that this
should be possible for Ho on a solid solution of LaF_3-CeF_3.

Mat. Res. Soc. Symp. Proc. Vol. 103. ©1988 Materials Research Society

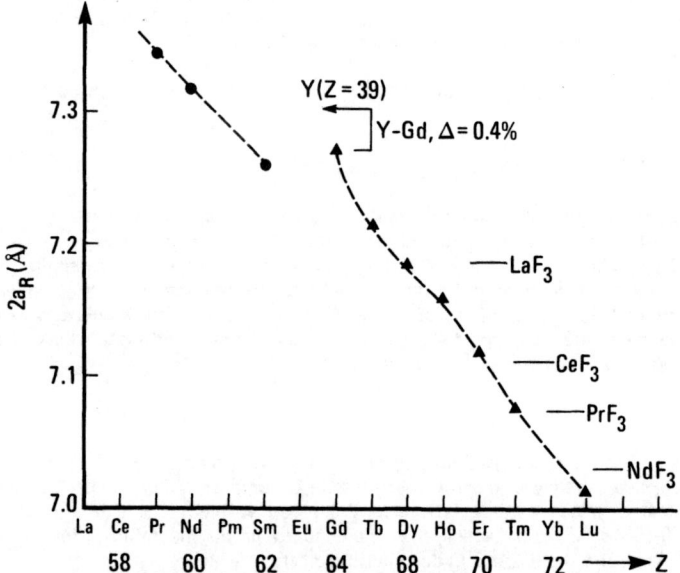

Figure 1 - Plot of the doubled , basal plane lattice parameter of the
rare earths as a function of atomic number. The basal plane
lattice parameters of the light rare earth trifluorides
are indicated by the horizontal lines.

Table I lists the free energies of formation[4] of the rare earth trifluorides.

TABLE I		
FREE ENERGIES OF FORMATION OF RARE EARTH FLUORIDES		
RF_3	$- \Delta G_f^\circ$ (k cal mole-1)	
	300K	900K
LaF_3	388.0	353.3
GdF_3	387.7	352.4
DyF_3	385.8	350.5
HoF_3	387.1	351.6
ErF_3	386.2	351.0
TbF_3	389.4	354.2

These energies are amongst the highest of all compounds and it may be noted that the differences between the values are very small. We have a situation , therefore, in which there is a negligible or negative thermodynamic driving force for interfacial reactions such as $Dy + LaF_3 = DyF_3 + La$.

EPITAXIAL GROWTH OF LaF$_3$/Dy/LaF$_3$ STRUCTURES.

As a prototype system we chose initially to study epitaxy of Dy on LaF$_3$ The basal plane (0001) surface of LaF$_3$ was generated by epitaxy of LaF$_3$ on the (TTT) As face of GaAs. In the initial experiments , a ~1000Å film of NdF$_3$ was grown on the GaAs surface at 500°C followed by a ~ 1000Å film of LaF$_3$. NdF$_3$ has a lattice constant intermediate between GaAs and LaF$_3$ and serves as a step grading film. The distribution of misfit dislocations in this structure has yet to be explored , however , in view of the in-plane lattice misfit of 2.1% between NdF$_3$ and LaF$_3$ we expect that the surface region of the LaF$_3$ will possess a lattice constant characteristic of relaxed bulk material and that the film will contain a high density of dislocations. In subsequent experiments a strained layer superlattice of alternating LaF$_3$- NdF$_3$ was interposed between the initial NdF$_3$ film and the top LaF$_3$ film to help prevent the propagation of misfit dislocations. In future experiments the LaF$_3$ film may be grown directly onto a buffer film of In$_x$Ga$_{1-x}$As with x selected[5] so that the buffer film , LaF$_3$ and Dy are all lattice- matched and free of coherency strain or misfit dislocations.

Wafers of GaAs were polished chemo-mechanically by the supplier (Morgan Semiconductor Inc) on the (TTT) As face, They were vapor degreased in iso propyl alcohol and etched in a solution of NH$_4$OH :H$_2$O$_2$:H$_2$O , 1:1:200 for 2 minutes. They were then loaded into an MBE machine (VG 80M-VG Semicon Ltd. U.K.) and cleaned by heating to ~550°C for a few minutes in a background pressure of 2x10^{-11} Torr. The RHEED (reflection high energy electron diffraction) pattern was observed during heat cleaning to determine at what point the surface assumed a sharp, well-defined (2x2) reconstruction indicative of a clean, well-ordered surface. Low energy Auger electron spectroscopy studies showed that the surface composition was identical with the MBE-grown , Ga-stabilized (TTT) surface reported by Ranke and Jacobi[6]. Typically this surface contained ~1-5% of a monolayer of C as a surface impurity. NdF$_3$ and LaF$_3$ films were grown (from effusion sources containing single crystal pieces of the fluorides) sequentially onto this surface at a substrate temperature of 500°C . The growth rates were ~70Å/m. Dy was then deposited , from an electron gun evaporation source , at a rate of ~40Å/m at a substrate temperature of 300°C. The final layer of LaF$_3$ was grown onto the Dy surface at a temperature of 300°C.

Basal plane epitaxy of Dy on LaF$_3$ and LaF$_3$ on Dy was confirmed by in situ LEED (low energy electron diffraction) and ex situ X-ray diffraction (texture analysis) . Figure 2 shows LEED patterns recorded at various stages in the growth of an epitaxial LaF$_3$/Dy/LaF$_3$/NdF$_3$/GaAs structure. A full description and indexing of the LEED patterns will be presented elsewhere. Here we simply indicate the main features which are indicative of basal plane epitaxy. The LEED pattern of the heat-cleaned (2x2) surface of the GaAs showed streaks and features (see pattern recorded at 56eV) between the first order beams (most intense spots) and the zero order beam indicative of the onset of microfacetting. The LaF$_3$ surface exhibited a LEED pattern consistent with the a = 7.186Å (Z =6) bulk unit cell. It should be noted that the sample charged to the primary beam potential within 10-20s resulting in loss of the LEED patterns. The patterns shown in Figure 2 were recorded within the first 5s of exposure to the primary beam. The Dy surface exhibited a LEED pattern consistent with the expected epitaxial relation : orientational registry between the Dy basal plane and the hexagonal basal plane cell of LaF$_3$ .The LEED pattern recorded from the final LaF$_3$ overlayer was similar to that of the LaF$_3$ underlayer. X-ray texture photographs confirmed the epitaxial relation indicated by the LEED study.

GaAs(111) As
47 eV

GaAs(111) As
56 eV

LaF₃/NdF₃/GaAs
56 eV

Dy/LaF₃/NdF₃/GaAs
56 eV

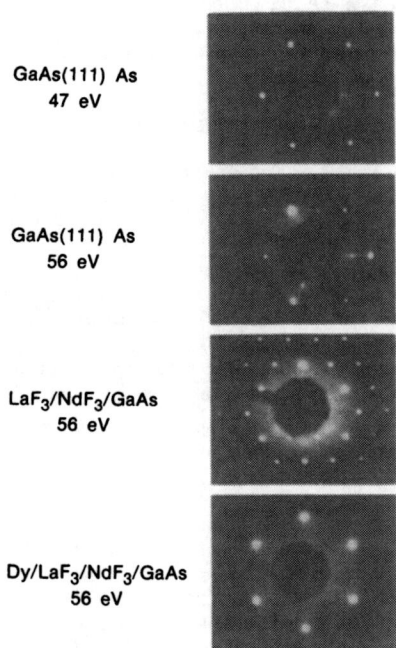

Figure 2 - LEED patterns recorded of the GaAs (TTT) surface before
and at intermediate stages in epitaxy of a
$LaF_3/Dy/LaF_3/$
NdF_3 structure.

MAGNETIC CHARACTERIZATION OF THE $LaF_3/Dy/LaF_3$ STRUCTURES.

The buried epitaxial film of Dy was characterized magnetically using SQUID magnetometry over the temperature range 6-300K. The initial structures , in which the LaF_3 was step-graded to the GaAs using a single NdF_3 film , as well as one structure which used a 4-period strained layer superlattice (~10 unit cells of LaF_3 / ~10 unit cells of NdF_3 per period) between the LaF_3 and the NdF_3 film were studied. Figure 3 shows the results for the latter structure. The magnetic moment of the Dy film (corrected for the susceptibility of the GaAs substrate) is plotted as a function of temperature. Two values of the applied field were used , 5 and 10kOe. As in the case of bulk Dy crystals[7] the transitions from the paramagnetic to helically ordered state (occurring at 178K) and from the helical to ferromagnetic state (occurring at ~ 100K-150K) are strongly field dependent. Previous studies [8] had reported the ferromagnetic transition at ~50K (in a field of 2.5kOe) for films of comparable thickness (~300Å) to ours. Even if the effect[7] of our higher applied fields is taken into account this is a considerably lower transition temperature than we observe. However, unlike the present case , the Dy film in Ref.8 was subjected to considerable coherency strain resulting from the ~1.6% lattice misfit between Dy and Y. In the present case the Dy film is closely lattice-matched to the adjacent LaF_3 films and has a lower coherency strain by a factor of ~40.

In the case of the Dy films grown without the superlattice buffer film, the magnetic transitions occurred at similar temperatures to bulk Dy crystals . However the transitions were superimposed on a Curie-Weiss tail indicating the presence of some disordered Dy. We speculate that this may be due to a high defect density in the Dy film in this case.

Future experiments will be directed at growing much thinner Dy films to investigate the effect of reduced dimensionality on magnetic ordering. In addition we plan to study the effect of strain on ordering by tailoring the lattice constant of the fluoride films adjacent to the Dy. Furthermore, these buried metal structures are well suited for magneto-transport (Hall effect) studies of quantum size effects and spin-orbit coupling in rare earth and related alloys.

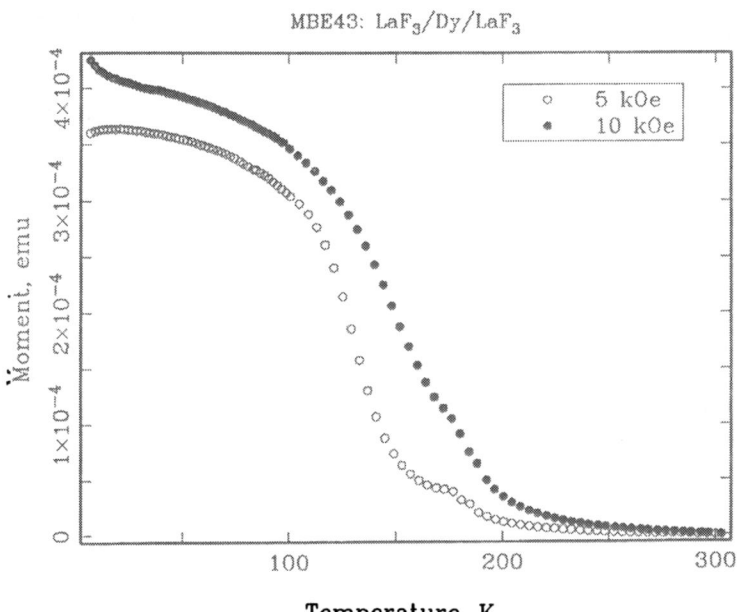

Figure 3 - Temperature dependence of magnetic moment of a buried
epitaxial film of Dy. The measurements were made for two
different values of applied field along the a axis.

CONCLUSIONS

We have prepared two new epitaxial systems by MBE techniques: basal plane epitaxy of the rare earth metal Dy onto LaF_3 films and vice versa. These systems are one member of a new family of basal plane epitaxy systems of rare earth metals (and rare earth -transition metal alloys) on rare earth trifluorides.

The ability to adjust the lattice constant of the fluoride films on either side of the metal will permit the effects of strain and dimensionality on magnetic ordering of the rare earths, to be separated. Furthermore, these buried metal structures are well suited for magneto-transport (Hall effect) studies of quantum size effects and spin-orbit coupling in rare earth and related alloys.

REFERENCES.

1. R.W. Wyckoff, Editor CRYSTAL STRUCTURES , Vol. 2, Second Edition, Interscience, 1964, pp. 60-61.

2. R.F.C. Farrow, S. Sinharoy, R.A. Hoffman, J.H. Rieger, W.J. Takei, J.C. Greggi Jr., S.Wood, T.A. Temofonte, in LAYERED STRUCTURES, EPITAXY AND INTERFACES , Edited by J.M. Gibson and L.R. Dawson (Mat. Res. Soc. Symp. Proc. 37 , Pittsburgh, Pa. 1985.)

3. C. Fontaine, M. Berrabah, J. Nejjar and A. Munoz-Yague J. Cryst. Growth 81 , 547 (1987)

4. L. B. Pankratz, THERMODYNAMIC PROPERTIES OF HALIDES , Bulletin 674, United States Department of the Interior, Bureau of Mines.

5. R.F.C. Farrow , S.S.P. Parkin , V.S. Speriosu , C.H. Wilts , R.B. Beyers , P. Pitner , J.M. Woodall , S.L. Wright , P.D. Kirchner, G.D. Pettit , Proc. Mat. Res. Soc. Fall 1987 Symposium on EPITAXY OF SEMICONDUCTOR LAYERED STRUCTURES.

6. W. Ranke and K. Jacobi , Surf. Sci. 63 , 33 (1977).

7. B. Coqblin, THE ELECTRONIC STRUCTURE OF RARE-EARTH METALS AND ALLOYS: THE MAGNETIC HEAVY RARE EARTHS, Academic Press, 1977, pp. 437-440.

8. M. Hong , R.M. Fleming , J. Kwo , J.V. Waszczak , L.F. Schneemeyer , J. P. Mannaerts , Doon Gibbs , and J. Bohr , J. Appl. Phys. 61 , 4052 (1987) also J. Kwo in THIN FILM GROWTH TECHNIQUES FOR LOW DIMENSIONAL STRUCTURES, NATO ASI Series B: Physics, Vol.163 , Edited by R.F.C. Farrow , S.S.P. Parkin , P.J. Dobson , J.H. Neave , A.S. Arrott. Plenum Press , New York (1987).

THE EFFECT OF INTERFACIAL DISORDER ON THE X-RAY DIFFRACTION OF
SUPERLATTICES

J.P. LOCQUET, D. NEERINCK, W. SEVENHANS, Y. BRUYNSERAEDE
Laboratorium voor Vaste Stof-Fysika en Magnetisme, Katholieke Universiteit
Leuven, B-3030 Leuven, Belgium

H. HOMMA
Materials Science Division, Argonne National Laboratory, Argonne, Illinois
60439, U.S.A.

IVAN K. SCHULLER
Physics Department-B019, University of California - San Diego, Lajolla,
California 92093, U.S.A.

ABSTRACT

We have generalised our x-ray diffraction results from amorphous/
crystalline multilayers, to include random interfacial disorder of a
gaussian type. A general relation is obtained which can be applied to both
crystalline/crystalline and crystalline/amorphous multilayers. This
gaussian fluctuation or "roughness" can strongly reduce the long-range
atomic order along the growth direction of the multilayer. Using classical
structure factor calculations, we simulate the evolution of x-ray patterns
as a function of the fluctuation amplitude, the superlattice wavelength, and
the interatomic distances. Applying this model to the crystalline/
crystalline case we fit the experimental Nb/Cu data, deduce a fluctuation
amplitude of about 0.4 Å, and relate it to the lattice mismatch between Nb
and Cu. For crystalline/amorphous systems (Pb/Ge) this amplitude can be
significantly larger (2 Å).

I. Introduction

The structural analysis of superlattices and multilayers is fundamental
in the understanding of their novel physical and metallurgical properties.
Especially x-ray diffraction has been used [1,2] in determining the chemical
composition and modulation wavelength of these materials. However, since
structural details cannot be obtained directly from an inversion of the
diffraction pattern, theoretical models are introduced which are then fitted
to the experimental data.

Recently, a variety of models for compositionally modulated structures
have been developed [3-14] . The "step model" assumes an abrupt composition
profile and uses the bulk lattice spacing distance for each material,
whereas the "strain model" assumes lattice-spacing variations due to
in-plane coherency strains [4]. These one-dimensional models have been
successfully used in semiconducting [4] and metallic superlattices [5-8] to
derive peak intensities and positions.

In more realistic models one takes into account fluctuations at the
interfaces, on the thickness of the layers as well as lateral fluctuations
(in the plane of the layers) [3,9-16]. The fluctuations can be continuously
distributed [11-15] (for instance the distance at the interface between
atoms of material A and B) or can be discretely distributed [9,10,15] (for
instance the number of atoms in a layer). At high angle (large q), the
latter distribution of width c^{-1} equal to an atomic distance gives rise to a
slight reduction in diffraction peak intensity and a disappearance of the
secondary peaks [15]. In a previous study [14] we showed that a continuous
distribution explains the total loss of high angle superlattice peaks in
crystalline/amorphous systems.

A number of different mechanisms which cause thickness fluctuations include : i) imperfections in the deposition process and ii) geometric constraints at the interfaces. The difference in lattice parameter and in symmetry of the constituent planes are accomodated by distorting the layers or the interfaces. This can be achieved by the introduction of in-plane coherency strain, or by the creation of misfit dislocations. The former mechanism occurs in multilayers with a small lattice mismatch (<1%) (Nb/Al [8], Nb/Ta [18]), while the latter mechanism is present in multilayers with an important lattice mismatch (Nb/Cu [5,19], Pb/Ag [7,10], Fe/Mg [13], Mo/Ni [17], Pd/Au [11]). Furthermore, when the thickness of the layers increases, a transition from a coherent to an incoherent structure is observed in multilayers with a small lattice mismatch.

In this paper, we report on the effect of continuous fluctuations of the interface distance on the line-broadening in multilayers. Using a one dimensional kinematical diffraction model, a relation is derived which can be used for crystalline/crystalline (Nb/Cu) as well as crystalline/amorphous multilayers (Pb/Ge). The width of the high-angle diffraction peaks can be explained and important structural information is obtained.

II. Theoretical Model

The structure factor for a multilayer consisting of M crystalline blocks of material A (lattice spacing d_a, scattering power f_a, number of planes N_a) and B (lattice spacing d_b, scattering power d_b, number of planes N_b), separated by an interface distance a_i, is given by:

$$F(q) = \sum_{n=0}^{N_a-1} f_a \exp(iqnd_a) + f_b \exp[iq((N_a-1)d_a+a_1)] \sum_{m=0}^{N_b-1} \exp(iqmd_b)$$

$$+ \exp[iq((N_a-1)d_a + (N_b-1)d_b+a_1+a_2)]\{ \sum_{n=0}^{N_a-1} f_a \exp(iqnd_a) +$$

$$f_b \exp[iq((N_a-1)d_a+a_3)] \sum_{m=0}^{N_b-1} \exp(iqmd_b) \} + \dots \qquad (1)$$

Following reference [2] we assume that the interface distance is not constant but fluctuates around an average value $\hat{a}$ following a continuous gaussian distribution of width c^{-1}. The distribution function of every interface distance a_i is given by:

$$p(a_i) = (c/\sqrt{\pi}) \exp[-c^2(a_i-\hat{a})^2] \qquad (2)$$

Integrating $F(q)F^*(q)$ over all real values a_i gives the average diffracted intensity:

$$I(q) = M (A^2 + B^2 + 2AB \exp(-q^2/4c^2) \cos(q\Lambda/2))$$

$$+ 2 \sum_{m=1}^{M-1} (M-m) \{ (A^2+B^2) \exp(-2mq^2/4c^2) \cos(2mq\Lambda/2)$$

$$+ AB (\exp(-(2m+1)q^2/4c^2) \cos((2m+1)q\Lambda/2)$$

$$+ \exp(-(2m-1)q^2/4c^2) \cos((2m-1)q\Lambda/2) \} \qquad (3)$$

where $A = f_a [\sin(N_a qd_a/2)]/[\sin(qd_a/2)]$

$B = f_b [\sin(N_b qd_b/2)]/[\sin(qd_b/2)]$

$\Lambda = (N_a-1)d_a + (N_b-1)d_b + 2\hat{a}$

For $c^{-1} = 0$, eq. (3) reduces to the step model, while for $c^{-1} = \infty$ it reduces to the scattering of two independent blocks of material A and B without any trace of superstructure. For crystalline/amorphous systems, eq. (5) of ref. 14 is recovered when $f_b = 0$.

Using eq. (3), the high-angle x-ray diffraction pattern of a crystalline/crystalline multilayer is calculated for different values of the distribution width c^{-1} (see Fig. 1). The distribution width c^{-1} is expressed as a percentage of the average interface distance $\hat{a}$, conventionally taken to be $(d_a + d_b)/2$. An increase of c^{-1} gives rise to a decrease of the peak intensities and an increase of the linewidth.

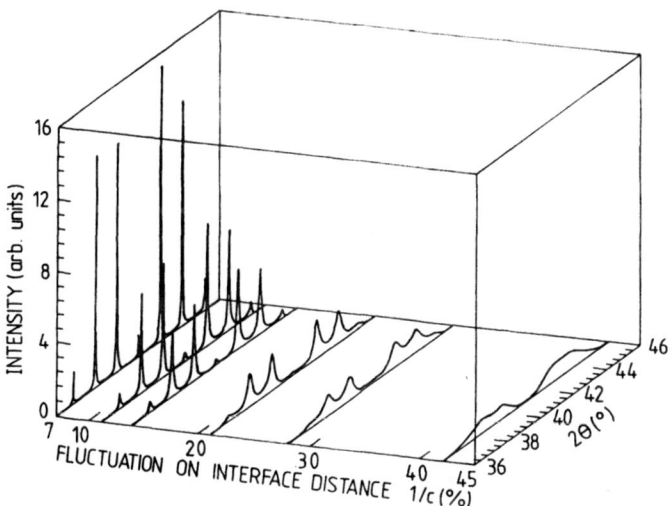

Fig. 1. Evolution of simulated high angle spectra for different values of the fluctuation amplitude c^{-1}, for d_a = 2.33 Å, d_b = 2.08 Å, $N_a = N_b = 24$, $\hat{a} = (d_a + d_b)/2$.

III. Discussion

It is well known that the coherence length, the distance over which the x-rays are coherently scattered, can be calculated from the full width at half maximum (FWHM) of an experimental diffraction line using the Scherrer equation :

$$\xi = 0.9\ \lambda_x\ /\ (\ \text{FWHM}(2\theta)\ \cos(\theta_B)\ ) \tag{4}$$

with λ_x the x-ray wavelength, and θ_B the Bragg angle of the diffraction peak.

In the case of crystalline/amorphous multilayers (Pb/Ge), it is assumed that the crystalline regions are separated from each other by "non-scattering" layers and the coherence length varies from the finite size length of one crystalline layer (Pb) to the total multilayer thickness.

To date, all diffraction experiments on crystalline/amorphous multilayers exhibit only one broad peak at high angle. In a previous paper [14] it was shown that small fluctuations (2 Å) of the thickness $\bar{a}$ of the amorphous layer can explain this fact (Fig. 2). This value is close to the nearest-neighbour distance in a-Ge (2.5 Å). Unfortunately this model only gives a lower limit for the fluctuation amplitude. An even better determination can be obtained if an amorphous material is layered between large d-spacing material. The reason for this is because the coherence between atoms spaced at 5 Å for instance is less affected by a fluctuation amplitude of 2 Å than it is from atoms spaced with a distance of say 2.87 Å (Fig. 3).

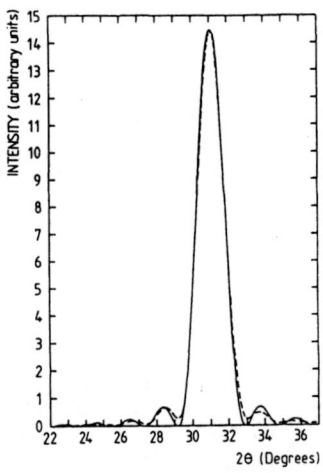

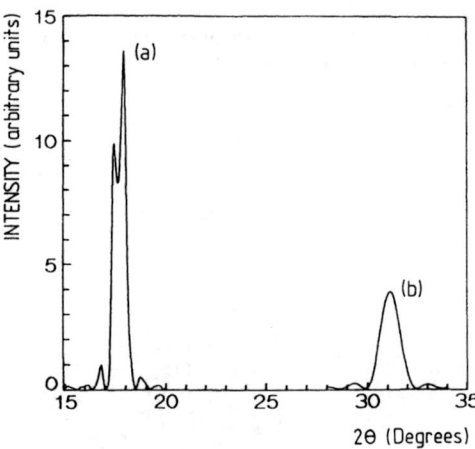

Fig. 2. Experimentally measured x-ray spectrum (dashed line) for a Pb (49 Å) / Ge (59 Å) multilayer fitted with eq. (5) of ref. 14 with $c^{-1} = 0.04\ \bar{a}$ (solid line).

Fig. 3. Simulation of high angle spectra with eq. (5) of ref. 14 for (a) d = 5 Å, $c^{-1} = 0.07\ \bar{a}$, N = 25, $\bar{a} = 28$ Å, M = 13, and (b) d = 2.87 Å (Pb), $c^{-1} = 0.07\ \bar{a}$, N = 25, $\bar{a} = 28$ Å, M = 13.

For crystalline/crystalline superlattices (Nb/Cu), the coherence length observed at high angles, is much larger than the finite-size thickness of each crystalline block.

In order to further quantify our model we introduce the concept of the "number of coherent scattering modulation wavelengths", η . This number is obtained by normalizing the coherence length by the modulation wavelength ($\eta = \xi\ /\ \Lambda$). We studied the dependence of η on the modulation wavelength and on the fluctuation amplitude. A doubling of Λ, for a constant c^{-1}, simply doubles the coherence length, and keeps η constant. This is plausible as the amount of disorder at the interface remains constant. Increasing the fluctuation amplitude, for a constant Λ, lowers the coherence length and reduces η. Figure 4 shows that for zero fluctuation amplitude, η approaches the value of the step model while for an infinite fluctuation amplitude η tends toward 0.5 (if $N_a d_a = N_b d_b$). This figure can directly be used to analyse experimental x-ray data : extracting the coherence length from experimental spectra gives the corresponding fluctuation amplitude.

Figure 5 shows the fluctuation amplitude as a function of the modulation wavelength, for Nb/Cu multilayers [5]. The observed fluctuation amplitude is much smaller than in the Pb/Ge case. For small modulation wavelengths the observed value is of the order of the lattice mismatch between Nb and Cu (0.4 Å).

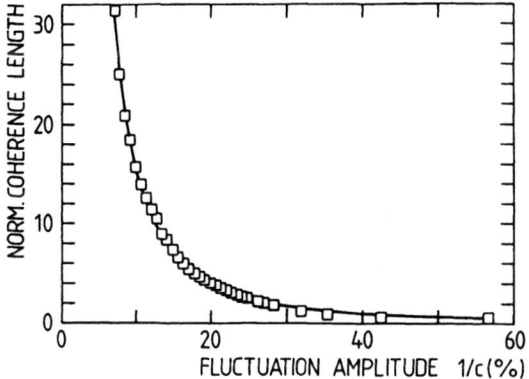

Fig. 4. The number of coherent scattering modulation wavelengths η as a function of fluctuation amplitude using eq. (3) for Nb/Cu multilayers.

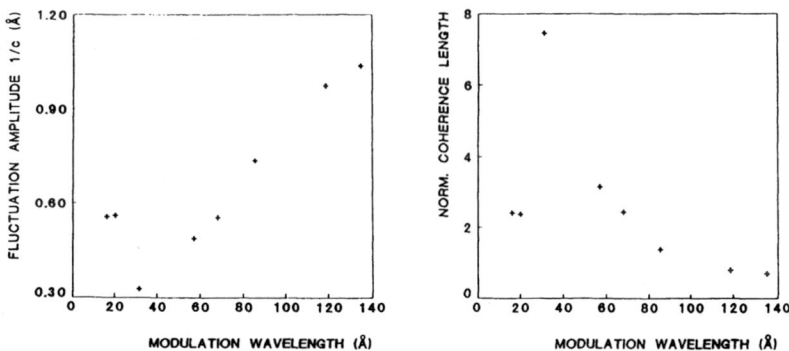

Fig. 5. Fluctuation amplitude c^{-1} versus modulation wavelength for Nb/Cu multilayers.

Fig. 6. Normalised coherence length versus modulation wavelength for the same samples as in Fig. 5.

Fig.6 shows the dependence of the normalized coherence length on the modulation wavelength, and indicates that the increase of the fluctuation amplitude as Λ increases is not an artefact of our model. This increase is a common feature in most superlattices as will be shown elsewhere [20]

Hilliard [21] calculated the energy needed to create a misfit dislocation at the interface, as well as the energy needed for epitaxial rearrangement. He found that the energy for a dislocation is inversely dependent on the thickness of the layers, whereas the energy for an epitaxial rearrangement is independent of the layer thickness. The actual configuration at the interface is determined by the balance of these energies. The increase of the fluctuation amplitude as Λ increases can thus be explained by stating that the amount of dislocations increases with Λ.

As there are many causes [16] for line-broadening it is difficult to unambiguously conclude that most of the disorder observed in diffraction patterns from multilayers is due to the above described mechanism. However the close agreement between the values of the lattice mismatch and the observed fluctuation amplitude strongly supports this model.

IV Conclusion

We developed a model which accounts for the line broadening in crystalline/amorphous and crystalline/crystalline multilayers by assuming that all disorder is due to fluctuations of the interface distance. The value of the fluctuation amplitude is 2 Å for crystalline/amorphous while for crystalline/crystalline multilayers it is much smaller and of the order of the lattice mismatch (0.4 Å).

Acknowledgements

This work was supported by the Office of Naval Research contract number N00014-83-F-0031, the U.S. Department of Energy, under contract number W-31-109-ENG-38 (at ANL), and DE-FG03-87ER45332 (at UCSD), and the Belgian Inter-University Institute for Nuclear Sciences (I.I.K.W.). International travel was provided by NATO grant number RG85-0695. J.P. Locquet is a Research Fellow of the I.I.K.W. and D. Neerinck is a Research Assistant of the National Fund for Scientific Research (Belgium).

References

1. For an early review, see D. de Fontaine, in Atomic Arrangements Studied by X-Ray Diffraction, edited by J.B. Cohen and J.E. Hilliard (Gordon and Breach, New York, 1966)
2. For a recent review, see D.B. McWhan, in Synthetic Modulated Structures, edited by L.L. Chang and B.C. Giessen (Academic Press, New York, 1985), Chap. 3
3. S. Hendricks and E. Teller, J. Chem. Phys. **10**, 147 (1942)
4. A. Segmüller and A.E. Blakeslee, J. Appl. Cryst. **6**, 19 (1973)
5. I.K. Schuller, Phys. Rev. Lett. **44**, 1597 (1980)
6. K.E. Meyer, G.P. Felcher, S.K. Sinha and I. K. Schuller, J. Appl. Phys. **52**, 6608 (1981)
7. M. Jalochowski, Thin Solid Films, **101**, 285 (1983)
8. D.B. McWhan, M. Gurvitch, J.M. Rowell and L.R. Walker, J. Appl. Phys. **54**, 3886 (1983)
9. D. Chrzan and P. Dutta J. Appl. Phys. **59**, 1504 (1986)
10. M. Jalochowski and P. Mikolajczak, J. Phys. F **13**, 1973 (1983)
11. P.F. Carcia and A. Suna, J. Appl. Phys. **54**, 2000 (1983)
12. N. Nakayama, K. Takahashi, T. Shinjo, T. Takada and H. Ichinose, Jap. J. Appl. Phys. **25**, 552, (1986)
13. Y. Fuji, T. Ohnishi, T. Ishihara, Y. Yamada, K. Kawaguchi, N. Nakayama and T. Shinjo, J. Phys. Soc. Japan **55**, 251 (1986)
14. W. Sevenhans, M. Gijs, Y. Bruynseraede, H. Homma and I.K. Schuller, Phys. Rev. B **34**, 5955 (1986)
15. B.M. Clemens and J.G. Gay, Phys. Rev. B **35**, 9337 (1987)
16. D.B. McWhan, Proc. Nato ASI on "Physics, Fabrication and Applications of Multilayered structures" June 1987
17. M.R. Kahn, C.S.L. Chun, G.P. Felcher, M. Grimsditch, A. Kueny, C.M. Falco and I.K. Schuller, Phys. Rev. B **27**, 7186 (1983)
18. S.M. Durbin, J.E. Cunningham and C.P. Flynn, J. Phys. F, **12**, L75, (1982)
19. W.P. Lowe, T.W. Barbee, Jr., T.H. Geballe and D.B. McWhan, Phys. Rev. B **24**, 6193 (1981)
20. J.P. Locquet, D. Neerinck, L. Stockman, Y. Bruynseraede, H. Homma en I.K. Schuller, to be published
21. J.E Hilliard, in Modulated Structures, edited by J.M. Cowley, J.B. Cohen, M.B. Salamon and B.J. Wuensch, (A.I.P. Conf. Proc.), 407 (1979)

CRYSTALLIZATION AND MELTING IN MULTILAYERED STRUCTURES

W. SEVENHANS, H. VANDERSTRAETEN, J.P. LOCQUET, Y. BRUYNSERAEDE,
Laboratorium voor Vaste Stof-Fysika en Magnetisme, Katholieke Universiteit Leuven, B-3030
Leuven, Belgium
H. HOMMA,
Materials Science Division, Argonne National Laboratory, Argonne, Illinois 60439, U.S.A.
IVAN K. SCHULLER
Physics Departement-B019, University of California-San Diego, LaJolla, California 92093, U.S.A.

ABSTRACT

The stability of Pb/Ge and Pb/C multilayers has been studied over a broad temperature range
by x-ray diffraction experiments. In the Pb/Ge system an amorphous to microcrystalline phase
transformation of the Ge-layers was already observed at $\simeq 100\,^{\circ}\text{C}$. This transition destroys the
modulation structure and improves the Pb(111) texture. In the Pb/C multilayers, the layered
structure was still present at temperatures higher than the melting temperature of Pb. Contrary
to recent publications, no depression of the melting temperature of the two-dimensional Pb layers
could be observed.

1 Introduction

Metal/metal and metal/semiconductor multilayers and superlattices have received considerable
attention in recent years [1]. This has been motivated by the development of novel preparation
techniques which allow microscopic control of the thickness and structure of the layers, and by
the prediction of unusual superconducting and magnetic properties [2]. More recently interesting
phenomena where predicted and observed in the recrystallization [3] and melting [4] behaviour
of those multilayered structures [4,5,6].

Amorphous Si and Ge in contact with certain metals appear to crystallize at temperatures
much lower than the usual bulk crystallization temperature [7,8]. The precise mechanism for
the nucleation and growth of a crystalline film starting from an amorphous layer is however not
well understood.

In an interesting experiment [4] it is claimed that the melting temperature of Pb in Pb/Ge
multilayers is suppressed and the transition changes from first to second order as the Pb layer
thickness is reduced along the (111) growth direction. Ion-shadowing and blocking measurements
[9] reveal a reversible order-disorder transition at the (110) surface of a Pb crystal well below
its melting point T_m. This transition is however not present in the Pb (111) surface [10]. The
orientation dependence of the degree of positional disorder appears to be directly correlated
with the large anisotropy in surface free energy.

In this paper we report on a detailed study of the crystallization of a-Ge and the melting of
Pb in Pb/Ge and Pb/C multilayers prepared by electron-beam evaporation. Both phenomena
where studied using x-ray diffraction, transmission electron microscopy (TEM) and electron
diffraction (ED).

2 Experimental techniques

The Pb/Ge and Pb/C multilayers were condensed on liquid nitrogen cooled sapphire substrates
in a load-locked molecular beam epitaxy (MBE) apparatus equipped with two electron beam
guns [11]. The evaporation rates were controlled using a quadrupole mass-spectrometer in
feedback mode. After warming up to room temperature, the sample was brought into air and
cut into pieces of $\simeq \frac{1}{12}" \times \frac{1}{2}"$ enabling us to carry out different experiments on the same sample.

X-ray diffraction measurements were performed in air and vacuum (10^{-6} Torr) on a Rigaku
Dmax II diffractometer (2 kW Cu K_α x-ray gun) equipped with a high-temperature stage. The

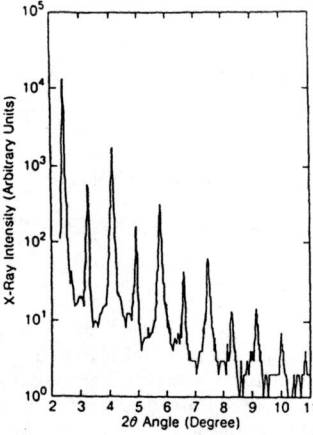
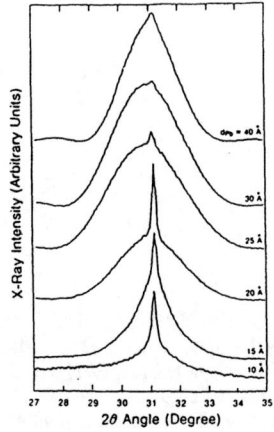

Figure 1: Small angle x-ray diffraction from a Pb (50Å)/Ge (50Å) multilayer. The fact that the even order peaks are smaller in intensity than the odd order ones shows that the modulation is almost square with minimal interdiffusion.

Figure 2: Evolution of the Pb (111) diffraction peak for different thicknesses of Pb in Pb/Ge(50Å) multilayers. For $d_{Pb} \geq$ 30Å, a typical broad peak shows up.

crystallinity and purity of the layers was checked by independent Debye-Scherrer measurements on samples scraped off the substrate. Transverse cross-sections for TEM and ED were prepared using standard thinning techniques. TEM and ED were performed on a Jeol 100 CX electron microscope equipped with high temperature capabilities.

3   Results and discussion

3.1   Initial structure

The multilayers are examined by $\theta - 2\theta$ small angle x-ray diffraction revealing the well developed layered structure, as shown in Fig. 1. The presence of a large number of small angle peaks clearly illustrates the quality of the resulting multilayer. At wide angles, no high order multilayer reflections [1] could be observed. Instead, a broad peak corresponding to the Pb (111) texture as well as secondary fringes were visible. As reported earlier [12] the wide angle spectrum is due to continuous thickness variation (interface roughness) of the amorphous Ge which limits the perpendicular coherence length to the thickness of an individual Pb layer. The evolution of the wide angle x-ray spectrum with Pb layer thickness is illustrated in Fig. 2. The smaller the individual Pb layer thickness the wider the Pb (111) peak becomes. The sharp peak on top of the broad one originates from the edge of the sample where the Pb layers can connect. This hypothesis was checked by removing the edges photolitographically.

3.2   Crystallization

After an examination at ambient temperature the substrate is heated while the small and wide angle x-ray scans are recorded. Far below the melting temperature of Pb, the a-Ge crystallizes, the Pb texture improves and the layered structure disappears [13]. The latter is clearly illustrated in Fig. 3 where the intensity of the first three observed small angle peaks of a Pb/Ge (49Å/59Å) multilayer is plotted versus substrate temperature.

To what extent the layered structure is destroyed is shown in Fig. 4. At room temperature the layered structure is clearly visible in the transmission electron micrograph (Fig. 4a) taken

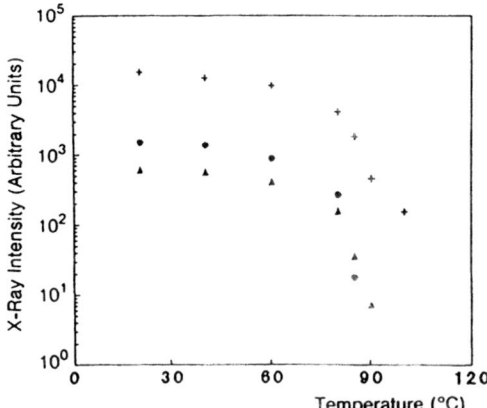

Figure 3: Evolution of the intensity of the first three observed (+ 13, △ 14, ● 15) small angle peaks of a Pb/Ge (49Å/59Å) multilayer as a function of temperature by keeping a fixed temperature for 5 minutes.

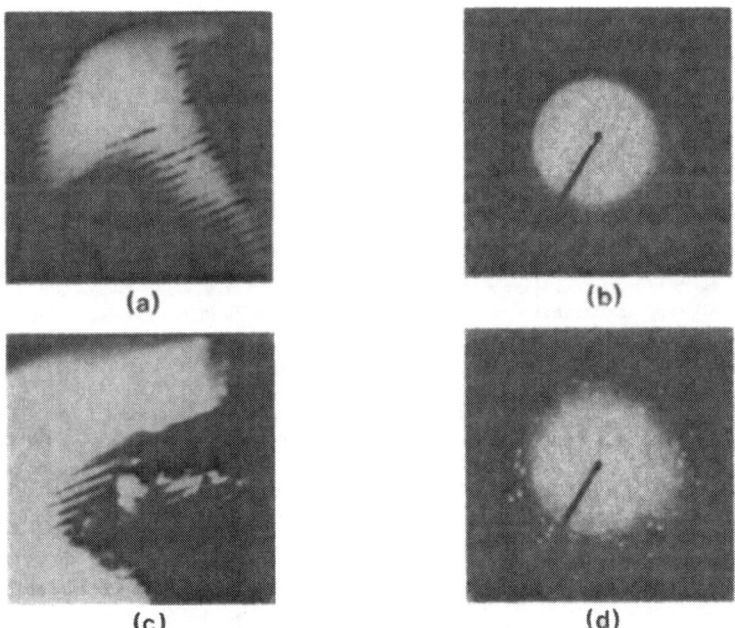

Figure 4: a) TEM of a Pb/Ge(120Å/80Å) multilayer clearly showing the layered structure. b) ED showing Pb (111) crystalline spots with no trace of crystalline Ge. c) TEM picture of the sample shown in a) after annealing. Note that the layered structure is absent. d) ED of the annealed sample showing crystalline Ge spots.

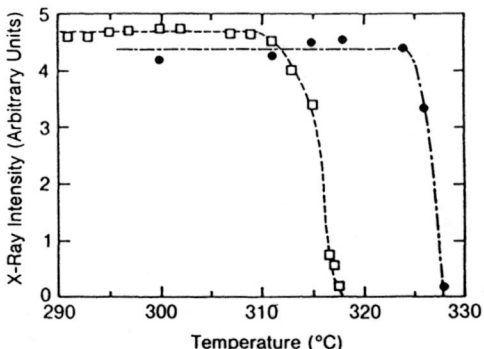

Figure 5: Pb (111) x-ray peak intensity as a function of temperature (by staying at a particular temperature for 5 minutes) for a Pb/Ge (25Å/50Å) multilayer near the Pb melting point. Data were taken in air (open squares) and in vacuum (closed circles).

from a transverse cross section of a Pb/Ge (120Å/80Å) multilayer. Moreover, no trace of crystalline Ge next to the Pb (111) spots shows up in the electron diffraction picture (Fig. 4b). During annealing the layered structure disappears (Fig. 4c) and crystalline spots of the Ge are showing up (Fig. 4d). This behaviour is seen in all the examined Pb/Ge multilayers. The temperature T_x at which the crystallization of the a-Ge occurs increases with increasing Ge layer thickness and decreases with decreasing Pb layer thickness [13]. A direct relationship between the crystallization temperature and the initial mosaic spread regardless of the Pb and Ge layer thickness was also found. The latter suggest that the crystallization is driven by the interface structure. Another explanation for the extreme low crystallization temperature and its functional behaviour with varying Pb and Ge layer thickness may be the enhanced kinetics of diffusion at the metal/semiconductor interface [14]. In the Pb/C multilayer no phase transformation was observed below the melting temperature T_m of bulk Pb.

3.3 Melting

According to Willens et al. [4] the onset of the melting of Pb in Pb/Ge multilayers decreases significantly and the transition becomes second order as the Pb layer thickness is decreased. This is in agreement with theoretical predictions [3]. The melting experiments conducted on our Pb/Ge multilayers, prepared in a similar way, however do not show any behaviour of that kind i.e. the melting transition of the Pb in these multilayers is very sharp (Fig. 5) and the melting temperature corresponds within 2 °C to the value of bulk Pb. This is not suprising in view of the fact that the multilayer is destroyed (Fig. 4c) at a temperature much lower than T_m. Indeed, as the layered structure disappears, the broad Pb (111) diffraction peak sharpens drastically [14] indicating that the crystal dimension of the Pb in the multilayer growth direction increases dramatically (a factor of 50 for the thinnest Pb layers). Therefore, no two dimensional melting behaviour of Pb can be expected in these multilayers.

A reasonable explanation can be given for these contradicting results. The melting of the Pb is in both cases measured by monitoring the decrease of the Pb (111) diffraction peak intensity. This intensity decrease may however be induced by other mechanisms.

Indeed, alloying of Pb and Ge as well as oxidation of Pb results in an intensity drop of the Pb (111) diffraction peak due to the fact that the amount of pure Pb decreases. Since Pb and Ge do not alloy in the bulk [15] we performed the melting experiments in air to allow oxidation. A decrease of T_m as reported in [4] was indeed observed. A typical transition of the Pb (111) peak of a Pb/Ge (20Å/50Å) multilayer heated in air is shown in Fig. 5. Additional Debye-Scherrer powder diffraction experiments on samples scraped off the substrate clearly showed that Pb

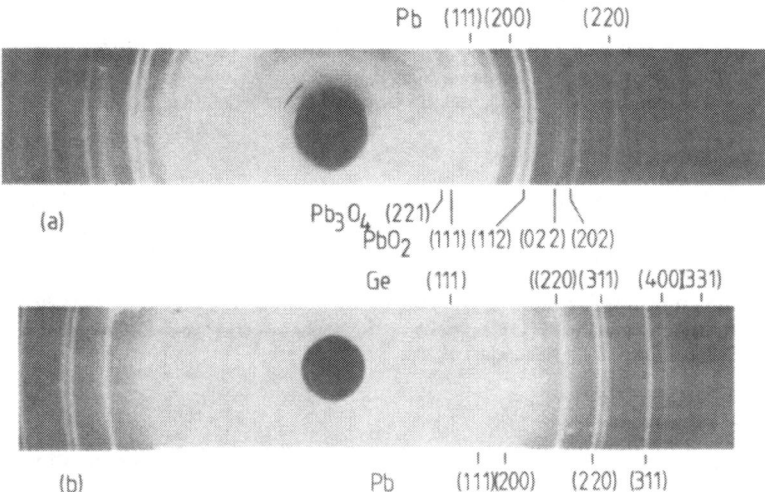

Figure 6: Debye-Scherrer powder diffraction pattern of a Pb/Ge multilayer heated in a) air; the Pb-oxides are clearly visible. b) vacuum; no oxides are visible.

oxidizes when the multilayer is heated in air (Fig. 6a), whereas only pure Pb and Ge was found when heated in vacuum (Fig. 6b).

Destruction of the multilayered structure due to crystallization does not occur in Pb/C multilayers. Partly due to the very large difference of the melting temperature of Pb and the sublimation temperature of C the quality of the Pb/C multilayers was not as good as that of the Pb/Ge multilayers. Nevertheless, very interesting results were obtained. Whereas in Pb/Ge multilayers the layers were destroyed far below the melting temperature of Pb they clearly survived in the Pb/C multilayers even far beyond the melting temperature of Pb. In Fig. 7 the intensity of the Pb (111) as well as the second and third order (l=2, 3) multilayer diffraction peaks versus temperature is shown for a Pb/C (25Å/46Å) multilayer. Again a sharp transition in the Pb (111) intensity is noticed at 328 °C, indicating the absence of two-dimensional melting

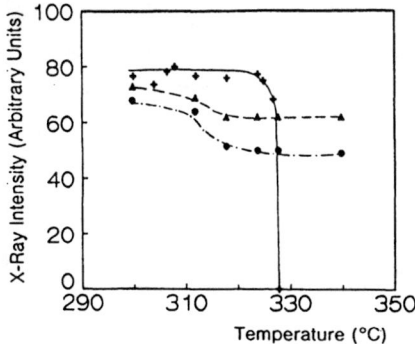

Figure 7: Evolution of the intensity of the Pb (111) and the second and third order small angle diffraction peaks (+ Pb(111), △ l2, ● l3) as a function of the temperature.

effects. This is in accordance with the experiment of Frenken and VanderVeen [6,9]. However, after the Pb was melted the second and third order multilayer peaks only dropped to 60 % of its initial value, indicating that the liquid Pb and solid C is still stacked in layers, even up to 340 °C. To our knowledge, this has never been observed before. Moreover, when the multilayer was cooled to room temperature its intensity recovered to $\simeq$ 80% of its initial value.

4 Conclusions

The amorphous to crystalline phase transition of Ge in Pb/Ge multilayers was extensively studied. It is shown that the crystallization temperature decreases with decreasing amorphous Ge thickness and increases with decreasing thickness of the metallic component. Due to the broken up structure of the Pb/Ge multilayers at T > 200°C, no two-dimensional melting was observed. The layered structure survived in Pb/C multilayers far beyond the melting temperature of Pb. The sharp transition in the Pb(111) intensity at 328 °C indicates the absence of two-dimensional melting effects down to Pb thicknesses of 20Å.

Acknowledgements

This work was supported by the Office of Naval Research contract number N00014-83-F-0031, the U.S. Departement of Energy, under contract number W-31-109-ENG-38 (at ANL), and DE-FG03-87ER45332 (at UCSD), and the Inter-University Institute for Nuclear Sciences (I.I.K.W.). International travel was provided by NATO grant number RG85-0695. H. Vanderstraeten and J.P. Locquet are Research Fellows of the I.I.K.W.

References

[1] For a recent review see Synthetic Modulated Structures, edited by L.L. Chang and B.C. Giessen (Academic Press, New York, 1985)

[2] D. Allenda, J. Bray and J. Bardeen, Phys. Rev. B 7, 1020 (1973)

[3] J.M. Kosterlitz and D.J. Thouless, J. Phys. C 6, 118 (1973), J. Phys. C 7, 1046 (1974)

[4] R.H. Willens, A. Kornblit, L.R. Testardi and S. Nakahara, Phys. Rev. B 25, 290 (1982)

[5] G. Devaud and R.H. Willens, Phys. Rev. Lett. 57, 2683 (1986)

[6] J.W.M. Frenken and J.F. van der Veen, Phys. Rev. Lett. 54, 134 (1985)

[7] T. Oki, Y. Ogawa and Y. Tujiki, Japan. J. Appl. Phys. 8,1056 (1969)

[8] S.R. Herd, P. Chaudhari and M.H. Brodsky, J. Non-Cryst. Solids 7, 309 (1972)

[9] J.W.M. Frenken, P.M.J. Marée and J.F. van der Veen, Phys. Rev. B 3, 7506 (1986)

[10] J.F. van der Veen, Proc. of the 18th CMD of the EPS, Pisa (1987)

[11] W. Sevenhans, J.P. Locquet and Y. Bruynseraede, Rev. Sci. Instrum. 57, 937 (1986)

[12] W. Sevenhans, M. Gijs, Y. Bruynseraede, H. Homma and I.K. Schuller, Phys. Rev. B 34, 5955 (1986)

[13] H. Homma, I.K. Schuller, W. Sevenhans and Y. Bruynseraede, Appl. Phys. Lett. 50, 594 (1987)

[14] W. Sevenhans, J.P. Locquet, Y. Bruynseraede, H. Homma and I.K. Schuller, to be published

[15] M. Hansen, Constitution of Binary Alloys, (McGraw-Hill, New York, 1958) and supplements

SOLID-PHASE AMORPHIZATION IN LAYERED METAL/SILICON THIN FILMS

MENACHEM NATHAN
Martin Marietta Laboratories, 1450 S. Rolling Rd., Baltimore, MD 21227

ABSTRACT

A general scheme for determining which metal-Si systems undergo solid-phase amorphization (SPA) upon rapid thermal annealing is presented and used to investigate Ni-Si, Ti-Si, V-Si, Co-Si and Cr-Si reactions. SPA occurs only in the first three systems. With the glaring exception of Co-Si, the results agree with the thermodynamic predictions of SPA in systems in which the free energy of a glassy phase is significantly lower than the free energy of the separate components. The amorphization may also be influenced by the diffusing species and contamination. Following SPA, the first crystalline compound is determined by nucleation kinetics.

INTRODUCTION

The rapid thermal annealing/transmission electron microscopy (RTA/TEM) technique[1] has been successfully used in recent years to investigate various aspects of thin-film interfacial reactions, including silicide phase sequences[2-4], solid-phase amorphization[2,4], metal-metal reactions[5] and metal-oxide reactions[6]. The basics of the technique include sequential deposition of bilayer or multilayer films on electron microscope grids followed by rapid thermal annealing of the grids and examination in the electron microscope. In general, interfacial reactions create new crystalline phases which are detected and identified with electron diffraction. However, in some cases, extensive interdiffusion and solid-phase amorphization occur prior to interfacial compound nucleation. Diffraction data prove to be extremely useful for detecting such changes, except in the case where the thickness of the interfacial amorphous layer is very small (~ 1-2 nm).

Thermally induced SPA is known to occur in both metal-metal and metal-Si reactions[7]. SPA leads to a metastable intermediate phase and involves the destruction of long-range order in the original components. The driving force for SPA is a reduction of the system free-energy, but there are strong indications that other factors, such as the suppression of nucleation of crystalline compounds by low-level contamination, are sometimes necessary to obtain the amorphous phases. It follows that thermodynamically based predictions for amorphization are not very reliable, and that one would benefit greatly from of a fast technique that could establish the existence of SPA in a particular system. RTA/TEM answers this requirement very well.

Five metal-Si systems in which heats of mixing, $-H_{mix}$, of a theoretical equiatomic liquid-like (or amorphous) state were calculated by Bene[8], were chosen for SPA studies. They included four systems (Ni-, Co-, Ti- and V-Si) in which $-H_{mix}$ is significantly more negative than the free energy $-G$ of the separate components at the reaction temperature, and which therefore favor SPA, and the Cr-Si system, in which the difference is small and SPA is thus unlikely.

EXPERIMENTAL

Trilayer a-Si/M/a-Si (a-Si = amorphous Si) films-on-grid with up to ten metal/Si thickness ratios varying roughly from 0.3 to 1.5 (generally, Si layers of constant thickness, 60 Å; and metal layers, of variable thickness, 20 to 200Å) were prepared in a single E-beam evaporation run, using the setup shown in Fig. 1. Rows of Si- to M-rich films were then rapidly thermal annealed in a tungsten halogen lamp system, Fig. 2. The annealing temperature/time combinations were chosen so that the reaction was stopped at various stages. The details are given elsewhere[9]. After each annealing, all films were examined in the electron microscope. The simultaneous deposition and annealing of films with a wide range of metal/Si ratios enables one to determine[9]: a) the preence and extent of SPA; b) an approximate diffusion coefficient of the fast-diffusing species (generally the metal in Si); c) the first crystalline compound(s) and its (their) dependence on stoichiometry; and d) nucleation kinetics. The method may also be very valuable for preparing both metastable and equilibrium thin-film phase diagrams, because it can clearly show the reaction products in an A-B diffusion couple at various stoichiometries, temperatures and times.

RESULTS and DISCUSSION

Of the five systems, only three -- Ni-Si, Ti-Si and V-Si -- showed SPA, whereas Co-Si and Cr-Si did not under any combination of RTA conditions. Representative diffraction patterns of trilayer films before and after annealing are shown in Fig. 3. In the Ni-, Ti- and V-Si reactions, there is a noticeable crystalline-to-amorphous pattern change. However, in Co-Si and Cr-Si, the metal reflections give way to those of a crystalline silicide, without any trace of an amorphous precursor. Moreover, this silicide (CoSi and $CrSi_2$ respectively) is unique and appears at all stoichiometries. This point is emphasized because of the prevailing confusion as to the identity of the first

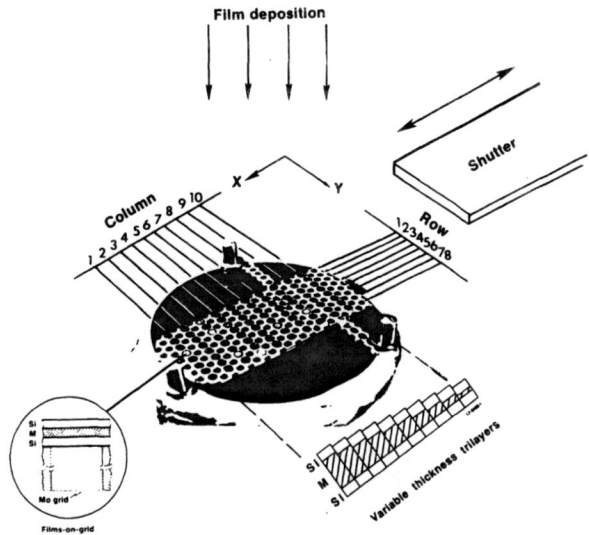

Fig. 1. Set-up for deposition of up to ten groups of films-on-grid with different metal/Si ratios. Molybdenum grids are positioned on a Ni grid carrier in a matrix of 10 columns and 8 rows. The shutter travelling in the x-direction controls the thicknesses in each column. Inset shows a cross-section of the trilayer geometry.

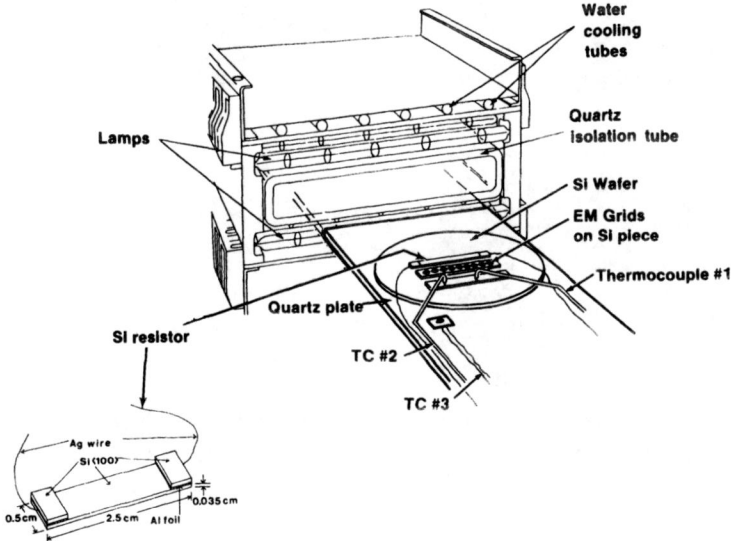

Fig. 2. RTA apparatus and geometry for the simultaneous annealing of ten grids in one row.

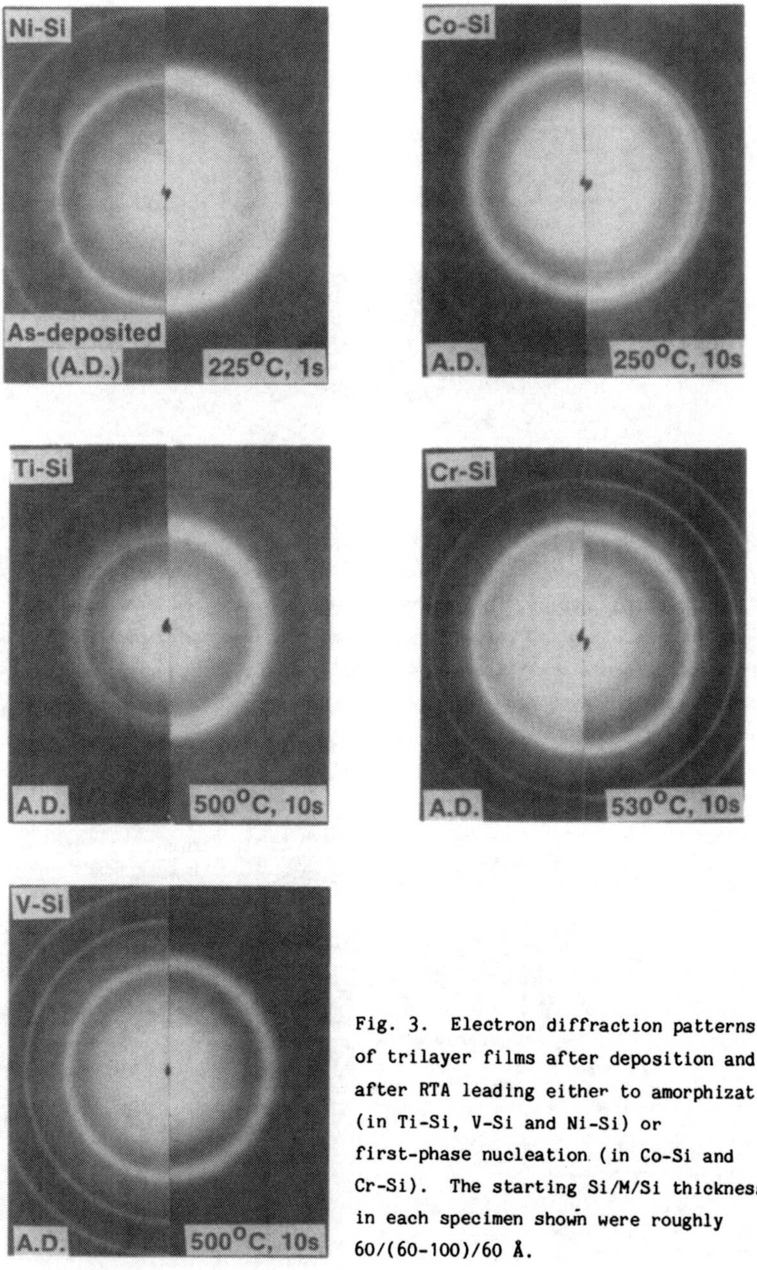

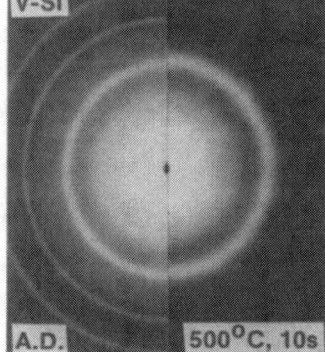

Fig. 3. Electron diffraction patterns
of trilayer films after deposition and
after RTA leading either to amorphization
(in Ti-Si, V-Si and Ni-Si) or
first-phase nucleation (in Co-Si and
Cr-Si). The starting Si/M/Si thicknesses
in each specimen shown were roughly
60/(60-100)/60 Å.

phase in Co-Si reactions[9], and because CoSi and $CrSi_2$ form even when the stoichiometry can yield other silicides such as Co_2Si or Cr_3Si.

The extent of SPA was largest in Ti-Si reactions and smallest in V-Si. When the experiment was repeated with much thicker Si/M/Si films (300/100-1000/300 Å), the amorphous region reached a few hundred angstroms in the Ti-Si couples, but only a few tens of angstroms in V-Si. The amorphization did not depend on the presence of impurities. Auger sputter profiling of thin (60/60/60 Å) Ni-Si films, as well as thick (300/200/300 Å) Ti-Si films, showed no presence of C and O above background levels in either species.

Regarding the $|H_{mix}-G|$ criterion, SPA occurred when the difference was large, e.g., in Ti-Si (-12 kcal/g at), Ni-Si (-6 kcal/g at), and V-Si (-6 kcal/g at), and did not occur when it was small, e.g., in Cr-Si (-3 kcal/g at). A major exception was Co-Si which, despite the quite large difference (-5 kcal/g at) did not amorphize. Given the well-known similarities between Ni-Si and Co-Si, this is quite surprising. I suggest that a likely reason is the difference in the moving species. It is well established that Ni always diffuses in Ni-Si couples, whereas Si diffuses when CoSi forms[10](and we find unequivocally that CoSi is indeed the first nucleating phase). One can visualize more easily the occurrence of SPA through the diffusion of the metal into amorphous Si than the opposite a-Si-into-metal case. Unfortunately, in many systems the moving species are measured in silicide reactions leading to only one particular phase, and are generally unknown in the very early diffusion and reaction stages, when the interfacial phase can be different (amorphous and/or metastable). An instructive example is the Ti-Si system: the moving species - Si - has been clearly determined in $TiSi_2$ formation only[10]. However, the sequence of phases in Ti-amorphous Si reactions is Ti+Si ⟶ amorphous TiSi ⟶ Ti_5Si_3; for both phases the moving species is undetermined. In the few silicide systems in which the moving species has been determined for more than one phase, the picture is mixed[10]: In some, such as Ni-Si and Rh-Si, the species is the same for all phases (Ni and Si, respectively). In others, such as V-Si, Co-Si, Pd-Si, Pt-Si and Ir-Si, either the metal or Si (or both) may diffuse, depending on the phase formed. To date, there are no reliable data on the moving species through amorphous silicides in any M-Si system.

Following amorphization, two new interfaces, one metal-rich, the other Si-rich, are introduced, and they provide two different chemical environments for silicide nucleation. "Multiple" nucleation does occur, but not simultaneously: In Ti-Si, we first detect Ti_5Si_3 even in Si-rich stacks, followed by the metastable C49 $TiSi_2$[9]. In Ni-Si, the first phase is either the metastable $\theta-Ni_2Si$ or $\delta-Ni_2Si$, followed closely by NiSi[2], whereas in V-Si, VSi_2

forms first even in V-rich stacks, closely followed by V-rich silicides[9]. In non-amorphizing systems such as Co-Si and Cr-Si, there is only one first silicide, CoSi and $CrSi_2$, respectively (see Fig. 3). The first phase in Cr-Si and Ni-Si reactions appears to be kinetically determined, i.e., has the fastest nucleation kinetics. In Ti-Si, Vi-Si, and Co-Si, the first phase is the one which is energetically most favored, so the role of kinetics is not clear. Once formed, this phase may be limited in its growth to a critical thickness[12], and may even disappear. This is clearly seen in Co-Si reactions, where CoSi was found to be unstable when sandwiched between Co and Si[13], and where we obtain the sequence Co+Si —>CoSi —>Co_2Si unequivocally over a wide range of starting stoichiometries. The concept of a "kinetic barrier"[12] leading to a critical thickness should hold for the amorphous layer as well, and may explain the limited amorphization in V-Si reactions.

CONCLUSIONS

The occurrence of solid-phase amorphization has been confirmed in Ti-Si, Ni-Si and V-Si layered films reacted under rapid thermal annealing conditions. A criterion based on the decrease in the free energy of the system in going from the separate M-Si components to an amorphous (glassy) phase[8] seems to work well except for Co-Si reactions. The lack of SPA in some thermodynamically favored systems may be due to "wrong" moving species, specifically Si. The determination of these species for various crystalline and amorphous silicides will improve our understanding of the primary factors leading to SPA and subsequent first-phase nucleation.

ACKNOWLEDGEMENT

I would like to thank Mr. G. Mendenilla and Ms. K. Olver for their technical assistance.

REFERENCES
1. M. Natan, MRS Symposia Proceedings, 54, 115 (1985).
2. M. Natan, Appl. Phys. Lett. 49, 257 (1986).
3. M. Natan, J. Vac. Sci. Technol. B 4, 1404 (1986).
4. M. Natan, Le vide, les couches minces, 42, 17 (1987).
5. A. Katz and Y. Komem, unpublished.
6. M. Natan, MRS Symposia Proceedings, 74, 679 (1986).
7. See, for example, W. L. Johnson in Prog. in Mater. Sci. 30, 81 (1986).
8. R. W. Bene, J. Appl. Phys. 61, 1826 (1987).
9. M. Nathan (Natan), to be published in J. Appl. Phys.
10. M.-A Nicolet and S. S. Lau in VLSI Electronics: Microstructure Science, 6 edited by N. G. Einspruch and G. B. Larrabee (Academic Press, NY, 1983), pp. 329-464.
11. K. Holloway and R. Sinclair, J. Appl. Phys. 61, 1359 (1987).
12. U. Gosele and K. N. Tu, J. Appl. Phys. 53, 3253 (1982).
13. K. N. Tu, G. Ottaviani, R. D. Thompson and J. W. Mayer, J. Appl. Phys. 53, 406 (1982).

A SOLID STATE AMORPHISATION REACTION IN Ti-Si DIFFUSION COUPLES: THE PHASE FIELD

IVO J.M.M. RAAIJMAKERS, PIET H. OOSTING AND ALEC H. READER.
Philips Research Laboratories, P.O. Box 80.000,
5600JA Eindhoven, The Netherlands

ABSTRACT

The reactions in sputter deposited Si-Ti-Si diffusion couples were investigated with X-ray diffraction, Auger electron spectroscopy and cross-section transmission electron microscopy. Anneals at a temperature of 400°C resulted in the growth of an amorphous phase at the Si-Ti interfaces. Crystalline silicides were only found after an anneal at temperatures of 500°C or higher.

It was demonstrated that an amorphous layer of approximately 8 nm thickness sustained a concentration gradient from about 73% Si at the Si side to about 28% Si at the Ti side of the diffusion couple. The measured width of the phase field agreed with the width predicted from a calculated free energy versus composition diagram. Actually the phase field was found to be so wide, that it contains the stoichiometry of all equilibrium silicides. The consequences of our results for the explanation of silicide first phase nucleation were discussed.

INTRODUCTION

The reactions of Ti thin films with Si (amorphous, poly or mono crystalline) are being investigated because of the application of $TiSi_2$ as a low resistivity material in integrated circuits.

Holloway et al. /1/ recently published a study on the reactions in thin (5 nm) Ti-Si multilayers. They found that significant interdiffusion occurred over distances of the order of 5 nm at low temperatures (455°C). Crystalline silicides did only form at temperatures in excess of 500°C. Raaijmakers et al. /2/ published a similar study on the reactions in thicker (upto 100 nm) sputter deposited Ti and Si layers. Both authors arrived at the conclusion that the growth of crystalline titanium silicides was preceeded by the formation of an amorphous Ti-Si alloy at low temperatures.

The aim of the present work is to measure the composition of the amorphous titanium silicide. A first estimate of the composition of the phase /1,2/, averaged over the growing layer, showed it to be close to a 50/50 composition. It will be shown that the amorphous silicide has a very wide phase field. The representation of the phase as a stoichiometric line compound is far from the actual situation.

In this paper, we will start with an illustration of the growth of the amorphous alloy at 400°C. After that the composition of the alloy is derived from Auger depth profiles. The thus obtained values for the Si concentration at the phase boundaries are then compared with the values calculated from the Ti-Si free energy diagram. The effect of the wide phase field of the amorphous

Mat. Res. Soc. Symp. Proc. Vol. 103. ©1988 Materials Research Society

silicide on first phase nucleation in metal-Si diffusion
couples will be discussed.

EXPERIMENTAL

Substrates were thermally oxidised Si wafers. A
tri-layer stack of Si-Ti-Si was sputter deposited on these
substrates. The thickness of each Si-layer was always
targetted to be 3/2 times the thickness of the Ti layer.
Excess Si is therefore present as compared to the composition
of the disilicide. Three different Ti layer thicknesses
(d=20 nm, 60 nm, 100 nm) were used. Anneals were performed in
a vacuum furnace at a residual gas pressure not exceeding
10^{-6} torr. Samples were analysed by cross-section
transmission electron-microscopy (TEM), X-ray diffraction
(XRD) and Auger electron spectroscopy (AES) with ion sputter
depth profiling. Details of the sample preparation, anneals
and analyses were presented in ref./2/.

RESULTS AND DISCUSSION

In the as-deposited samples, the Ti layers were found
to be poly-crystalline, strongly textured with the {001}
planes parallel to the substrate. The Si layers were found to
be amorphous according to XRD and TEM.
At low temperatures the reaction starts with the formation of
an amorphous phase /1,2/. Cross-section TEM micrographs for
the unreacted sample and the sample reacted at 400°C during
1 h are presented in Fig.1. The micrographs show the
formation of a planar amorphous reacted layer of approximately
8 nm thickness at both Si-Ti interfaces. It was demonstrated
/2/, that during anneals at temperatures between 300°C and

(a)   (b)

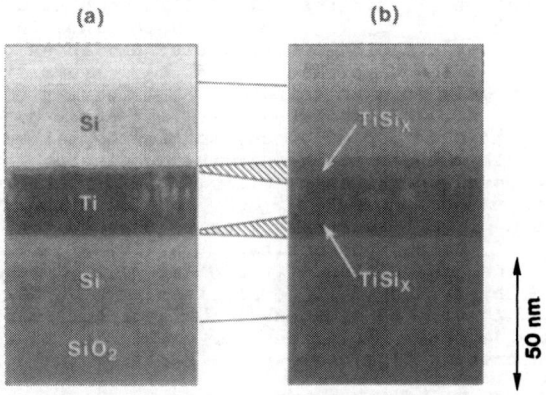

Fig.1: Cross-section TEM micrographs of a 30nm Si-20 nm
Ti-30nm Si diffusion couple. (a) as deposited; (b) annealed at
400°C during 1 h. The shaded areas illustrate the growth of
the amorphous alloy (TiSix) upon anneal.

450°C only this amorphous phase formed at the Ti-Si interfaces. If the anneal temperature was increased to 500°C, the nucleation and growth of crystalline TiSi₂ (ZrSi₂ or C49 structure) was observed.

In this paper the composition of the amorphous silicide is examined. Depth resolved Si concentrations will be derived from Auger measurements. Samples consisting of tri-layer stacks of different thicknesses (Ti thickness d= 20 nm, 60 nm and 100 nm) were annealed at 400°C during 1 h. Fig. 2 shows the Auger intensity ratio Y_{Si}/Y_{Ti} as a function of sputter time. Only the reacted region at the upper Si-Ti interface is shown. To facilitate comparison of the profiles of the samples with different layer thicknesses the sputter time axis was shifted until the centers of the reacted regions coincided. It is immediately apparent from the curves in Fig. 2 that within the present measurement accuracy the reacted regions are of the same width and composition. The curves in Fig. 2 show two clear inflexions which are most probably associated with the Si-silicide and the silicide-Ti interface. The slope of the steep parts of the curves (outside the two inflexions) are representative for the depth resolution of our measurement (approximately 4 nm). The gradually decreasing intensity ratio between the two inflexions is representative for the

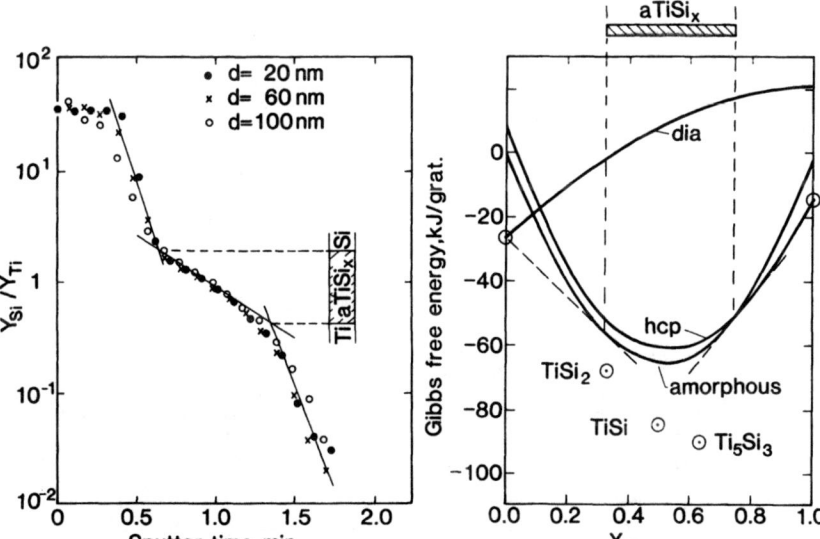

Fig.2: The Auger intensity ratio Y_{Si}/Y_{Ti} at the reacted interface as a function of sputtertime (Ti_{LMM} (390-410eV); Si_{LMM} (80-100eV)). Results for three different Ti thicknesses are shown. The thickness of the reacted amorphous layer is approximately 8 nm (TEM). The phase field of the formed compound is indicated by the shaded bar at the right hand side of the figure.

Fig.3 : Calculated free energy diagram (at a temperature of 400°C; dia = diamond; hcp = hexagonal close packed). Metastable two phase equilibria at the Si-silicide and the silicide-Ti interface are indicated by dashed lines. The phase field of the amorphous phase is indicated by the shaded bar at the top of the figure.

decreasing Si concentration with increasing depth in the
reacted amorphous layer. The intensity ratio changes from
$Y_{Si}/Y_{Ti} = 2.0\pm0.2$ at the Si-silicide interface to 0.45 ± 0.1
at the silicide-Ti interface. Using relative sensitivity
factors F_{Si}/F_{Ti}, the corresponding silicon concentrations
(x_{Si}) are found to be $x_{Si} = 0.73\pm0.02$ and $x_{Si} =$
0.28 ± 0.05 respectively. The higher x_{Si} was calculated with
$F_{Si}/F_{Ti} = 0.76$ as obtained from a $TiSi_2$ reference sample
($x_{Si} = 0.67$; $Y_{Si}/Y_{Ti} = 1.52$). The lower x_{Si} was
calculated with $F_{Si}/F_{Ti} = 1.45$, obtained from a Ti_5Si_3
reference sample ($x_{Si} = 0.37$; $Y_{Si}/Y_{Ti} = 0.87$).
Note that even substantial errors in the Auger intensity ratio at the in-
flexion points only cause fairly small errors in the concentration values.
   The very wide phase field of the amorphous alloy is a
direct consequence of the large heat of mixing of Ti and Si
/3,4/. The growth of an amorphous alloy is conveniently
illustrated in a free energy versus composition diagram.
Metastable equilibria are easily elucidated in such diagrams.
We have calculated the Gibbs free energy of amorphous and
crystalline alloys in the Ti-Si system using Kaufmans /3/
compilaton of thermo-chemical data. Generally the free energy
of amorphous solid alloys of elements which show large
negative heat of mixing is considerably smaller than that of
the (hypothetical) under-cooled liquid due to short range
ordering effects /4/. To obtain the free energy of the
amorphous solid alloy we have added an extra negative
contribution to the free energy of the under-cooled liquid.
The correction term was taken equal to $x_{Si}(1-x_{Si})$ H_a, where
x_{Si} is the Si concentration in the alloy and H_a the enthalpy
contribution due to short range order. An estimate for the
magnitude of H_a could be derived from measured heats of
crystallisation of amorphous Ti-Si alloys /5/. The value of H_a
is estimated to be 80kJ/grat. Fig. 3 shows the calculated free
energy curves of the amorphous, hcp and diamond states, and
the minimum of the free energy curves of the equilibrium
stoichiometric compounds. The diagram is calculated for a
temperature of $400^\circ C$. Metastable two phase equilibria are
shown as dashed lines in the diagram. From the free energy
diagram one derives for the phase field of the amorphous
alloy: $0.25 < x_{Si} < 0.75$. In the stages of reaction which
concern us here the growth of the amorphous phase is probably
controlled by diffusion of Si through the growing phase /2/.
Hence it is reasonable to assume that the situation at the
interfaces is close to metastable equilibrium. In that case
the position of the phase boundaries derived from the free
energy diagram correspond directly to the composition at the
interfaces. On the Ti side of the diagram, metastable
equilibrium prevails between the amorphous phase and a
super-saturated solid solution of Si in hcp Ti. The shift in
the Ti interplanar spacings to larger values as was described
in ref. /2/ is thought to be due to the super-saturation of
hcp Ti with Si
   The width of the calculated phase field is subject to
some errors, the largest being the uncertainty in the exact
shape and position of the free energy curve of the amorphous
phase. Moreover, the composition of the alloy on the Si-rich
side was derived using the assumption that metastable
equilibrium between diamond Si - amorphous silicide exists.
Our sputtered Si layers however were amorphous according to
XRD and TEM, and their free energy most certainly is higher

than the value for the diamond state. This effect would shift
the composition at the Si rich side to even higher
concentrations of Si. Although the exact boundaries of the
phase field are rather uncertain, it is believed that the
large phase field which was found experimentally agrees with
the Ti-Si free energy diagram.

We would like to note here that the phase field of the
amorphous phase is wide enough to contain the composition of
all equilibrium compounds. Clearly the assumption made by
Walser and Bené /6/ that the amorphous layer at the metal-Si
interface has a composition close to the deepest eutectic in
the binary system is not necessarily true in the Ti-Si system.

CONCLUSIONS

The growth of an amorphous silicide from sputter
deposited Ti-Si thin films during 400°C anneals was
demonstrated. This is in agreement with work published
by Holloway et al. /1/. It was shown that as long as the
supply of reactants to the growth interface is not exhausted,
the composition and amount of amorphous phase which forms is
independent of the initial layer thicknesses. It was found
that a large composition gradient exists over the growing
amorphous silicide, the silicide being very Si rich near the
Si-silicide interface ($x_{Si}= 0.73 \pm 0.02$)   and Si poor near
the silicide-Ti interface ($x_{Si}= 0.28 \pm 0.05$). The
experimentally determined phase field agreed with that derived
from the calculated free energy diagram. The composition of
the equilibrium crystalline compounds ($TiSi_2$, $TiSi$, Ti_5Si_3)
lies within the wide phase field of the amorphous alloy.
First phase nucleation in thin film diffusion couples could be
strongly related to the crystallisation behaviour of
this (inhomogeneous) amorphous alloy.

REFERENCE LIST

1. K. Holloway and R. Sinclair, J. Appl. Phys. 61, 1359
   (1987).
2. I.J.M.M. Raaijmakers, A.H. Reader and P.H. Oosting,
   accepted for publication in J. Appl. Phys.
3. L. Kaufman, Calphad 3, 45 (1979).
4. R. Boom, F.R. de Boer and A.R. Miedema, J. Less Common
   Metals 46, 271 (1976).
5. The heat of crystallisation of amorphous Ti-Si alloys was
   measured with differential scanning calorimetry
   (unpublished); see also M. Kemper and P.H. Oosting, J.
   Appl. Phys. 53, 6214 (1982).
6. R.M. Walser and R.W. Bené, Appl.Phys. Lett. 28, 624
   (1976).

ELECTRONIC AND MECHANICAL PROPERTIES OF DC SPUTTERED COMPOSITIONALLY
MODULATED METAL FILMS

J.S. DREWERY AND A.L. GREER
University of Cambridge, Department of Materials Science and Metallurgy,
Pembroke Street, Cambridge CB2 3QZ, United Kingdom.

ABSTRACT

We describe a recently constructed apparatus for the measurement of the
Young's modulus and Poisson's ratio of free-standing sputtered multilayer films. In comparison with more conventional methods, this is expected
to permit both measurement with lower uncertainties and interpretation which
is more straightforward. A separate apparatus for the accurate measurement
of the conductivity, magnetoresistance, and Hall coefficient of such systems
has been prepared which will enable the contribution of Fermi surface-
Brillouin zone interactions to the behaviour of these materials to be
assessed.

1. A NOVEL APPARATUS FOR ELASTICITY MEASUREMENTS

Previous measurements of elastic moduli in thin films have fallen into
two classes.

(1) Indirect determinations based on measurements of the velocity of sound
or on vibrating reed techniques. The former are adversely affected by
surface reflections. In the latter, distortion of the film leads to an
enhancement of the effective stiffness, and possibly to deviations from
harmonic behaviour; to take account of the distortion several resonances of
the film must be found [1].

(2) Direct determination of the stress-strain relationship in microtensile
apparatus or in a bulge test [2]. Here the strain is difficult to determine
because of slip where the thin film is gripped. The measurements are
complicated by the uncertainty in the length of the film over which the
strain is uniform and by possible non-rigidity of the mounting points.

We have constructed an apparatus less sensitive to the above problems,
independent of the movement of the grips, which detects the strain in the
sample by optical diffraction from a grating pattern on the sample surface.
The change in the angle of the diffracted spots of light from a gridiron
pattern grating can be used to study the in-plane Poisson's ratio, which has
not previously been measured.

1.1. The Apparatus

A diagram of the overall arrangement is given in fig. 1, and of the
sample mounting assembly in fig. 2. Two large piezoelectric tubes are used
to strain the samples. A large DC voltage, on to which an AC modulation can
be superimposed, is applied to the outside of the tubes. The inner surfaces
are earthed as are the end plates. This provides a screened environment in
the cavity. The stress in the sample is measured using a pair of piezoelectric rings whose end faces are connected to the inputs of a lock-in
amplifier (Brookdeal 9206) so that the voltage output is proportional to the
sum of the measured stresses. Stresses due to translational motion of the
sample are thereby eliminated. The connection from the sample to the
sensors is made via a light titanium shaft. The lower sensor is connected

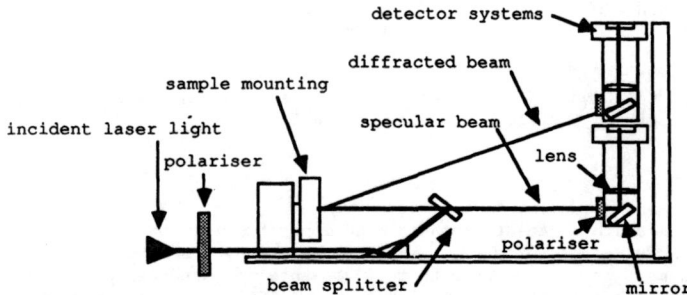

Fig. 1 The overall arrangement of the apparatus for elasticity measurements. The optical beam path is shown.

to the end plate of the driver by a fixed nylon rod. The upper sensor is mounted on a similar rod with two differential threads, in order to effect coarse adjustment of the sample mounting.

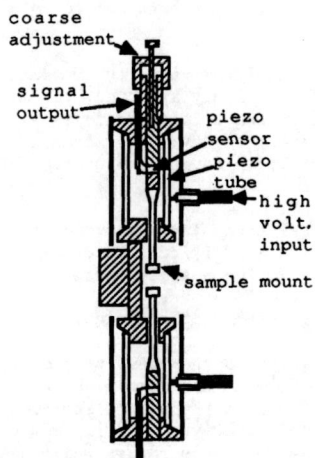

Fig. 2 Close-up of sample mounting.

The sample is illuminated by light from a Spectra Physics He-Ne laser beam whose angular fluctuations are within 20 microradians, attenuated by means of a rotatable polariser. Two prisms and a beam splitter are used to deflect the beam normally onto the sample. The beam splitter may be adjusted to aim the beam at different points on the sample surface.

The angular positions of the deflected beams are measured by two position-sensitive detectors (Micro Robotics Ltd.), mounted so that each is at the focal point of an 100 mm focal length convex lens. The angular variations of the beam are thereby converted into a displacement of the image of the beam on the two dimensional sensor array. In front of the lens, there is a further rotatable polariser which is adjusted to compensate for the difference in intensity between diffracted spots. An adjustable mirror is used to deflect the beam on to a convenient point on the sensors. The detector assemblies, on magnetic bases, may be positioned anywhere on a large vertical steel plate so that a number of diffracted spots may be studied. To allow for distortion of the sample mountings during stretching, the motion of the zero order (specularly reflected) beam is measured by one camera and compared with the motion of a suitable diffracted spot. With the gridiron grating, the movement of a diffraction spot off the symmetry axes of the diffraction pattern enables Poisson's ratio to be measured simply and directly.

The light-sensitive detectors are interfaced directly to a BBC microcomputer which also controls the lock-in amplifier and the high voltage supply. The computer displays the diffraction maxima detected on the computer screen to allow optimisation of detector position and light intensity, and calculates the centroid of the peaks.

1.2. Sample Mounting

It is desirable that at no point should samples be strained during the mounting procedure, that the stress in the sample be uniform and that there should be not tearing when stress is applied. A complex mounting procedure is followed as below.

Aluminium (1 μm thickness) is evaporated on to the film surface. The sample is then bonded, aluminised face down, to glass or sapphire, and the original substrate is removed either by dissolving in water (in the case of rock salt) or by dissolving the copper underlayer used to start epitaxy in a suitable etch. If the copper underlayer was not removed earlier it is now removed by ion-milling in a 900 V Ar beam. Photoresist is applied to the film and exposed using a mask as in fig. 3(a). The resist is developed and the superfluous portions of the film removed by a suitable etch. The resist is removed. Fresh resist is now applied to the film and exposed using a mask as in fig. 3(b). After development about 50 μm of copper is electroplated and the resist is removed to give thick pads at the end of the sample which can be used for mounting. Further resist is applied in a thin coat of not more than 1 μm thickness and exposed with a mask as fig. 3(c) which consists of a series of parallel lines at spacings down to 4 μm or a gridiron pattern with periodicity 8 μm. At present, after development the resist pattern is used as the diffraction grating but it is hoped to utilise etching processes to generate a thinner phase grating in future. Steel mountings are fixed using wax to a copper yoke and bonded to the copper pads of the sample using epoxy resin. The glass sample mounting is now removed by dissolving the aluminium overlayer in dilute sodium hydroxide solution. The steel mountings may now be screwed on to the main apparatus and the yoke removed by heating to melt the wax. This lengthy procedure is necessary to ensure reliable mounting.

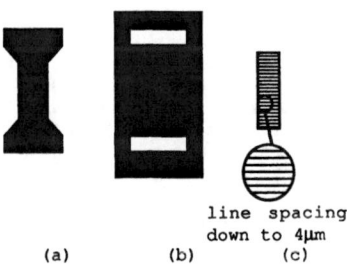

line spacing
down to 4μm

(a)   (b)   (c)

Fig. 3 Mask patterns used in lithography. The dark areas represent photoresist remaining after development.

1.3. Stress Measurement, Calibration and Sensitivity.

The derivative of the stress sensor output with driver voltage is measured by applying a modulation voltage of about 50 V at 93 Hz to a DC bias of up to 3 kV. The sensor output is measured by the lock-in amplifier. The derivative is integrated to give stress as a function of driver voltage. Calibration of the sensors may be carried out by measuring the response of each driver/sensor pair in the absence of a sample as a function of driver frequency f before and after addition of a small weight to the mounting points. The f^2 signal component is proportional to the mounting mass and the response to the small increase may thereby be determined. The maximum frequency which can be applied is limited by the effects of sound waves in the nylon mounting rods to about 2000 Hz.

For normal beam incidence and grating spacing d, the nth order diffracted beam occurs at an angle $\theta = \sin^{-1} (n\lambda/d)$. The change of angle with sample strain ε is $\delta\theta = \varepsilon.\tan \theta$. With 100 mm focal length lenses, the movement of the diffracted spot on the detectors is $L = 100 \ \varepsilon.\tan \theta$ mm.

The detector pixel spacing is 10 μm on a 256 x 128 grid. However, as the detectors are two-dimensional and there is a small uncertainty of the pixel sensitivity, the measurable displacement is 1 μm provided that about

1000 pixels are illuminated. The resulting resolution in strain is 2×10^{-5} for the 3rd order diffracted beam. Owing to the variation in sensitivity with tan θ and the need to subtract the movement of the zero order beam, it is desirable that not too high an order n should be taken; if the incident beam is not perfectly normal to the sample there could be a deviation from the expected diffracted angle leading to an error in tan θ and hence a sub-traction error.

The output of the stress sensors is approximately 0.02 V/N. The lock-in amplifier sensitivity is in practice of order 1 µV leading to a sensitivity of 50 µN. The maximum voltage which can be applied to the straining elements is 3 kV. This leads to a motion of 10 µm, or for a sample of 5 mm length a strain of approximately 0.2%. It is hoped to improve on this rather small value in future.

2. ELECTRONIC MEASUREMENTS

Precise measurements of the electronic properties of multilayers are of great interest both for characterisation and to investigate the possible effect of Fermi surface anomalies due to the superlattice on the film properties. In particular the Hall coefficient and magnetoresistance might be expected to show anomalous temperature dependences as the blurring of the Fermi wavevector due to scattering is reduced at lower temperatures. For non-magnetic films, disorder within the layers and at the interfaces may lead to an anisotropic diffusion coefficient for the electrons. Below about 30 K in a fairly disordered film, weak localisation effects [e.g. 3] should be observed and this anisotropy is mirrored in the magnetoresistance in magnetic fields normal and parallel to the film.

Measurements of the electronic properties are carried out in a cryostat mounted inside a superconducting solenoid with field up to 5 T. Suitable contact geometries are defined by lithographic etching of films, which are mounted on sapphire substrates prior to removal of the original substrates and the copper underlayer. The contact pads are coated with 0.5 µm of aluminium and the sample is connected into a carrier by ultrasonically welded 10 µm aluminium wires. The sample is mounted in the cryostat in such a way as to allow rotation in the field. This allows quick measurement of the Hall coefficient and the magnetoresistance anisotropy as the magnetic field need not be changed. Hall coefficients are measured using an AC method [4]. Resistivity and magnetoresistance are measured using an AC current; a nulling resistor is used to enable small changes of signal to be measured. Initial trials show that the method of making contacts results in very low contact noise compared with connections using silver paint.

3. CONCLUSION

The above apparatuses have been completed very recently and no results are as yet forthcoming. We intend to study a number of compositionally modulated systems such as Cu-Ni with both [100] and [111] epitaxy. It is hoped that the results will clarify the behaviour of the supermodulus effect, and provide unambiguous corroboration, or otherwise, of the role of Fermi surface-Brillouin interactions in these materials.

REFERENCES

R.C. Cammarata, Ph.D. thesis, Harvard University (1985).
D. Baral, J.B. Ketterson and J.E. Hilliard, J. Appl. Phys. 57, 1076 (1985).
P.A. Lee and T.V. Ramakrishnan, Rev. Mod. Phys. 57, 287 (1985).
R.H. Friend and N. Bett, J. Phys. E 19, 958 (1980).

ELLIPSOMETRIC, MAGNETO-OPTIC, AND MAGNETIC PROPERTIES OF SPUTTERED RARE EARTH-TRANSITION METAL MULTILAYERS[+]

THOMAS E. TIWALD,[*] JOHN A. WOOLLAM,[*] Z. S. SHAN,[**] and D. J. SELLMYER[**]
[*]Department of Electrical Engineering, University of Nebraska, Lincoln, NE 68588-0511
[**]Department of Physics, University of Nebraska, Lincoln, NE 68588
[+]Research supported by NSF Grant DMR8605367.

ABSTRACT

Ellipsometric and magneto-optical properties of amorphous Dy (3.5 Å) and amorphous Fe (25 Å to 12.5 Å) multilayers were investigated over the spectral range from 3000 Å to 8000 Å in magnetic fields to 0.21 Tesla. In this range of layer thickness the magnetic anisotropy is vertical. Kerr rotations, θ_k, were weakly spectrally dependent, and as large as 0.1 degrees. The diagonal and off-diagonal elements of the optical dielectric response function were determined over the full spectral range, and were found to be dependent on iron layer thickness.

THEORY

Ellipsometry is a measure of the change of state of polarization of a light wave reflected from a material under study [1]. The measured ellipsometric parameters are ψ and Δ defined from:

$$\rho = R_p/R_s = \tan\psi(\exp j\Delta) \tag{1}$$

where R_p and R_s are the complex Fresnel reflection coefficients for polarized light, which are angle of incident and materials dependent. The effect of the material on the polarization state is represented by the wavelength dependent complex dielectric response tensor [2,3].

$$\tilde{\varepsilon} = \begin{pmatrix} \tilde{\varepsilon}_{xx} & j\tilde{\varepsilon}_{xy} & 0 \\ j\tilde{\varepsilon}_{xy} & \tilde{\varepsilon}_{xx} & 0 \\ 0 & 0 & \varepsilon_{xx} \end{pmatrix} \tag{2}$$

where the $\tilde{\varepsilon}_{xy}$ component is magnetically induced, and $\tilde{\varepsilon}_{xx}$ is the complex dielectric function measured ellipsometrically in zero field.

The magneto-optical effects ($\tilde{\varepsilon}_{xy}$) are found from the real part (Kerr rotation) and imaginary part (Kerr ellipticity) of

$$\frac{\rho(+) - \rho(-)}{\tan\theta[\rho(+) - \rho(-)] - 2\cot\theta} \tag{3}$$

where $\rho(\pm)$ are defined from Eqn. (1) with an applied magnetic field up (+), or down (−), along the axis normal to the surface, and θ is the azimuth of the electric field relative to the plane of incidence for the incoming linearly polarized wave.

RESULTS

Zero Field Ellipsometric Analysis

The spectral dependence of ψ and Δ were measured for four multilayer samples, as well as for thin (amorphous) pure iron and pure dysprosium layers deposited on quartz substrates using U.S. Gun plasma sputtering [4]. The optical dielectric function was determined assuming the sample to be optically nontransparent and consisting of a single layer on the glass substrate. Plots of the real part $(\tilde{\varepsilon}^{\,l}_{xx})$, and imaginary part $(\tilde{\varepsilon}^{\,ll}_{xx})$ of the dielectric function are shown in Figs. 1 and 2 for the four multilayer samples. The dependences of these parameters on thickness are shown in Fig. 3 $(\tilde{\varepsilon}^{\,l}_{xx})$ and in Fig. 4 $(\tilde{\varepsilon}^{\,ll}_{xx})$.

Magneto-optical Measurements

The real and imaginary parts of ε_{xy} were measured, and found to uniformly increase in magnitude for increasing wavelength. These data were then converted (because of the technological importance for magneto-optical recording) to Kerr rotation and Kerr ellipticity, as shown (for the rotation) in Fig. 5. Fig. 6 shows the dependence of rotation and ellipticity on Fe layer thickness, at a wavelength of 6000 Å and a field strength of 0.21 Tesla.

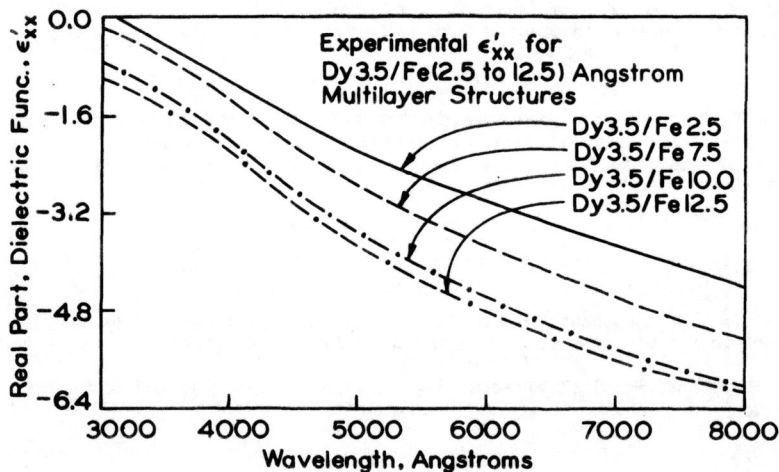

Fig. 1.   Real part ε^{l}_{xx} of the (diagonal) dielectric function vs wavelength of light.

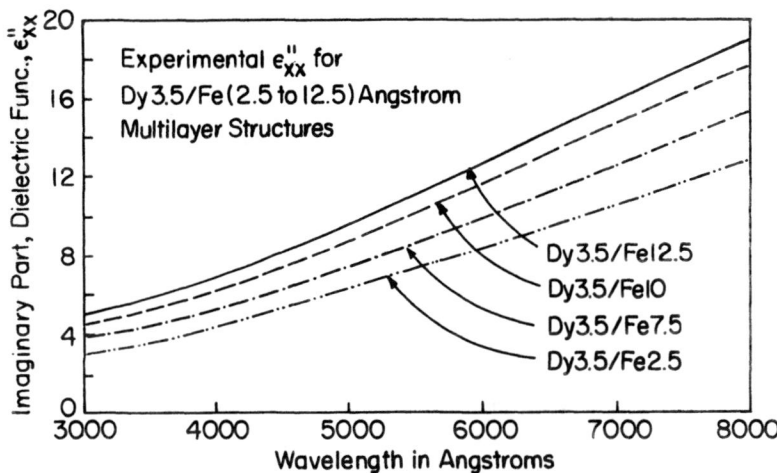

Fig. 2.   Imaginary part ε''_{xx} of the (diagonal) dielectric function vs wavelength of light.

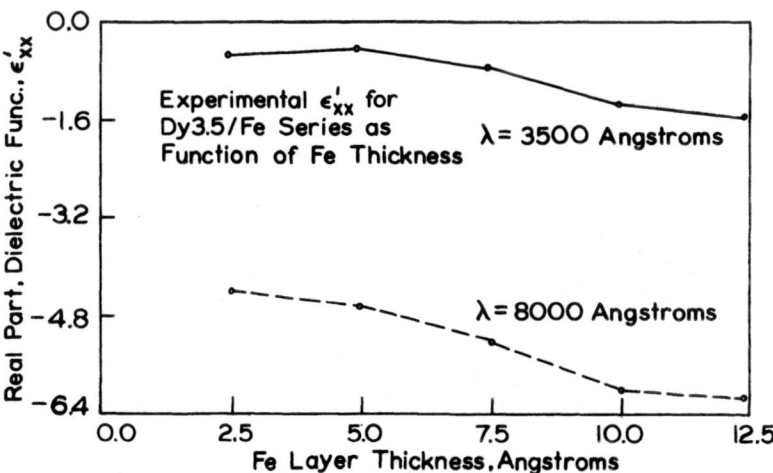

Fig. 3.   Dependence of ε'_{xx} on Fe layer thickness.

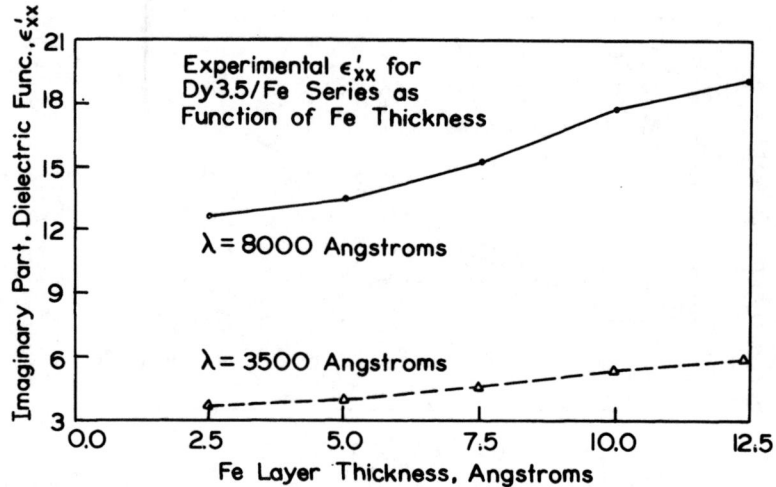

Fig. 4.   Dependence of ϵ_{xx}^{II} on Fe layer thickness.

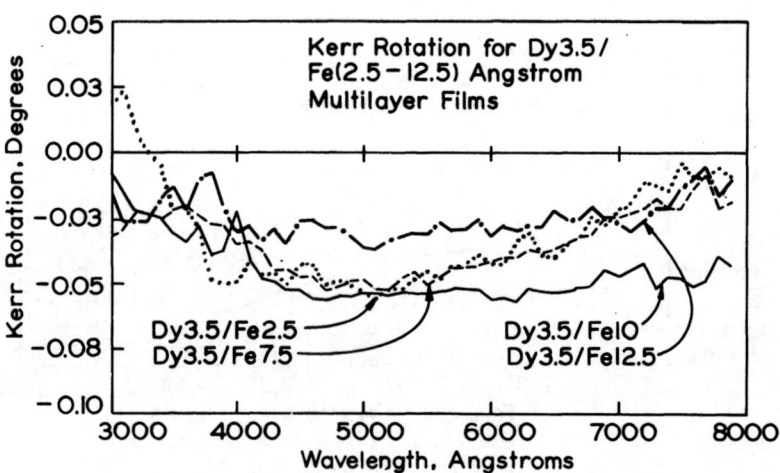

Fig. 5.   Kerr rotation vs wavelength.

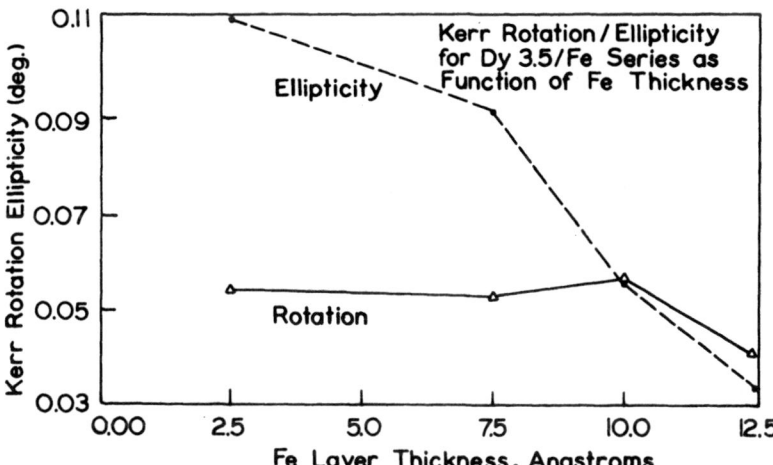

Fig. 6.   Kerr rotation and ellipticity vs Fe layer thickness.

DISCUSSION

The zero field optical data could be modeled in three ways: 1) assume a multilayered sample and use bulk Fe and bulk Dy optical constants, 2) assume a multilayered sample with measured (thin film) Fe and Dy optical data, and 3) assume bulk or thin film optical constants in a homogeneous mixture. All three approaches were tried and none gave satisfactory fits to the measured ψ and Δ data. Thus the measured optical constants are not simply due to a combination of effects from noninteracting layers or to an alloy. A new optical material has been created by the multilayer process. Although both the diagonal and off-diagonal components of $\tilde{\varepsilon}$ are affected by multilayering, we feel that it is the diagonal component which is dominating the multilayered material Kerr rotation results.

The rare earth-transition metal alloy materials system is of special interest for magneto-optic recording. Specifically the Tb/Fe system is under intense international study [5]. Magneto-optic effects are frequently dominated by the transition metal [3], and thus it is likely that the most significant element in our system is the iron, the thickness of which was the variable in the present study. It is likely that our multiple gun sputtering system produces anisotropic chemical short-range order. This appears to lead to magnetic anisotropy which is vertical (as it must be for optical recording), and a Kerr rotation which is reasonably large.

REFERENCES

1. R. M. A. Azzam and N. M. Bashara, <u>Ellipsometry and Polarized Light</u>, North Holland Publishing, New York, NY (1976).
2. M. J. Freiser, IEEE Trans. Mags. <u>MAG-4</u>, 152 (1968).
3. M. Parker, <u>Magneto-optics of Thin Magnetic Films</u>, Reference T29, SPIE Symposium Proceedings (1987), to be published.
4. See for example, C. M. Falco, I. K. Schuller, in <u>Novel Materials and Techniques in Condensed Matter</u>, G. Crabtree and P. Vashishta, eds. North Holland Publishing, New York, NY (1982).
5. See Proceedings of MMM Conferences, published annually in the Journal of Applied Physics.

"MAGNETIC EXCHANGE INTERACTIONS IN Fe/TbFe BILAYERS"

E.E. Marinero, G.S. Sprokel and H. Notarys, IBM Almaden Research Center, 650 Harry Road, San Jose, Ca. 95120-6099.

ABSTRACT

We report on magnetic exchange coupling phenomena between Fe and TbFe thin films. The structures are grown utilizing D.C. magnetron sputtering from separate Tb and Fe targets and their magnetic characteristics are measured utilizing VSM and Longitudinal Kerr Rotation. In addition, Auger depth profiling and x-ray diffraction studies were conducted to study film composition and structure.

Uniaxial alignment of the Fe moment in the plane of the substrate is revealed by the longitudinal Kerr measurements. The coercivity of the Fe layer varies from 19 Oe to 130 Oe depending on the azimuthal angle between the normal to the plane of the bilayer and the direction of the external field. The hysterisis loops vary in shape from rectangular in the alignment direction to S-shape in the direction orthogonal to this direction.

These effects are discussed in terms of magnetic exchange coupling interactions between the ferromagnetic layer and the ferrimagnetic alloy.

INTRODUCTION

Exchange coupling between ferromagnetic and antiferromagnetic materials has been extensively studied in recent years and some devices based on this phenomenon are currently available (1-3). In this work we study the magnetic interaction between a ferromagnetic thin film (Fe) and a ferrimagnetic alloy (TbFe). Thin films of Fe possess in-plane magnetization which is determined by shape anisotropy, whereas amorphous alloys of the rare-earth and the transition metals exhibit perpendicular anisotropy which has a structural origin. If the magnetization properties of the surfaces of these thin films were identical to the bulk, no exchange coupling should occur at the interfaces on account of the orthogonallity of the respective magnetic moments. However, as reported by Seredkin et al (4) for the case of NiFe/TbFe and by Hellman et al (5) for NiFeMo/TbFe, exchange interactions are observed for bilayers of these materials. In the case of NiFeMo/TbFe, the origin of the interaction is attributed to a small tilt of the TbFe axis with respect to that of the NiFeMo. This is assigned to a nonperpendicular incident angle of deposition and depends critically on the incident angles of the Tb and Fe vapor beams.
In this work we examine exchange interactions between Fe and TbFe contacted layers and attribute the origin of the coupling to surface magnetic effects at the interface between the ferro and the ferrimagnetic layer.

EXPERIMENTAL

Bilayers of TbFe and Fe were fabricated by co-sputtering from individual Fe and Tb targets utilizing DC magnetron sources in a diffusion pumped system in which the substrates are continuously rotated over the metal targets. Typical deposition rates of 0.2 nm/s are utilized and the sputter-gas pressure is typically 5 milliTorr. The bilayers were grown in the following fashion: the TbFe alloy was first grown on glass substrates to a thickness of 100 nm and then the Tb source was shielded to allow the growth of the Fe overlayers (6 to 26 nm). An Al2O3 dielectric overcoat (100 nm thick) was deposited by reactive sputtering on top of the Fe layer to provide protection against oxidation and corrosion reactions.

Magnetic measurements of these bilayers were conducted via vibrating sample magnetometry (VSM) and Kerr rotation measurements. To study the properties of the Fe layer utilizing Kerr studies, the longitudinal Kerr effect is exploited (6). In this, the internal magnetic field of the layer in question is parallel with the plane of incidence of the light beam as well as its plane of polarization. As shown in figure 1 the light is reflected obliquely from the magnetized surface and it is subsequently sent through a Wollaston prism and the s and p polarization components measured utilizing Si photodiodes. In this manner as the external applied field is scanned, a hysteresis loop of the layer of interest is obtained. In the figure, the sample surface is in the XY plane, the Fe moment is assumed to be along the Y-axis and the plane of incidence is the YZ axis. In-plane magnetic anisotropy can be readily examined by rotating the sample about the Z-axis. In our work the angle of incidence for the He-Ne laser was chosen to be 70 deg. A Varian electromagnet with suitably modified pole pieces was used to provide applied field from -2.5 to 2.5 KGauss. The data acquisition and the electromagnet were controlled via a PC-AT.

In addition to their magnetic properties, we also examined the structural properties of these layers by utilizing x-ray diffraction as well as their compositon depth-profiles by means of Auger spectroscopy combined with Ar ion depth profiling. The latter reveals that the Fe/ TbFe interface is free of contaminants and that the Fe concentration changes from 65% to 100% over a 1 nm region of the interface.

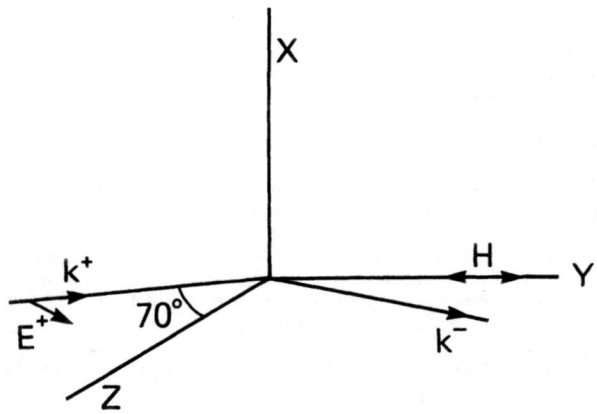

FIGURE 1. Kerr rotation coordinate system. The sample surface is the XY plane, the plane of incidence is YZ and the external field is along Y. The sample is rotated around Z-axis for in-plane anisotropy studies.

Al$_2$O$_3$/Fe (6 nm)/TbFe/G

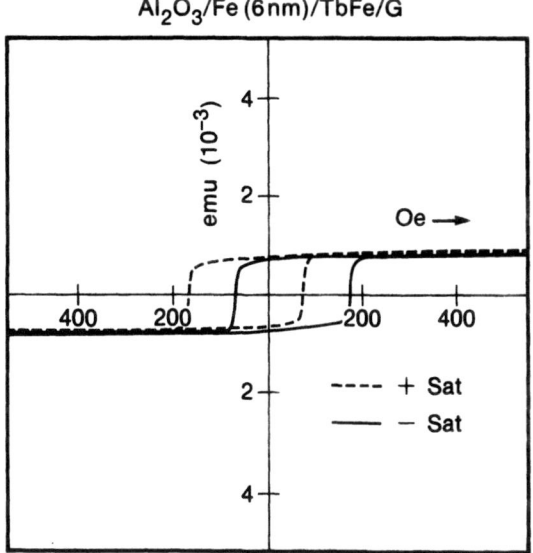

FIGURE 2. Exchange coupling in Fe/TbFe bilayers. The in-plane Fe moment was measured after saturating the underlying TbFe with fields of 17 kOe perpendicular to the film surface.

RESULTS AND DISCUSSION

The exchange coupling between the Fe and the TbFe layers is shown in figure 2. The amorphous alloy was saturated along its easy axis utilizing the VSM magnetic field of up to 17 kOe and then the sample was rotated 90 degrees and the magnetization of the in-plane moment was measured. By reversing the TbFe magnetization moment, (solid curve) and acquiring again the Fe hysterisis loop one can readily determine the magnitude of the exchange bias field. For the structure shown in the figure (6 nm Fe overlayer) the exchange bias field is 110 Oe. This varies inversely proportional with the thickness of the Fe overlayer.

Similar results are obtained via the longitudinal Kerr effect and are shown in figure 3. In this, in addition to the effect of exchange coupling, we also display the hysterisis loop obtained when the sample is rotated by 90 degrees about the Z-axis (dashed curve). This S-shaped curve with a very different coercivity indicates that the moment of the Fe layer is not isotropic and that a hard axis is present indicative of alignment of the Fe overlayer. This is in marked contrast to identical measurements done with pure Fe thin films in which the in-plane magnetization is found to be isotropic.

The variation of the Fe coercivity vs azimuthal angle is shown in figure 4 for the Fe/TbFe structure with Fe thickness = 6nm. It can be seen that the coercive force ranges from approximately 40 Oe to 140 Oe. It should be noted that the magnitude of the Kerr rotation for these measurements remained constant. For thicker Fe overlayers the anisotropy is still present although the coercive force angular dependence is not as marked as for the 6 nm layer.

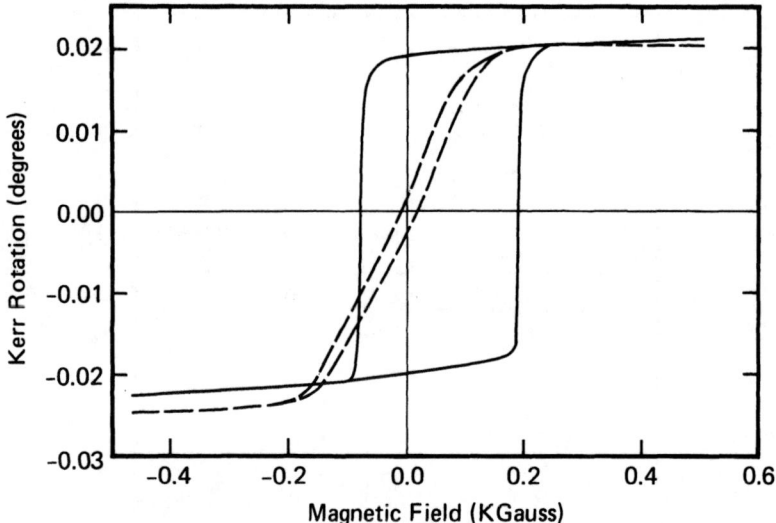

FIGURE 3. Anisotropy in the Fe moment of magnetization as determined by the longitudinal Kerr effect. The TbFe is first saturated along its easy axis. The dashed curve is the result of rotating the sample 90 degrees with respect to the position of the solid curve.

Polar Kerr measurements obtained through the Fe and TbFe layers revealed that whereas the hysterisis loop measured from the alloy side is as expected, that obtained from the Fe side exhibits a non-saturating behavior over the range of the external applied field. This can be explained by the fact that although the polar Kerr rotation originates purely from the TbFe alloy, a contribution at high fields from the Fe layer is obtained as the in-plane moment of magnetization is aligned in the perpendicular direction.

X-ray analysis indicates that the Fe overlayer in these structures is polycrystalline and the TbFe amorphous. The extent of amorphicity in these alloys has been determined in separate experiments utilizing EXAFS and TEM and within the resolution limits of these techniques, one can conclude that these materials exhibit no microcrystallinity. It is conceivable that these Fe layers grown uninterruptedly over the amorphous alloy may exhibit preferential crystal growth alignment and that this may be responsible for the in-plane anisotropy observed in our work. Further measurements are being conducted to test this observation.

Finally, we briefly address the origin of the exchange coupling in these contacted layers. Characterization of the easy-axis of magnetization of TbFe alloys utilizing Torque magnetometry reveals no significant tilt that would explain the exchange interaction as proposed by Hellman et al (5) in the case of NiFeMo/TbFe.

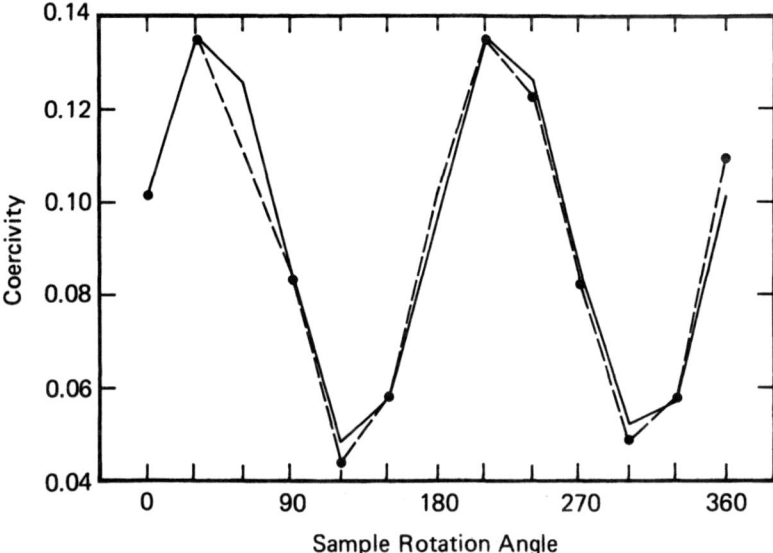

FIGURE 4. In-plane anisotropy of the Fe overlayer. The sample is rotated around the Z-axis and the results indicate an uniaxial alignment direction in the plane of the film.

Spin polarized photoemission studies by Allenspach et al (7) of Fe (100) and by Bona et al (8) of TbFe alloys, show that the surface magnetization properties in these materials differ substantially from the bulk. This leads to surface magnetic structures that can significantly reduce the coercivity of the surface layers. In the case of Fe/TbFe one can expect the magnetization of the Fe layer to lie in-plane and be determined by shape anisotropy. Thus at the interface, the magnetization of the TbFe first layers are modified by the Fe overlayer. As a consequence, a deviation of the magnetization direction from the easy-axis of TbFe results which permits the exchange interaction. One can expect in agreement with Hellman et al (5) that this effect will be obviated when the films are purposedly grown with canted axis of magnetization. However, for atomically clean interfaces, this is not a necessary requirement as the results described in this paper indicate. A more comprehensive discussion of this model is given in ref 9.

CONCLUSION

Exchange interactions between Fe layers (with in-plane magnetization) and TbFe alloys (perpendicular magnetization) are observed leading to exchange bias fields of typically 110 Oe for a 6 nm Fe overlayer on a 100 nm thick TbFe. Longitudinal Kerr measurements reveal that the Fe moment is uniaxially aligned with respect to the surface of the film and this may be the consequence of preferential crystalline growth of the Fe overlayer. The origin of the exchange interaction is attributed to surface magnetic effects that render the magnetization of the interface layers in TbFe to assume a different orientation from that of the bulk leading therefore to non-orthogonal magnetization axis.

REFERENCES

(1) W. Meiklejohn, JAP, supplement to Vol 33, No 3,1328 (1962).
(2) R. Hempstead, IEEE Trans. on Mag. 14, 521, (1978).
(3) C. Tsang, N. Heiman and K. Lee, J. Appl. Phys, 52, 2471, (1981)
(4) V.A. Seredkin, G.I. Frolov and V.Y. Yakovchuk, Sov. Tech. Phys. Lett. 9, 621, (1983).
(5) F. Hellman, R.B. van Dover and E.M. Gyorgy, Appl. Phys. Lett., 50, 296, (1987).
(6) C.C. Robinson, JOSA, 53, 681, (1963).
(7) R. Allenspach, M. Taborelli, M. Landolt and H.C. Siegmann, Phys. Rev. Lett. 56, 953 (1986).
(8) G.L. Bona, F. Meier, M. Taborelli, H.C. Siegmann, A.E. Bell, R.J. Gambino and E. Kay, J.Mag. and Mag. Mat. 54-57, 1403, (1986).
(9) H.C. Siegmann, P.S. Bagus and E. Kay, Z. Physik, in press

Ellipsometric Characterization for Multilayers Containing Magneto-Optic TbFeCo Films

M. Ruane   A. Jain

Boston University, College of Engineering, 44 Cummington Street,
Boston, MA 02215 USA

1 ABSTRACT

Multilayered structures containing sputter-deposited films of amorphous TbFeCo can exhibit magneto-optical Kerr rotation and are leading candidates for erasable optical storage media. Ellipsometric characterization of optically active multilayered media is desirable during media development, testing and production, but traditional ellipsometry does not account for the presence of optical activity. A novel ellipsometer is described that can characterize both the dielectric overcoat and the optically active layers. Reflectivity measurements are collected for different incident angles and polarizations using a differential detector while the magnetic reversibility of the active film layer is exploited to enhance the magneto-optical signal from the TbFeCo. A multilayer film model is used to process observations and estimate media characteristics. The model explicitly accounts for the optical activity of the TbFeCo layer, and is parameterized in terms of the index of refraction and thickness of the overcoat, and the real and imaginary parts of the elements of the dielectric tensor of the active layer.

A series of TbFeCo films with varying composition, rf-sputter deposited on quartz substrates with Al_2O_3 dielectric protective overcoats, is characterized. Both Tb-rich and Fe-rich samples were deposited, with room temperature coercivities ranging from 1.6kOe to about 4kOe, and film-side Kerr rotations between 21 minutes and 28 minutes. Reflectivity data versus angle of incidence, and estimated dielectric tensor elements for the TbFeCo layers are presented. A standard figure of merit, based on the off-diagonal tensor elements, is used to compare competing media in terms of their optimal performance in a multilayered structure.

2 INTRODUCTION

Erasable magneto-optical media have been the subject of intense development because they offer high density data storage, optical write/read operations and ease of handling. A variety of amorphous binary and ternary rare-earth transition-metal (RE-TM) alloys have been studied as possible media. Appropriate compositions support perpendicular anisotropy and the thermomagnetic writing of perpendicular reverse magnetized domains. Readout uses the polar Kerr effect.[1-3] In a previous paper, we reported on the characterization of binary TbFe alloys that were sputter-deposited on quartz substrates and coated with glass.[4]

Compared to TbFe binary films, amorphous TbFeCo alloys have exhibited greater resistance to corrosion and are more likely candidates for practical magneto-optical disk systems. This paper describes extensions of our earlier characterization efforts to include ternary TbFeCo films with Al_2O_3 overcoat layers. Media will be described in terms of their dielectric tensor elements, which can be used directly in system modeling and calculation of magneto-optical system performance measures[5].

The measurement system resembles an ellipsometer, but exploits the Kerr magneto-optical effect of the films to allow measurement of off-diagonal elements of the dielectric tensor. Standard ellipsometric measurements could also be performed with the apparatus if desired. The films were sputter-deposited in argon on 0.5mm thick, 1in diameter quartz substrates using a rotating substrate carrier that passed under a sequence of 8in diameter targets. TbFeCo film thickness is 900Å, and a 4000Å overcoat of Al_2O_3 was applied in situ to prevent oxidation and provide mechanical protection. A Kerr loop tracer was used to determine Kerr rotation and the squareness of the sample hysteresis curves; film composition was found using Rutherford Back Scattering on carbon substrates that were on the substrate carrier during the deposition process.

A useful Figure of Merit is given by the right-hand side below:[5]

$$r_\perp \leq \frac{|\epsilon_{xy}|}{2Im\{\epsilon_z\}} \tag{1}$$

This performance measure provides an upper bound on the ratio of useful magneto-optical signal to incident laser power available from a film. Realization of the upper bound requires an optimized multilayered structure. Here ϵ_z and ϵ_{xy} are elements of the dielectric tensor and $r_\perp$ is the Kerr magneto-optical reflectivity.

3   EXPERIMENTAL MEASUREMENTS

A perpendicularly magnetized magneto-optical medium is described by the dielectric tensor:

$$\underline{\epsilon} = \begin{pmatrix} \epsilon_x & i\epsilon_{xy} & 0 \\ -i\epsilon_{xy} & \epsilon_y & 0 \\ 0 & 0 & \epsilon_z \end{pmatrix} \tag{2}$$

where the plane of the film is taken as x-y and the magnetization direction is z. Deposition conditions should create isotropic amorphous films in x-y, such that $\epsilon_x = \epsilon_y$; ϵ_z will be assumed to be equal to ϵ_x, thereby neglecting the effects of any possible growth-induced anisotropy. An earlier paper[6] presented the details of our reflectivity model relating n_o, the index of refraction of the dielectric overcoat layer, the complex Fresnel reflectivities, $r_\parallel^{(p)}$ and $r_\parallel^{(s)}$ and the magneto-optical reflectivity, $r_\perp$, as functions of angle of incidence, θ. The reflectivity model of the film gives the characteristic matrix for the film-overcoat interface.

3.1   Experimental Apparatus

Figure 1 shows the measurement apparatus. A polarized HeNe laser ($\lambda = 632.8nm$) is incident on a polarizing beam splitter (PBS) mounted in a rotating holder. The plane of polarization of the output from the rotating PBS can be oriented at angle β, measured CCW from the vertical axis. The side-exiting beam of the PBS is incident on a PIN detector and is used to normalize for intensity changes. The beam incident upon the sample is linearly polarized with p, s, or a mixture of polarization ($\beta=90$deg, $\beta=0$deg or $\beta=45$deg) A lens of focal length 200mm focusses the beam on the sample with spot size at normal incidence less than 0.1mm.

The sample is held in a vertical position under a hemispherical lens of index $n=1.51$. A thin film of index matching oil ($n=1.51$) completes the optical path, and prevents interference phenomena in the thin air layer between the lens and the overcoat. The incident focussed beam enters the hemispherical lens-overcoat combination at normal incidence, is refracted slightly at the interface between the Al_2O_3 ($n=1.76$) and matching oil, strikes the sample, and exits by a similar path. Refraction effects of the overcoat layer are thereby mitigated and angles of incidence as high as $60°$ can be reliably measured. Measurements at angle of incidence θ requires rotating the sample table and signal arms by $+\theta$ and $+2\theta$ from normal incidence. A small pulsed electromagnet, with peak field strength of about 8kOe and pulse duration of 150 msec, reverses the sample's magnetization direction.

The signal arm contains a rotating quarter-wave plate with angle ς, and a detector unit consisting of a second polarizing beam splitter and two large area PIN detectors on the output axes of the PBS. The PBS and detectors are mounted in a rotator whose angle is η.

PIN outputs S_1 and S_2 are filtered for noise, amplified, normalized for fluctuations in laser output, and added or subtracted to give system observations. A single final normalization factor, σ_0, is necessary to account for the losses in the optics, and is taken without a sample, using total internal reflection from the rear of the hemispherical lens. A computer-based data acquisition system collects all measurements at each angle of incidence.

3.2   Measurement Procedures

Seven distinct functions are read at each angle of incidence. The observed function depends on the set of angles $[\beta, \varsigma, \eta]$ of the rotators. These functions provide magnitudes for $\mathbf{r}_p$, $\mathbf{r}_s$, and $\mathbf{r}_\perp$ and relative, but not absolute, phase information. The functions are shown in Table 1.

3.3   Estimation of $\underline{\epsilon}$ Elements

Even without noise and measurement errors, the set of functions $[f_1, \ldots, f_7]$ are related to the elements of the tensor $\underline{\epsilon}$ in a complicated nonlinear fashion. The multivariate gradient algorithm of Levenberg-Marquardt[7,8] was used to estimate the set of $\underline{\epsilon}$ parameters $[\text{Re}\{\epsilon_z\}, \text{Im}\{\epsilon_z\}, \text{Re}\{\epsilon_{zy}\}, \text{Im}\{\epsilon_{zy}\}]$. Estimation proceeds incrementally to account for differences in relative magnitudes of the dielectric tensor elements. First the real and imaginary parts of ϵ_z and the refractive index of the overcoat, n_0, are estimated, using only functions f_1, f_2, and f_3 and assuming $\epsilon_z = \epsilon_y = \epsilon_z$. Next, the diagonal elements and n_0 are held fixed and ϵ_{zy} is estimated using functions $f_4 \ldots f_7$. To explore the effects of initialization of the gradient algorithm, parameters are repeatedly estimated over a grid of starting points, and the statistics of the estimates are collected automatically. For ϵ_z and ϵ_{zy}, standard deviations of estimates determined over the grid of starting points were consistently less than 2% of the mean estimates; the grid range was at least $\pm 100\%$ of the center values.

4 Results for TbFeCo Samples

Compositional and Kerr loop tracer data is given in Table 2 for the four samples that are characterized below. Each sputtering run had different Ar pressure and bias voltages, giving rise to different compositions. The column marked **Loop** indicates the sense of the hysteresis loop; Fe-rich films have positive loops since the net magnetization corresponds to the TM sublattice magnetization and the associated Kerr signal. All samples had square loops.

The estimates for ϵ_z and ϵ_{zy} in Table 3 are comparable to published values for TbFe films of similar composition.[4,9] Since Co is present in relatively small amounts (¡10%), dielectric tensor values were expected to be similar to those of TbFe samples. For example, $Tb_{15.8}Fe_{72.4}Ar_{11.8}$ was found[4] to have ϵ_z = -2.9 +i10.1 and ϵ_{zy} = -0.068 + i0.115. and a figure of merit of 6.59×10^{-3}.

Referring to Table 3, it is clear that high Kerr rotation alone does not guarantee a high performance medium. Runs A and C, with the lower Kerr rotation data, had similar Figures of Merit near 0.0045, exceeding those of Runs B and D. Given similarly well designed multilayered structures, the films of Runs A and C should produce a stronger magneto-optical readout signal. Reflectivity curves for each film were calculated using estimated dielectric tensor elements, and are shown in Figure 2. The upper curves show the similar $r_\perp$ and r_s behavior of all the films. The lower curves indicate the different strengths of the Kerr response in the films, and can be used to compare candidate films or to examine the falloff of readout signal for the most oblique rays of a high numerical aperture lens.

5 CONCLUSIONS

An ellipsometric device has been used to develop estimates of the dielectric tensor for a series of TbFeCo amorphous films exhibiting magneto-optical Kerr effect. Estimation of ϵ_z and ϵ_{zy} from reflectivity measurements of overcoated magneto-optical films has been used to characterize and compare representative TbFeCo media suitable for magneto-optical storage applications. For our limited set of samples, $Tb_{23.4}Fe_{57.6}Co_{8.5}Ar_{10.5}$ had the highest figure of merit. Samples with high Kerr rotation will not necessarily have the best figure of merit, which was about 4.5 for the best samples.

References

[1] P. Chaudhari, J.J. Cuomo, and R.J. Gambino, *Appl. Phys. Lett.* **22**, 337 (1973)

[2] Y. Mimura, N. Imamura, and T. Kobayashi, *IEEE Trans. Magn.* **12**, 779 (1976)

[3] Y. Togami, *IEEE Trans. Magn.* **18**, 1233 (1982)

[4] M. Ruane, A. Jain, R. Rosenvold, and M. Mansuripur, presented at the 1986 MRS Winter Meeting, Boston, MA, 1986, (unpublished).

[5] M.Mansuripur, *Appl. Phys. Lett.*, **49**, 19 (1986)

[6] M. Ruane, M. Mansuripur and R. Rosenvold, *Appl. Optics* **25**, 1946 (1986)

[7] K.M. Brown and J.E. Dennis, *Numerische Mathematik* **18**, 289 (1972)

[8] International Mathematical Subroutine Library. Ed. 9, IMSL, Inc., Houston, Texas (1982)

[9] G.A.N. Connell and D.S. Bloomberg, in Mott Festschrift, (Plenum, New York) (1985)

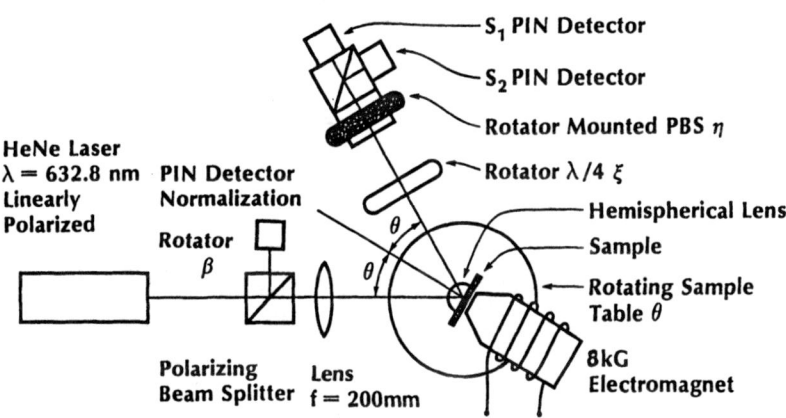

Figure 1: Oblique Incidence Reflectivity Measurement Apparatus

FUNCTION	β	η	ς	OBSERVATION
f_1	90°	0°	0°	$r_p{}^2$
f_2	0°	0°	0°	$r_s{}^2$
f_3	45°	45°	45°	$r_p \, r_s \cos (\phi_p - \phi_s)$
f_4	90°	0°	45°	$r_p \, r_\perp \sin (\phi_p - \phi_\perp)$
f_5	90°	45°	45°	$r_p \, r_\perp \cos (\phi_p - \phi_\perp)$
f_6	0°	0°	45°	$r_s \, r_\perp \sin (\phi_s - \phi_\perp)$
f_7	0°	45°	45°	$r_s \, r_\perp \cos (\phi_s - \phi_\perp)$

Table 1: Observations for Different Rotator Settings

Group	Tb%	Co%	Fe%	Ar%	Loop	Kerr Rot.	Coercivity
Run A	20.9	8.5	62.1	8.5	Tb-rich	24.1 min	3.8kOe
Run B	22.9	9.4	58.4	9.3	Tb-rich	28.0 min	1.9kOe
Run C	23.4	8.5	57.6	10.5	Tb-rich	21.8 min	1.6kOe
Run D	17.2	7.4	60.3	15.2	Fe-rich	26.6 min	2.3kOe

Table 2: RBS Sample Compositions, Hysteresis Loop Shapes, Kerr Rotation Angles in Minutes of Arc, and Coercivity, by Run (A, B, C, D)

ESTIMATE	Run A	Run B	Run C	Run D
$Re(\epsilon_z)$	-3.89	-2.81	-3.48	-2.59
$Im(\epsilon_z)$	7.71	5.51	7.96	5.06
$Re(\epsilon_{zy})$	-0.053	-0.021	-0.048	-0.013
$Im(\epsilon_{zy})$	0.046	-0.00075	0.054	0.0043
Fig Merit x10^3	4.5	1.9	4.6	1.4

Table 3: Dielectric Tensor Estimates for TbFeCo Samples

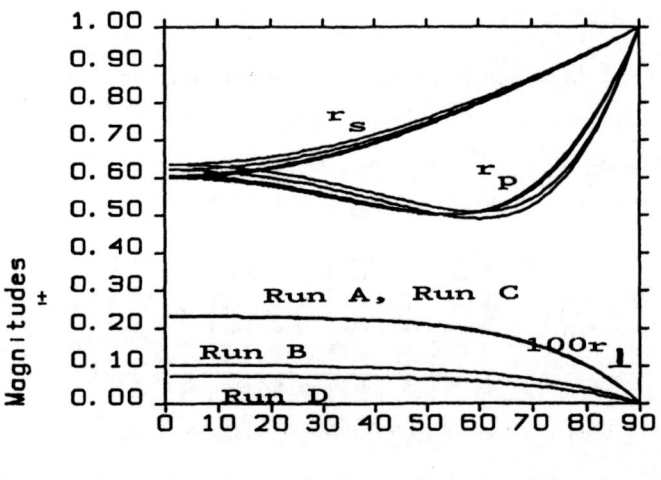

Figure 2: Reflectivity versus Angle of Incidence, based on Model of Film/Overcoat Interface and Estimated ϵ_z and ϵ_{zy} Values- Runs A, B, C, D

FABRICATION AND EVALUATION OF TRANSMISSIVE
MULTILAYER OPTICS FOR 8 keV X RAYS

R.M. Bionta, A.F. Jankowski, and D.M. Makowiecki, Lawrence Livermore National Laboratory, P.O. Box 5503, M.S. L-278, Livermore, CA 94550.

ABSTRACT

We have made and tested several sliced multilayer structures which can function as transmissive x-ray optical elements (diffraction gratings, zone plates, and phase gratings) at 8 keV. Our automated multilayer sputtering system is optimized to sputter layers of arbitrary thickness for very large total deposits at high deposition rates. Diffraction patterns produced by the multilayer devices closely match theoretical predictions. Such transmissive optics have the potential for wide application in high resolution microscope and spectrometer systems.

Introduction

Zone plates are focusing devices constructed of alternating transparent and opaque layers [1]. Constructive interference occurs at the focal point of the zone plate when the zones are properly spaced. Using electron beam and holographic lithography, zone plates have been constructed that operate at x-ray energies below 1 keV [2-6]. Unfortunately, for higher x-ray energies of interest in the industrial laboratory, namely 5 to 20 keV, zone plates cannot be fabricated by the holographic and electron beam lithography techniques, because such energies require large aspect ratios and very fine zones. To date, the only way to manipulate x-ray energies above 5 keV is by the use of grazing incidence mirrors or near-grazing incidence mirrors coated with multilayer coatings to increase the reflectivity.

We have investigated a different technique for fabricating zone plates for use in the 5 to 10 keV regime. Ultimately we plan to make zone plates by sputtering alternating layers of opaque and transparent materials onto a thin wire core, then slicing perpendicular to the core axis to produce many zone plates. This technique shows promise for making x-ray optical elements that can be used in industrial crystallography, microprobe, and radiography equipment. In a previous publication [7] we reported on the good agreement between the measured performance of an Al/Ta diffraction grating and our numerical simulation. Details of the test bed used to evaluate the gratings and zone plates may be found elsewhere [8]. In this report we concentrate on the fabrication techniques used to produce diffraction gratings and linear zone plates.

In the past 40 years, we have seen impressive advances in the multilayer technology used to enhance the reflectivity of grazing and near-grazing incidence mirrors [9-12]. Even so, transmissive multilayer optics is sufficiently novel to require considerable improvement in the techniques of multilayer sputtering. In general, transmissive optics require much thicker layers, ruling out techniques such as molecular beam epitaxy which is mainly useful for making very thin layers. Also, multilayer transmissive optics usually require many more layers than reflective optics. Because the number of layers often exceeds 10,000, to make transmissive multilayer optics requires computer control of the process. Futhermore, since the total thickness of the sputtered multilayer determines the aperture of the lens, very thick coatings are required. This means that the sputtering system must run for several days in order to achieve thicknesses approaching 500 μm, requiring extensive computer-controlled monitoring and logging of the sputtering process. In addition, absolute thickness control is required to achieve proper constructive interference. And finally, to make a wide variety of optics with different focal lengths requires a sputtering system capable of sputtering layers of arbitrary thickness. This means that there must be some way of individually specifying the thickness of several thousand layers.

Sputtering

For simplicity, we have avoided the problems of sputtering on a wire core by sputtering onto a flat substrate and then slicing the coatings into thin slabs. This technique allows us to produce linear zone plates which focus x rays from a point source into a line, much as a cylindrical lens does in ordinary optics. The sputtering system, shown in Figure 1, is optimized to produce transmissive linear optics. The system has two sputtering sources positioned 90° apart with respect to a rotating mandrel. Two substrates, measuring 5 x 20 mm, are mounted on the mandrel, and are alternately positioned in front of each source. Cylindrical shutters cover the opposing source during deposition, and the substrate is heated to ~300°C by a quartz lamp inside the mandrel. Two quartz crystal microbalances track the deposition rate during the process. The sputtering guns, shown in Figure 2, are of our own design. They employ neodymium iron boron magnets to produce very high field strengths. The magnets are configured into a small cylinder located in the center of the gun and a ring magnet surrounding it. The field strength produced by these magnets are sufficiently high that targets up to 0.75 inch thick may be employed. These very thick targets allow us to continuously sputter for several days in order to build up very thick coatings. The guns are powered by 10 kW power supplies in a DC mode.

Deposition is under the control of an HP9000 series computer (CAD/CAM), which is used to design and build the multilayer. To use the lens-making software, the user enters the focal length of the lens, its diameter, and the materials to be used. Then the program calculates the thickness of each layer and the material type. The thicknesses and materials are written into a file stored on a floppy disk. Both zone plates with variable spacing and diffraction gratings with constant spacing can be built. The software has provisions for inverting and concatenating the lens files to make more complex designs. The sputtering software is divided into an initialization section, a sputtering section, and a wrap-up section. During initialization, the user has the option of changing the preprogrammed sputtering parameters before the user inserts the diskette containing the lens file. These parameters include sputtering power, pressure, warm up time, and tooling factor, which is defined as the ratio of the thickness on the substrate to the thickness measured by the crystal.

To begin sputtering, the program first turns on the sputtering gas and purges the system with argon for about an hour. Then the program goes into the layer loop. For each layer, the program reads the layer thickness and material from the diskette, sets the pressure of the sputtering gas, brings up the sputtering gun to the desired sputtering power, moves the substrate to face the sputtering gun, and then opens the shutter. During deposition the substrate is made to oscillate back and forth in front of the sputtering gun (ensuring a uniform layer thickness). The computer continuously reads the value of the thickness on the crystal until the desired value is reached, taking into account the crystal tooling factor. When the desired thickness is reached, the computer closes the shutter, rotates the substrate to a neutral position, and powers down the gun. After all layers are complete, the program goes into the wrap-up mode, during which it shuts down the high voltage on the guns, turns off the argon, and prints out a summary sheet of the sputtering run.

Since the sputtering system is intended to run for many days without supervision, it is necessary to have emergency monitoring and shut-down procedures programmed into the computer. This is accomplished by having the computer read the current in the sputtering guns during deposition of the layers. The current is read once every second. If the current

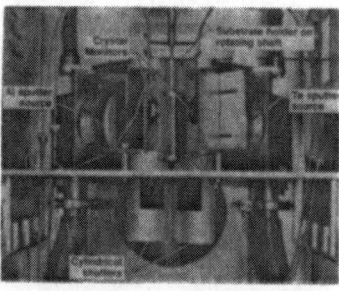

Figure 1. Automated sputtering system for multilayer deposition.

Figure 2. Sputtering source guns and targets (eroded and new).

exceeds a predetermined value (10 amps) for more than 4 seconds, the computer shuts off all power supplies and stops the process. Another fault check involves the measurement of the amount of time it takes to sputter a layer of a given thickness. If for a given layer the deposition time is more than 600% of the time that it should take based on the thickness of the current layer and the sputtering rate achieved on previous layers, the computer shuts down the operation. This is useful if the shutter sticks or the crystal becomes broken. In addition, the computer monitors the deposition process by recording all deposition parameters, including the substrate temperature, the sputtering rate, the current, the voltage, and the pressure, once each minute onto a log file on the disk. After the deposition run is over, plots of these recorded parameters are helpful in understanding what happened during the run.

The crystal tooling factors are obtained in special calibration runs. A cross-section of a calibration run, shown in Figure 3, consists of 4 μm of Ta, followed by 20 μm of Al, 20 μm of Ta, 4 μm of Al, then 4 μm of Ta. For calibration purposes we measure the actual thickness of the 20 μm layers and correct the crystal tooling factor for any discrepancies. The accuracy of the calibration is limited by our ability to measure the 20-μm-thick layers. In particular, it is limited by the sharpness of the interface between adjacent layers. In practice, we can achieve thickness measurement errors of less than 0.5 μm, which with a thickness of 20 μm gives a calibration error of ≤ 2.5%. Since the crystals are further away from the sputtering source than the substrate, the tooling factors are large, and generally vary between 8 and 15. The tooling factors are very sensitive to the distance between the target and the crystal. Figure 4 shows a plot of the tooling factor as a function of distance from the target on a log scale. The variation of tooling factor of distance in this sputtering system is seen to be an exponential, with a slope of about 37 mm. Thus an error of 1 mm in crystal position corresponds to a 3% error in calibration.

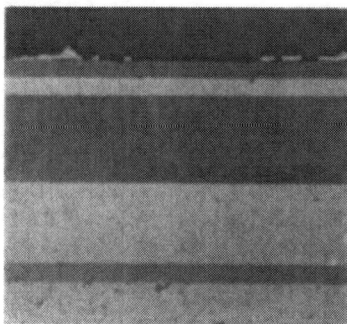

Figure 3. Cross section of Al/Ta calibration run.

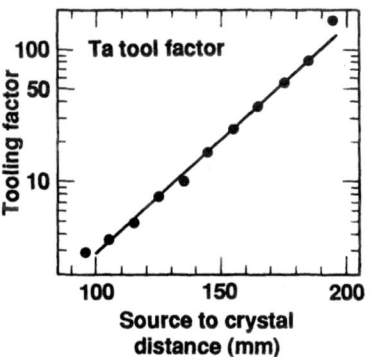

Figure 4. Typical plot of crystal tooling factor (defined as the ratio of the deposition rate on the substrate to the deposition rate on the crystal) vs. distance from sputtering source.

With a 0.75-inch-thick Al target and 1 kW power, we can achieve deposition rates of 100 Å per second. The influence of the strong magnetic field can be seen in Figure 5, which shows a log-log plot of the sputtering voltage versus the sputtering current for the Al gun. The plot shows the results for several layers as the gun voltage is raised and lowered for each layer. For magnetron operation, the current should be proportional to some power of the voltage, which would produce a straight line in the log-log plot. The slope of the line is related to the efficiency of the system for retaining the plasma. In the beginning of the run, when the target is thick, the slope is very shallow. It takes a large change in voltage to change the current. Towards the end of the run, the slope is very steep, because the target has eroded away considerably and the sputtered surface is now much closer to the magnetic field.

Another effect of target erosion is shown in Figure 6, which shows the sputtering rate versus time for a six day run at constant power. It can be seen that over a period of six days, the sputtering rate for Al drops by more than a factor of two as the target erodes.

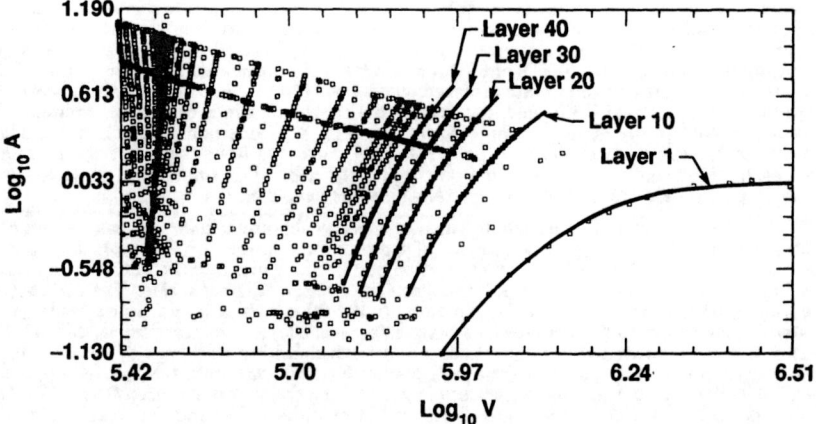

Figure 5. Current voltage relationship for magnetron sputtering of Al.
Data was recorded one per minute during deposition as well as
every tenth layer during the power up of the sputtering gun.

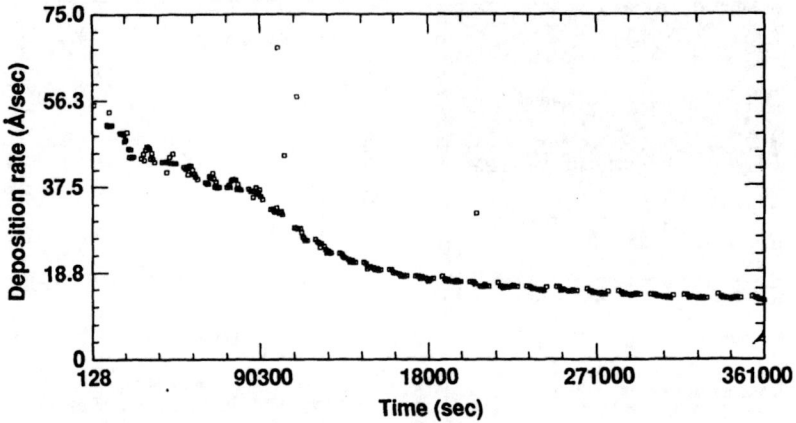

Figure 6. Reduction of Al sputtering rate during 6-day run due to target erosion.

Several improvements had to be made to the system in order to obtain extremely smooth layers of Al and Ta. Figure 7 shows several Al/Ta multilayers sputtered under various conditions. Our first coatings were done with the mandrel continuously rotating at 60 rpm and an argon pressure of 5 mTorr. Continuously rotating the mandrel would ultimately allow us to make as-deposited blended coatings by co-sputtering the two materials simultaneously. Unfortunately, rotating the substrate caused most of the coating to be deposited at oblique angles, promoting columnar defects as shown in Figure 7a. Changing to normal incidence coating by positioning, then holding the substrate in front of each gun produced smoother coatings as shown in Figure 7b. The coatings still had a few gross defects which were traced to fine particles of Ta that drifted onto the substrate. We found that at 5 mTorr the Ta coatings were highly stressed and tended to flake off of the shutters and surfaces of the chamber, causing very fine debris to land on the substrate and introduce defects. To overcome this problem we made test runs of Ta at various pressures in order to determine the pressure at which the Ta coatings had the least amount of stress. This was done by sputtering Ta films on very thin Ta substrates at various pressures and observing the stress-induced curling of the substrates. At low pressures, between 5 and 30 mTorr, the coatings were compressive and tended to curl inward. Above 60 mTorr, the coatings were tensile and tended to curl outward. At 50 mTorr, however, the coating seemed to have very little stress and did not curl the substrate, indicating that 50 mTorr was the pressure we should use to sputter Ta. Unfortunately, at 50 mTorr the Al deposition rate is very low, so it is necessary to sputter Al at 5 mTorr and Ta at 50 mTorr. We subsequently modified our control system so that the computer had control of the pressure. Figure 7c shows a thick multilayer coating made under these conditions, where the Al layers are sputtered at 5 mTorr and the Ta layers are sputtered at 50 mTorr. There is a considerable improvement in the quality of the multilayer as compared to Figure 7b, chiefly due to the fact that the Ta sputtered at 50 mTorr sticks very well to all of the surfaces inside

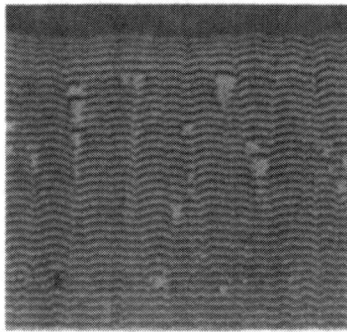

7a. Rotating substrate produces columnar defects.

7b. Nonmoving substrate, with gross defects due to particle contamination.

7c. Coatings sputtered at optimal pressures, with columnar defects in aluminum.

7d. Substrate temperature of 300°C results in extremely smooth coatings.

Figure 7. Al/Ta multilayers sputtered under various conditions.

of the chamber, and no debris is introduced onto the substrate. The remaining defects in Figure 7c are due to the columnar growth that is a property of Al coatings. This growth can be minimized by sputtering Al at a high substrate temperature, around 300°C [13]. To accomplish this, we located a quartz heater behind the substrate with a temperature control system that keeps the temperature of the substrate at 300°C. The resulting coatings were extremely smooth, as shown in Figure 7d.

Finishing

To make even thicker structures, we sputter two coatings at once onto two separate substrates and then bond them together in the middle. This is possible because the diffraction gratings are periodic, and the zone plates are symmetric about their axis. We use a solid state bonding technique to bond the two coated substrates together. The last layer of each coating is Al. The two substrates are placed facing each other in a vacuum chamber where they are brought up to a temperature of 550°C and then pressed together at a pressure of 3000 psi. Figure 8 shows a zone plate before and after bonding. The bonds are strong and durable, and the bonded Al layers behave as if they are one coating.

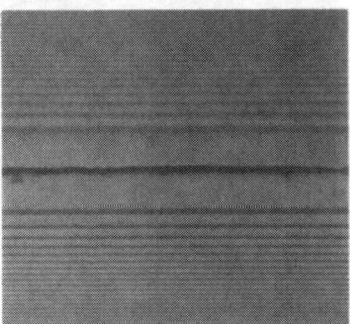 

Figure 8. Lens bonding process. Al/Ta zone plate before bonding (a) and after bonding (b).

After bonding, the coated substrates are sliced into wafers perpendicular to the layers. These wafers are about 0.050 inch thick. The sliced wafers are polished on one side and then glued down to a thinning fixture, where they can be polished from the other side. The fixture has boron carbide shims which stop the polishing at the desired thickness. Figure 9 shows a thinned mounted lens.

X-Ray Testing

The thinned lenses are tested in our x-ray test bed, which has a microfocus x-ray source with a copper anode. The x-ray source emits x rays at 8 keV. The lenses are positioned according to the thin lens law to image the x-ray source onto our scanning pinhole detector, which enables us to record the image. Diffraction gratings are mounted midway between the source and detector where the interference pattern is observed. Figure 10 shows the diffraction pattern produced by one of our diffraction gratings in this apparatus.

Future Work

Several improvements in our process are envisioned in order to make improved x-ray optics. First, the 2.5% calibration error is too large to produce high quality zone plates, which require an absolute calibration of < 1%. To achieve this, we plan to install an interferometer in the sputtering system which will register the thickness change of the coating during the deposition process. To achieve very-high-efficiency focusing devices requires the use of blazed structures having a continuously varying concentration of the two materials across each zone, in a saw-tooth like pattern. This will involve co-depositing the two materials simultaneously. We plan to use the interferometer to determine the thickness of the coating, and use the crystals to set the ratios of the two materials. Finally, we are constructing a sputtering system that will deposit onto thin strands of optical fiber in order to make circular lenses. The circular lenses will be very similar to transmissive lenses in visible optics, and will avail a wide assortment of optical techniques to the x-ray regime.

The authors are indebted to the technical staff of 'O' Division's X-Ray Microscopy Project who made this work possible, including: E. Ables, K. Cook, D. Coufal, P. Gabriele, H. Highstone, L. Kennedy, K. Miller, H. Olson, L. Ott, R. Tilley, R. Vital, and T. Viada, as well as the useful advice given by T. Barbee and L. Wood.

This work was performed under the auspices of the United States Department of Energy by Lawrence Livermore National Laboratory under Contract W-7405-Eng-48.

Figure 9. Thinned mounted lens.

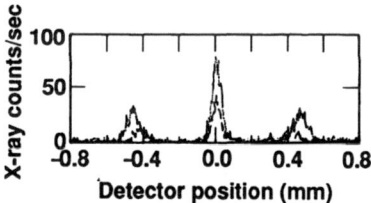

Figure 10. X-ray diffraction pattern produced by a diffraction grating.

REFERENCES

1. J.L. Soret, Arch. Sci. Phys. Nat. 52, 320 (1875).
2. A.V. Baez, J. Opt. Soc. Am. 42, 756 (1952); 51, 405 (1961).
3. G. Schmahl and D. Rudolph, Optik 29, 577 (1969).
4. J. Kirz, J. Opt. Soc. Am. 64, 301 (1974).
5. N.M. Ceglio, A.M. Hawryluk, and M. Schattenburg, J. Vac. Sci. Technol. 1, 1285 (1983).
6. R. Tatchyn, in X-Ray Microscopy (Springer, Berlin, 1984) 40.
7. R.M. Bionta, Appl. Phys. Lett., 51, 725, (1987).
8. R.M. Bionta, A.F. Jankowski, and D.M. Makowiecki, in X-Ray Microscopy, Springer-Berlin, 1988.
9. J. DuMond and J.P. Youtz, J. Appl. Phys. 11, 357 (1940).
10. J. Dinklage, J. Appl. Phys. 38, 3781 (1967).
11. E. Spiller, Appl. Phys. Lett. 20, 365 (1972); Appl. Opt. 15, 2333 (1976).
12. J.H. Underwood, T.W. Barbee, and D.C. Keith, Proc. Soc. Photo-Opt. Instrum. Eng. 184, 123 (1979).
13. J.A. Thornton, Ann. Rev. Mater. Sci. 7, 239 (1977).

LIGHT SCATTERING ANALYSIS OF STRAINED-LAYER SUPERLATTICES

G. P. SCHWARTZ, G. J. GUALTIERI, AND W. A. SUNDER
AT&T Bell Laboratories
Murray Hill, NJ 07974

ABSTRACT

Raman scattering has been used to examine strained-layer GaSb/AlSb superlattices for a variety of growth imperfections. Analysis of zone-folded acoustic phonon spectra permit a determination of the superlattice period, the individual layer widths, some estimate of the interfacial widths, and evidence for variations in layer thicknesses.

INTRODUCTION

During the last one and a half decades molecular beam epitaxy (MBE) and variants thereon have been extraordinarily successful in growing structures with superlattice periods on the order of tens of angstroms. Both lattice-matched and strained-layer systems are now routinely grown for a variety of device-related applications. As with any materials growth technique, there exists a variety of possible imperfections. Those intrinsic to the materials system and growth parameters include interfacial diffusion and the nucleation of misfit and related defects in strained-layer systems, while extrinsic imperfections include uncertainties in the superlattice period and the individual layer widths. In the present study we have examined the zone-folded acoustic phonon mode spectra of a series of strained-layer GaSb/AlSb superlattices using Raman spectroscopy. The ability of this technique and its limitations to address the issues of intrinsic and extrinsic imperfections in superlattices are the focus of this work.

EXPERIMENTAL

The superlattice samples were grown on (001) GaSb substrates at temperatures between 490 and 505C by MBE using elemental Ga and Sb sources. Raman spectra were obtained using 5145Å excitation incident at Brewster's angle in a surface reflection geometry. The electric vector of the incident radiation was polarized in the plane of incidence. The axes are denoted as x(100), y(010), z(001), x′(110), and y′(110), and standard scattering notation is employed.

RESULTS AND DISCUSSION

Five types of growth irregularities found in periodic $(AB)_n$ superlattices will be considered. These include (i) uncertainty in the period $d \equiv d_A + d_B$, (ii) uncertainty in the individual layer widths d_A and d_B, (iii) interfacial diffusion, (iv) distribution of d_A or d_B widths within the structure, and (v) the presence of misfit and related dislocations.

The frequencies ω_m and Raman intensities I_m of mth order zone-folded acoustic phonons constitute the basis for analyzing defects of types (i)-(iii).

The frequency dispersion relation is given by [1]

$$\cos(qd) = \cos(\omega d_A/v_A) \cos(\omega d_B/v_B)$$

$$-\frac{1}{2}\left[(Z^2+1)/Z\right] \sin(\omega d_A/v_A) \sin(\omega d_B/v_B), \tag{1}$$

where q is the phonon wavevector, $v_{A,B}$ the acoustic velocities, and Z is the acoustic impedance defined by $\rho_B v_B/\rho_A v_A$ ($\rho_{A,B}$ layer density). In the approximation that $v_A=v_B$ and $\rho_A=\rho_B$ Eqn. 1 can be replaced with

$$w_m = V_{SL} \mid 2\pi m/d \pm q \mid, \tag{2}$$

where V_{SL} represents an average acoustic velocity in the superlattice. At the same level of approximation, the Raman intensities of the zone-folded doublets are given by [1]

$$I_m \propto \mid p_A - p_B \mid^2 \times \left[\sin^2(m\pi d_A/d)/(m\pi)^2\right], \tag{3}$$

where $p_{A,B}$ are the elasto-optic constants for the two materials.

In our growth of GaSb/AlSb superlattices, variations in the period of 10% relative to the calibrated growth rates are often encountered. Refinement of the period obtained via Raman scattering is illustrated in Fig. 1(A,B) for a superlattice with nominal GaSb and AlSb widths of 25 and 98Å respectively (d=123Å). Comparison of the measured frequencies with the dispersion relation of Eqn. 1 leads to d = 128±3Å, or a correction of ~4.1%. Recent studies using x-ray scattering [2] have shown that the two techniques yield periods which agree to within 2-3%. The determination of d via Raman scattering is relatively insensitive to the individual values used for d_A and d_B because the acoustic velocities are of similar magnitude for GaSb and AlSb. In order to keep the most intense (lowest order) doublet frequencies in a frequency range accessible to commercial spectrometers ($\omega>5cm^{-1}$), it is necessary to restrict the period d to be less than 250-300Å. On the other hand, measurements of d in superlattices with a total thickness of less than 1000Å are readily accomplished.

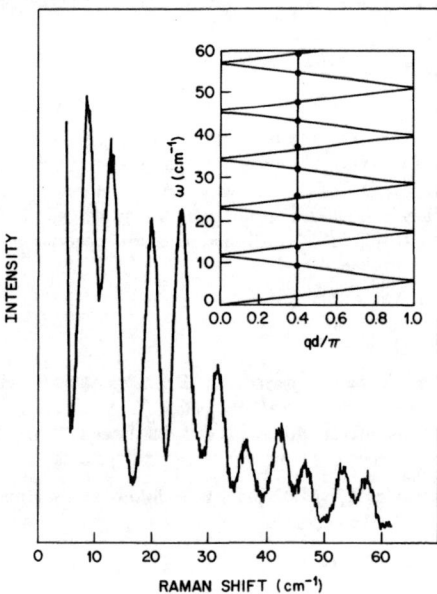

Fig. 1

Raman spectrum of the zone-folded longitudinal acoustic phonons of a GaSb/AlSb superlattice grown with a nominal period of 123Å. The insert shows the measured frequencies (solid circles) matched to an elastic continuum model dispersion curve with a refined period of 128Å.

The relative intensities of the folded acoustic mode spectra can be used in conjunction with Eqn. 2 to determine d_A and d_B separately. The analysis is most sensitive when $d_A = d_B$, in which case the even m doublets should be missing from the spectra [3]. Since the modulation scales with $\sin^2(m\pi d_A/d)$, when $d_A \ll d$ relatively large percentage deviations in d_A are required to substantially affect the intensities. This is illustrated in Fig. 2 for Lorentzian broadened simulations in which $d = 128\text{Å}$ and d_{GaSb} varies by ~30%. Note that for $d_A = 26\text{Å}$ the 5th doublet is extinguished, whereas for $d_A = 14\text{Å}$ the intensities of the m=1 and 2 doublets would be nearly equal. Neither condition occurs in the data of Fig. 1, thus providing some bounds on d_A. A major analysis limitation, however, is presented by the fact that the intensity of both peaks in a given mth order doublet are not necessarily equal in the experimental spectrum. Equation 3 is an approximation which achieves its simplicity of use at the cost of neglecting the Fourier coefficients of the derivative of the displacement amplitude [3]. The ambiguity of assigning an amplitude to a doublet with unequal peak amplitudes can introduce considerable uncertainty into the determination of individual d_A and d_B layer widths. He, Rouhani, and Sapriel [4] have recently derived a closed form expression for I_m under the less restrictive constraint that $\rho_A v_A = \rho_B v_B$. Their expression for the Raman intensity is given by

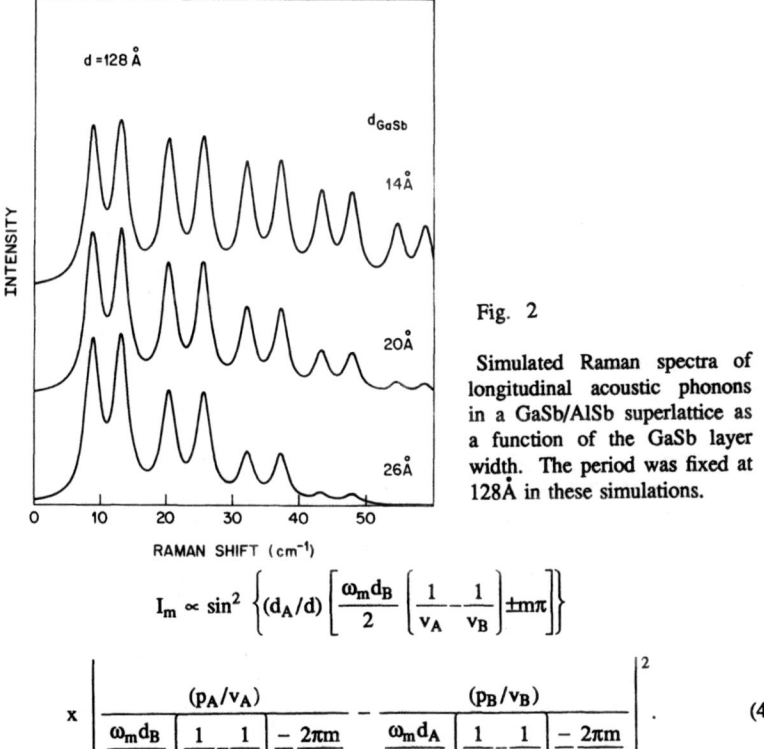

d = 128 Å

d_{GaSb}

14Å

20Å

26Å

INTENSITY

0   10   20   30   40   50

RAMAN SHIFT (cm⁻¹)

Fig. 2

Simulated Raman spectra of longitudinal acoustic phonons in a GaSb/AlSb superlattice as a function of the GaSb layer width. The period was fixed at 128Å in these simulations.

$$I_m \propto \sin^2\left\{(d_A/d)\left[\frac{\omega_m d_B}{2}\left(\frac{1}{v_A}-\frac{1}{v_B}\right) \pm m\pi\right]\right\}$$

$$\times \left|\frac{(\rho_A/v_A)}{\frac{\omega_m d_B}{d}\left(\frac{1}{v_A}-\frac{1}{v_B}\right)-2\pi m}+\frac{d}{d} - \frac{(\rho_B/v_B)}{\frac{\omega_m d_A}{d}\left(\frac{1}{v_B}-\frac{1}{v_A}\right)-2\pi m}+d\right|^2 . \quad (4)$$

Figure 3 shows the simulation of the Raman spectra for three values of the parameter $\alpha \equiv (d_{AlSb}/d)$ and $\rho_B/\rho_A \sim 0.4$. Since accurate values for the elasto-optic coefficients are not known for these materials, and adjustments of that parameter strongly influence the relative intensities of the ±m components, it remains difficult to specify α to better than ±0.05.

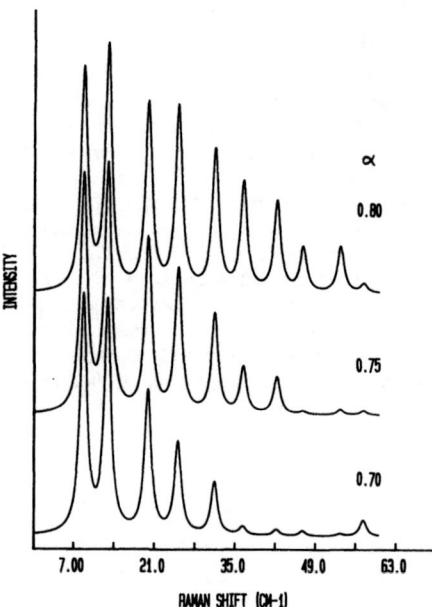

Fig. 3

Simulated Raman spectra using Eqn. 4 and parameter values listed in the text.

Jusserand *et al.* [5] have also related the intensity of folded acoustic modes to interface broadening due to differences in growth conditions. Earlier annealing studies [3] of GaAs-AlGaAs superlattices had shown that the frequencies and linewidths of folded acoustic modes were relatively insensitive to interfacial diffusion, whereas the absolute mode intensities were attenuated. To make a valid intercomparison between samples with the same nominal period it is critical to achieve similar growth morphologies, since any type of surface roughness will also attenuate the signal amplitude independent of interfacial diffusion. We have examined the question of interfacial diffusion by growing samples with thin GaSb layers of order 4 monolayers. The optic phonon spectra display narrow confined phonon modes with no discernible evidence for extensive alloy formation which would be expected if interdiffusion on the order of 2 monolayers accompanied the growth process.

Samples have also been grown and examined which contained intentional variations in one of the layer widths. The effect of weakly perturbing the symmetry leads to the activation of normally symmetry forbidden modes. As an example, Fig. 4 shows the Raman spectrum of a lattice grown with 15 periods of a repeat unit of the form (AB_-ABAB_+) where the B layer (GaSb) width $B_\pm \equiv B \pm \Delta$, with $d_A = 58.4$Å, $d_B = 42$Å, and $\Delta = 8.4$Å. In the structure the B-type layers were randomly sequenced with the sole constraint that the number of B_+ and B_- layers were equal. Panel (B) shows the simulated spectrum of an $(AB)_n$ lattice, while in panel (c) the spectrum is that of a periodic (AB_-ABAB_+) supercell with the B-type layers in a fixed B_-, B, B_+ sequence. Although the major peaks can still be identified with an "average" $(AB)_n$ lattice, the additional structure reflects the symmetry breaking associated with a variable layer width. It should be remarked that layer width variations are not the only possible mechanism for activating symmetry forbidden modes. Various other forms of disorder [6] have been cited, and in large period structures where $q > \pi/d$, Umklapp processes can also contribute [7].

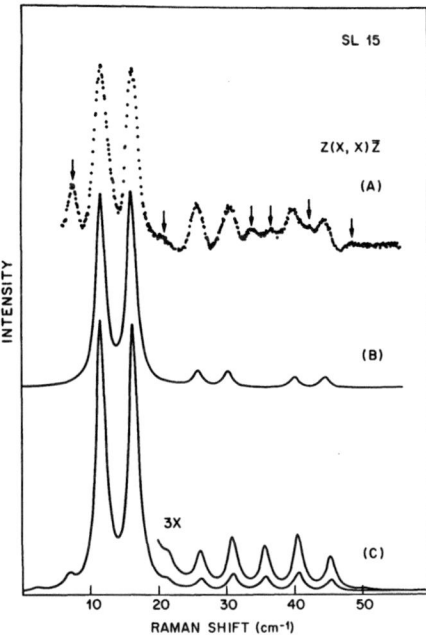

Fig. 4

Raman spectrum of (A) a superlattice with variable B-layer widths as described in the text. Panels (B) and (C) show simulated Raman spectra for $(AB)_n$ and $(AB_-ABAB_+)_n$ lattices with strict sequencing of the B-type layers.

In strained-layer materials systems such as GaSb/AlSb (0.65% lattice mismatch), there are also individual and total superlattice critical thicknesses which when exceeded will lead to the formation of misfit and related dislocations. One of our early interests was to examine the acoustic Raman spectra for evidence of such defect formation. Combined x-ray and Raman analyses in both the GaSb/AlSb [2] and Si/(SiGe) [8] systems on samples with misfits have shown that misfit dislocations do not necessarily give rise to symmetry forbidden modes, although the relaxation of the strain does shift the optic mode frequencies.

REFERENCES

[1] For a current review see M. V. Klein, IEEE J. Quantum Electron. *QE-22*, 1760 (1986).

[2] A. T. Macrander, unpublished.

[3] C. Colvard, T. A. Gant, M. V. Klein, R. Merlin, R. Fischer, H. Morkoc, and A. C. Gossard, Phys. Rev. B*31*, 2080 (1984).

[4] J. He, B. D. Rouhani, and J. Sapriel, Phys. Rev. B, in press. Preprint courtesy of the authors.

[5] B. Jusserand, F. Alexandre, D. Paquet, and G. Leroux, Appl. Phys. Lett. *47*, 301 (1985).

[6] J. Sapriel, J. Chavignon, F. Alexandre, and R. Azoulay, Phys. Rev. B*34*, 7118 (1986).

[7]   D. J. Lockwood, M. W. C. Dharma-wardana, J. M. Baribean, and D. C. Houghton, Phys. Rev. B35, 2243 (1987).

[8]   A. T. Macrander, G. P. Schwartz, and J. Bevk, unpublished.

HYDROGEN BONDING IN LOW-GAP a-Si,Ge:H,F SUPERLATTICES

J.P. CONDE, G. HAGER AND S. WAGNER
Department of Electrical Engineering, Princeton University, Princeton, NJ 08544

ABSTRACT

The integrated absorption of the Si-H, Ge-H and Si—H_2 IR stretching modes were measured for three series of superlattices where: a) the well to barrier thickness ratio was kept constant while the period was varied; b) the barrier thickness was kept constant while the well thickness was varied and c) the well thickness was kept constant while the barrier thickness was varied. The dark (σ_d) and photo (σ_{ph}) conductivities , the photoconductivity exponent γ and the activation energy of the dark conductivity E_a were measured perpendicularly to the plane of the layers for series a) and b). Structural changes induced by sandwiching were documented by IR absorption and correlated with the optoelectronic properties of the thin films.

INTRODUCTION

Since Abeles and co-workers suggested, in 1983, the concept of amorphous superlattices structures,[1] these materials have been intensively studied. The first studies focussed on the structure of interfaces between amorphous materials,[2-4] taking advantage of the large number of reproducible interfaces available in a superlattice structure. Since then the optical and electronical properties of these novel semiconductor materials,[5-8] their application to solar cells[9] and their properties as photoreceptors[6] have been the major topic of study.

Recent evidence[8, 10, 11] suggests that the thin layers (10~100Å) of amorphous semiconductor in multilayers structure are structurally different from those grown in the bulk. We present here a study of the hydrogen concentration and bonding in a-Si,Ge:H,F/a-Si:H,F superlattices. With this study we probe the structure of the thin layers in the superlattice, and we compare the structural information thus obtained with the electronic properties of the films.

EXPERIMENTAL PROCEDURES

The a-Si,Ge:H,F alloys and a-Si:H,F/a-Si,Ge:H,F multilayer structures were grown using a plasma-enhanced chemical vapor deposition technique.[12] The glow discharge was radio-frequency excited (13.56 MHz). The RF diode power density was 250 mWcm^{-2}. In all deposition runs the gas flow rates were 28 sccm SiF_4, 4.6 sccm H_2 and either 0 or 0.4 sccm GeF_4. The superlattice structure was obtained by periodically diverting the GeF_4 flow, using a computer controlled pneumatic valve without interrupting the discharge. The films were grown on c-Si substrates, polished to 1° wedges to reduce fringing in the infrared.

The thickness of the films was measured with a Dektak surface profiler. It agreed with the product of the growth rates of the individual layers, the layers deposition time and the number of layers. The infrared transmittance spectra were measured in a range between 400 and 4000 cm^{-1} using a Digilab-20C Fourier transform infrared spectrophotometer. From the spectra we determined the integrated intensity of the Si-H (2000 cm^{-1}), Si–H$_2$ (2100 cm^{-1}) and Ge-H (1875 cm^{-1}) infrared modes. Using conversion factors[13] we can calculate the number of bonds per cm^3.

Semitransparent Cr dots evaporated on the surface of the film were used as the top contact for perpendicular electronic measurements. σ_d and σ_{ph} were determined in the ohmic region of the I-V curves. σ_{ph} was measured using bandpass filtered illumination from a tungsten-halogen adjusted to yield an approximately uniform generation rate of 10^{21}cm^{-3}s^{-1}. The σ_{ph} dependence on the light intensity G is of the power law form $\sigma_{ph} \propto G^{\gamma}$. The dark conductivity activation energy E_a was measured in vacuum. σ_d was measured as a function of the temperature T over the range 10-100 ° C and E_a was computed from $\sigma_d = \sigma_{d,0} \exp(-E_a/kT)$.

RESULTS AND DISCUSSION

Fig. 1. shows a schematic energy band diagram of a a-Si:H,F/a-Si,Ge:H,F superlattice. d_s, d_g and d_r are the barrier, well and period thicknesses respectively. E_{gs} and E_{gg} are bulk values of the optical gaps of the well material of the barrier and well layers, respectively. E_g is the optical gap of the well material in the multilayer structure, where we can write $E_g = E_{gg} + W + W'$ if we interpret W and W' as the energies of the first quantum levels above the conduction band edge and below the valence band edge, respectively. U is the conduction band discontinuity (taken to be $0.8(E_{gs} - E_{gg})$)[14] and U' is the valence band discontinuity.

We study three series of a-Si:H,F/a-Si$_{0.45}$Ge$_{0.55}$:H,F superlattices differing in the geometry of the multilayer structure. In these samples E_{gs} is 1.7 eV and E_{gg} is 1.3 eV.

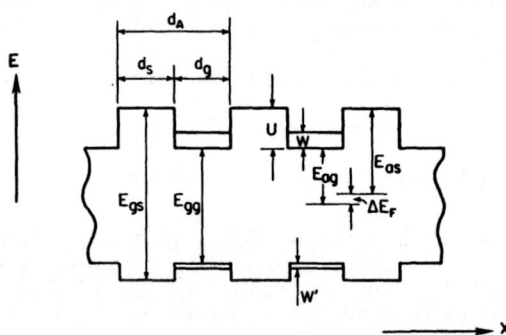

Fig. 1.   Energy band diagram of an a-Si:H,F/a-Si,Ge:H,F superlattice.

Constant composition superlattice series

In this series of samples we kept $d_s/d_g \sim 1$ and constant, and varied the period of the superlattice d_r from 240Å to 24Å . In such a series, the overall chemical composition of the films is kept constant and the variations will arise either from interface effects or structural changes.

In Fig. 2 we show the variation of the integrated absorption of the different kinds of bonds as a function of $1/d_g = 1/d_s$. This representation has the distinct advantage that $1/d_s$ is proportional to the number of interfaces in the films.

In Fig. 3 we show the electronic properties of these films: σ_d, σ_{ph}, E_a and γ. They show clearly the transition from electron transport over the barriers for wide barriers to tunneling through the barrier layer when the barrier becomes $\lesssim 50$Å. This behaviour has been extensively documented[7,8] and is characterized by a sharp increase in the σ_d, a decrease of E_a from values typical of a-Si:H,F for wide barriers to values typical of the low gap a-Si,Ge:H,F well layer when tunneling dominates. We report for the first time a dramatic decrease of γ from 0.6-0.7 typical of bulk alloys to 0.5 at the onset of tunneling.

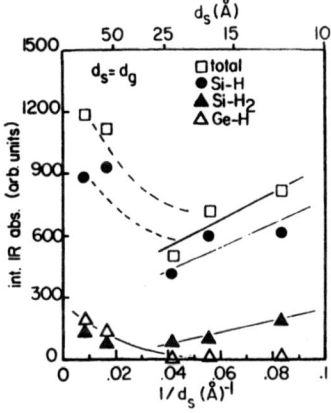

Fig. 2.

Integrated IR absorption of the H stretching vibrations for a superlattice series with constant overall composition and varying period d_r. In this paper $E_{gs} = 1.7$ eV and $E_{gg} = 1.3$eV.

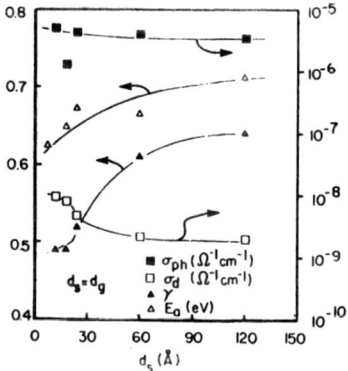

Fig. 3.

Dark conductivity σ_d, photoconductivity σ_{ph}, dark conductivity activation energy E_a and γ for the superlattice series of Fig. 2.

We tentatively suggest the following interpretation for the results of Fig. 2: as we go from essentially bulk layers (small $1/d_s$) to thin layers (small d_s) there is a reduction in H concentration in the superlattice that we attribute to structural modifications of the layer material. In addition, when the layers become very thin ($d_s < \sim 50$Å) the increased number of interfaces will play an important role : there is a linear increase in Fig. 2 in the H concentration bonded to Si for short period superlattices which results from increasing Si-H and Si—H$_2$ contribution combined with decreasing Ge-H.

If we compare the structural IR and the electronic results (Fig. 2 and 3) we see that the onset of tunneling, which has dramatic consequences on the optoelectronic properties, appears correlated with a structural effect. A linear increase with d_s^{-1} of the hydrogen bonded to Si points to induced changes in the barrier layer whereas the decrease to very low levels of the H bonded to the Ge points to structural modifications in the well layer itself. These multilayer-induced structural changes in the well may be related with the lower defect density[15] and enhanced stability[10, 11] in the alloy layer; in the barrier layer they relate possibly with the very high tunneling probability measured. The high tunneling probability has been attributed to an electron effective mass in the order of 0.2.[5, 7, 8] We have then a higher defective barrier layer (reflected in the higher Si–H_2 density and tunneling probability) and a less defective alloy (lower subgap defect density).

Constant barrier layer thickness superlattice series

In this series of samples, we keep d_s=120Å constant and vary the well layer thickness d_g from 120Å to 12Å. The barrier layer is kept very thick so that its bulk properties are preserved. In Fig. 4 we show the integrated IR absorption for the hydrogen stretch vibration and in Fig. 5 the electronic properties for the samples in this series.

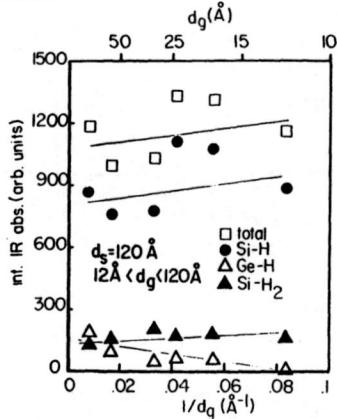

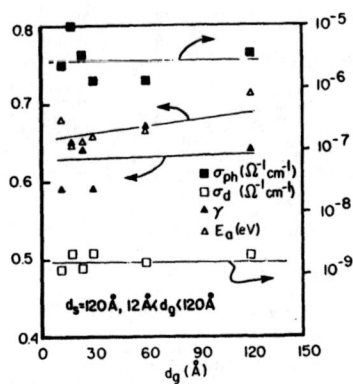

Fig. 4. Integrated IR absorption of the H stretching vibrations for a superlattice series with constant barrier thickness d_s=120Å and varying well thickness d_g.

Fig. 5. σ_d, σ_{ph}, E_a and γ for the superlattice series of Fig. 4.

Electronic transport is dominated by thermal emission over the thick a-Si:H,F barriers. The transport properties are then remarkably independent of the alloy layer characteristics.[7, 8] To analyze the bonded hydrogen behaviour in this series we have to take into account that the overall composition changes as we shrink the well layer width and that the change in the number of interfaces is much smaller than in the first series. We expect both the composition changes and the number of interfaces to be roughly proportional to $1/d_g$. Since the change in the number of interfaces is comparatively small in this series, the deviation from linearity of the plots can be interpreted as a multilayer-induced structural change. The results shown in Fig. 4 confirm that the hydrogen bonded to the silicon is mainly contributed by the barrier layers; in this series they dominate.

The approximate linearity with d_g^{-1} of Fig. 4 curves for hydrogen bonded to Si and the constancy of the electronic properties in Fig. 5 point that "quasi-bulk" a-Si:H,F has a stabilizing effect in the superlattice, keeping its structure as we vary the well layer thickness, and preventing the multilayer-induced changes that occurred in the previous series.

Constant well layer thickness superlattice series

Finally, we grew two series of superlattices in which we kept a fixed well layer thickness (in Fig. 6, $d_g=120\text{Å}$ and in Fig. 7 $d_g=60\text{Å}$) while varying the barrier layer thickness (in Fig. 6 from 120 to 18Å and in Fig. 7 from 60 to 15 Å). Once again we must be aware of the changes in composition overall along the series.

We show the integrated absorptions of the IR stretching bands as a function of $1/d_s$ to simplify the analysis. In both Fig. 6 and Fig. 7 the H bonded to Si is dominated by the barrier contribution whereas the H bonded to Ge is obviously controlled by the constant well layer thickness. The approximate linearity of the plots shows that the large well layer prevents multilayer-induced structural changes. The Ge-H bond integrated intensity does not change, which indicates that the alloy layer remains essentially bulk-like.

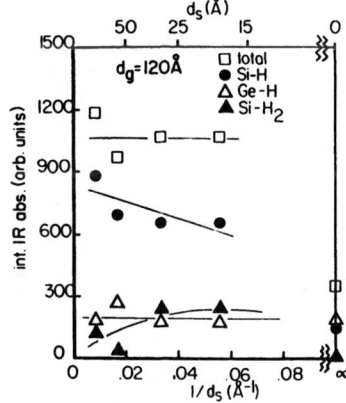 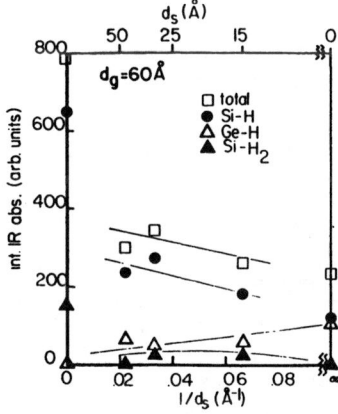

Fig. 6. Integrated IR absorption of the H stretching vibrations for a superlattice series with constant well layer thickness $d_g=120\text{Å}$ and varying barrier thickness d_s.

Fig. 7. Integrated IR absorption of the H stretching vibrations for a superlattice series with constant well layer thickness $d_g=60\text{Å}$ and varying barrier thickness d_s.

CONCLUSIONS

Infrared spectroscopy proved to be a powerful tool to study and understand the structural changes in amorphous materials caused by a superlattice structure. It is shown that both the well and barrier present properties different from the corresponding in bulk materials. These differences do not manifest themselves when one of the layers is thick enough to retain its own bulk characteristics.

The new optoelectronic properties shown by amorphous semiconductor superlattices put them into a new category of materials with both fundamental and practical interest. Structural measurements can help us understand how these new properties arise and which are the possibilities and limitations that should be expected from these amorphous based structures.

Acknowledgements

We acknowledge V. Chu, A. Maruyama, Y. Okada and D.S. Shen for help in the laboratory. J.P. Conde acknowledges an IBM Graduate Fellowship. This work was supported by the Electric Power Research Institute under Contract No. 2824-2.

References

1. B. Abeles and T. Tiedje, *Phys. Rev. Lett.*, vol. 51, p. 2003, 1983.
2. P.D. Persans, A.F. Ruppert, B. Abeles and T. Tiedje, *Phys. Rev.*, vol. B32, p. 5558, 1985.
3. Y. Okada, D. Slobodin, S.F. Chou, R. Schwarz, and S. Wagner, in *Mat. Res. Soc. Symp. Proc. Vol. 70*, ed. D. Adler, Y. Hamakawa and A. Madan, p. 289, Materials Research Society, Pittsburgh, 1986.
4. J.P. Conde, D.S. Shen, I.H. Campbell, P.M. Fauchet and S. Wagner, in *Mat. Res. Soc. Symp. Proc. Vol. 77*, Materials Research Society, Pittsburgh, 1986, to be published.
5. C.R. Wronski, P.D. Persans and B. Abeles, *Appl. Phys. Lett.*, vol. 49, p. 569, 1986.
6. H. Shirai, A. Tanabe, J. Hanna, S. Oda, T. Nakamura and I. Shimizu, *Japn. J. of Appl. Phys.*, vol. 25, p. L537, 1986.
7. J.P. Conde, S. Aljishi, D.S. Shen, V. Chu, Z E. Smith and S. Wagner, in *Mat. Res. Soc. Symp. Proc. Vol. 95*, ed. Y. Hamakawa and A. Madan, Materials Research Society, Pittsburgh, 1987, to be published.
8. J.P. Conde, S. Aljishi, D.S. Shen, M. Angell and S. Wagner, in *J. of Non-Cryst. Solids*, Proceedings of the 12th Int. Conf. on Liquid and Amorphous Semiconductors, Prague, 1987, to be published.
9. S. Tsuda, H. Haky, H. Tarui, T. Matsuyama, K. Sayama, Y. Nakashima, S. Nakano, M. Ohnishi and Y. Kuwano, in *Mat. Res. Soc. Symp. Proc. Vol. 95*, ed. Y. Hamakawa and A. Madan, Material Research Society, Pittsburgh, 1987, to be published.
10. I. Honma, H. Hotta, K. Kawai, H. Komiyama and K. Tanaka, 1987, to be published.
11. S. Miyazaki, Y. Ihara, M. Hirose, *J. Non-Cryst. Solids*, Prague, 1987, to be published.
12. J. Kolodzey, S. Aljishi, R. Schwarz, D. Slobodin, and S. Wagner, *Journal of Vacuum Science Technology*, vol. A 4 (6), p. 2499, 1986.
13. C.J. Fang, K.J. Gruntz, L. Ley and M. Cardona, *J. Non-Cryst. Solids*, vol. 35 &36, p. 255, 1980.
14. F. Evangelisti, *J. of Non-Cryst. Solids*, vol. 77&78, p. 969, 1985.
15. J.P. Conde, V. Chu, D.S. Shen, S. Aljishi, S. Tanaka and S. Wagner, *Tecnhical Digest of the International PVSEC-3, Tokyo, Japan, 1987*, p. 717, 1987.

INTERDIFFUSION IN SHORT-WAVELENGTH MODULATED MATERIALS STUDIED BY MONTE-CARLO SIMULATIONS

M. Atzmon[*]
Division of Applied Sciences, Harvard University, Cambridge, MA 02138.

Abstract

Interdiffusion in a two-dimensional compositionally modulated lattice has been studied by Monte-Carlo simulations. In the initial stages, the interdiffusion coefficient has been observed to change with time due to the development of short-range order simultaneously with the interdiffusion process. When the short-range order parameter approached its limiting value, the diffusion coefficient approached a constant value. The dependence of the interdiffusion coefficient on the modulation wavelength does not agree with the prediction of one-dimensional theories. For ordering alloy systems, the effective interdiffusion coefficient is positive, i.e., an initially present modulation decays in time, for all wavelengths.

Introduction

The basis for the theory of interdiffusion in compositionally modulated films was given by Hillert [1] and by Cahn and Hilliard [2]. Their treatment is based on expressing the free energy of a nonuniform system as a sum of the contribution of the bulk free energy and a gradient energy term:

$$F = \int [f_o(c) + \kappa (\vec{\nabla} c)] dV \qquad (1)$$

where $c(r)$ is the composition, $f_0(c)$ the local free energy per unit volume, and κ the gradient energy coefficient. The diffusion equation was derived by setting the diffusional flux proportional to the gradient of the chemical potential, which is obtained from the free energy. The composition is expressed as a Fourier series:

$$c - c_o = \sum_\beta c_\beta \cos (\beta \cdot r) \qquad (2)$$

where $\beta = 2\pi/\lambda$ and λ is the wavelength of the composition fluctuation. c_0 is the average composition. For small amplitudes, each Fourier component is assumed to evolve independently. For compositionally modulated films, the problem reduces to one dimension (see review by Greer and Spaepen [3]). The solution to the diffusion equation for one Fourier component is:

$$c = c_\beta \exp [- D\beta^2 (1 + \frac{2\kappa\beta^2}{f_o}) t] \cos (\beta \cdot r) \qquad (3)$$

where t is the time and: $f_o = \dfrac{\partial^2 f_o}{\partial c^2}$ The amplitude of the modulation A will evolve according

to the amplification factor R given by: $R = \dfrac{d}{dt} \ln A = - \bar{D} (1 + \dfrac{2\kappa\beta^2}{f_o})\beta^2 = - \bar{D}_\lambda \beta^2.$ (4)

$\bar{D}$ is the bulk interdiffusion coefficient and $\bar{D}\lambda$ is an effective interdiffusion coefficient which is a function of the modulation wavelength λ. Depending on the sign of R, the modulation amplitude can grow or decay. Within the approximations made in Ref. 3, the sign of $\bar{D}\lambda$ will be the opposite of the sign of the heat of mixing ΔH for $\Delta H \gg kT$. The continuum treatment of Cahn and Hilliard [2] does not take into account the atomistic nature of matter. Cook et al. [5] introduced a discrete analysis of the problem in order to describe systems with wavelengths comparable to atomic dimensions. They obtained the same result as eq. 4, with the term β^2 replaced by a dispersion relation given by

$$B^2 = \frac{2}{d^2} [1 - \cos \frac{2\pi}{\lambda} d] \qquad (5)$$

in one dimension, where d is the interplanar spacing. The effective, wavelength dependent diffusion coefficient will be denoted by $\bar{D}_B$ in this paper. In the limit $\lambda \gg d$, B becomes equal to $\beta = 2\pi/\lambda$. The qualitative predictions of a sign change in the amplification factor as a function of the wavelength remain the same. It should be noted that the discrete theory assumes, as does the continuum model, that perpendicular modes evolve independently for small amplitudes. It

* Present address: Department of Nuclear Engineering, The University of Michigan, Ann Arbor, Michigan 48109-2104.

is therefore a one-dimensional treatment. General reviews of atomic transport in inhomogeneous materials were given by several authors (e.g. Refs. 6-8).

For an ordering system ($\Delta H<0$) the contribution of the gradient energy is predicted to make the amplification factor negative below a critical wavelength. Because λ is then usually of the order of an atomic diameter, this is equivalent to the development of long-range order in one dimension. However, the lowest enthalpy state of an ordering system is associated with one-dimensional order only in specific crystallographic directions. Therefore, an ordering system approaching equilibrium will develop three-dimensional order, but this does not imply an increase in the amplitude of a pre-existing one-dimensional modulation. In order to examine the applicability of the one-dimensional theories to higher-dimensional modulated materials, a simulation was performed in two dimensions. The wavelength dependence of the interdiffusion coefficient was studied and compared to the previously described theories and to experimental results.

Details of the simulation

A two-dimensional array of 400x400 elements (A and B atoms) was constructed. A modulated structure with the amplitude A was created by giving the nth row the composition $c(n)=0.5+A \cdot \sin[(n-1/2)2\pi/\lambda]$, where λ is the modulation wavelength in units of atomic rows. Within each layer, the arrangement was chosen at random subject to the constraint on the average composition. The energy of the system was modeled as follows: for a positive heat of mixing, the contribution of a nearest-neighbor bond was -1 for like neighbors and +1 for unlike neighbors. For a negative heat of mixing, the signs were reversed. The two-dimensional short-range order (sro) parameter s was defined as the total number of like nearest-neighbor bonds minus the number of unlike nearest neighbor bonds, divided by the number of atoms. The neighbors were counted in a manner corresponding to either a simple cubic or hexagonal structure and with the modulation vector perpendicular to the <1,0> direction in either case. Diffusion was simulated by selecting a random atom and a random neighbor at each step. The change in energy ΔH upon exchanging the two atoms was calculated and they were exchanged with the probability $\alpha/(1+\alpha)$, where α is $\exp\{-\Delta H/kT\}$ and the temperature T is chosen to be below the ordering temperature. Periodic BOUNDARY conditions were applied. The output of the random number generator was scrambled following Ref. 9 in order to insure a practically random sequence. The modulation amplitude A was obtained by performing discrete integration :

$$A = \frac{2}{N} \sum_{n=1}^{N} c(n) \sin[(\frac{n-1/2}{\lambda})2\pi] \qquad (6)$$

where n is the serial number of a row, c(n) its average composition and N the total number of rows. The amplification factor was obtained from eq. 4 by monitoring the amplitude A as a function of time with the time given in normalized units of Γ^{-1}, where Γ is the frequency of attempted jumps per atom. The calculated diffusion coefficient was given in normalized units of $a^2\Gamma$, where a is the interatomic spacing.

Results and discussion

Fig. 1 shows the modulation amplitude ln(A) and the simultaneous development of the two-dimensional short-range order parameter s as a function of the normalized time for a hexagonal lattice with a negative heat of mixing. The initial amplitude is 0.05 in a. and 0.2 in b. The nonlinear behavior of ln(A) is a result of a continuous change of the diffusion coefficient as a result of the change of sro. As the sro develops, the lowering in energy reduces the driving force for homogenization and the diffusion coefficient decreases. The smaller the amplitude, the closer is each volume element to the average composition and the faster is the approach to the limiting value of the sro parameter. Therefore the curve in Fig. 1a becomes linear sooner than in Fig. 1b. The curvature of ln(A) is not caused by Brownian motion (see Cook [10]), because it is weaker for smaller amplitudes, whereas the relative effect of Brownian motion increases for decreasing amplitudes. An approximately constant slope and therefore constant diffusion coefficient is obtained after the sro parameter s has reached 0.95 of its limiting value -1. A more quantitative description of the time dependence of the inter-diffusion coefficient will require knowledge of the ordering kinetics and a corresponding time dependent expression of the chemical factor $(1+d(\ln\gamma)/d(\ln c))$ in Darken's equation [4], where γ is the activity coefficient of one of the components and c its concentration.

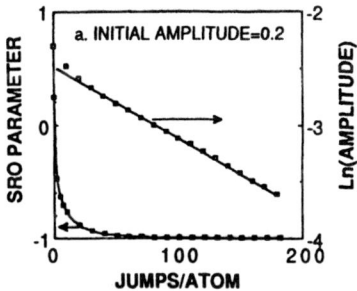
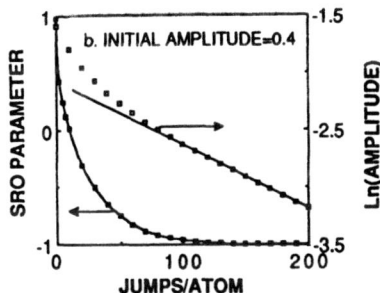

Fig. 1 The logarithm of the modulation amplitude as a function of the normalized time for a two-dimensional hexagonal lattice with a modulation wavelength of 40 layers at a temperature of 0.5. The two-dimensional short-range order parameter is also plotted. The initial amplitude is 0.2 in a. and 0.4 in b.

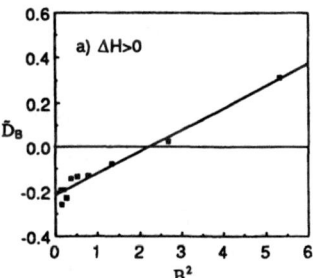
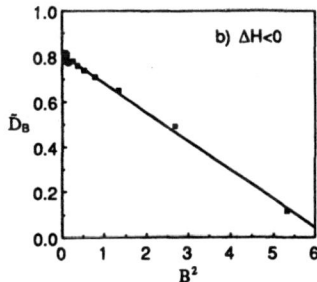

Fig. 2 The effective interdiffusion coefficient as a function of $B^2 = \dfrac{2}{d^2}[1 - \cos\dfrac{2\pi}{\lambda}d]$ for a two-dimensional hexagonal lattice at a temperature of 0.2. a. Positive and b. negative heat of mixing. $B^2 = 5\frac{1}{3}$ corresponds to the minimum attainable wavelength of two monolayers.

In Fig. 2, the interdiffusion coefficient in a hexagonal array is plotted as a function of the previously defined wavelength-dependent quantity B^2. Graph a. corresponds to $\Delta H > 0$ and b. to $\Delta H < 0$. It was not possible to obtain the interdiffusion coefficients after the short-range order parameter had approached its limiting value because the amplitude fell below the noise level. Therefore, their value was determined after an equal annealing time for all samples. In case a., the qualitative predictions of the previously discussed theories are satisfied: at long wavelengths, the modulation amplitude increases in time, whereas at short wavelengths , it decays due to the dominance of the gradient energy term. In case b., however, there is no corresponding behavior. The amplitude decays for all possible wavelengths. It is a result of the fact that the ordered equilibrium state, toward which the system transforms, is not a one-dimensional modulation.

Several experimental results are consistent with the simulation presented in this paper. A large number of authors report initial nonlinear behavior of the logarithm of the amplitude (see Ref. 11 and examples in Ref. 3). This has been attributed to relaxation phenomena of various types (especially in amorphous materials) and to nonlinear diffusion. The present study shows that part of the nonlinear behavior can also be explained as a result of ordering during interdiffusion within the simple model of nearest-neighbor interactions and equal intrinsic diffusivities. Even for modulation amplitudes of several percent, the development of order causes a composition and time dependence of the interdiffusion coefficient through a change in the driving force.

The author has reported the wavelength dependence of the interdiffusion coefficient in compositionally modulated amorphous Ni-Zr [11]. The results were consistent with the behavior of an ordering system and a critical wavelength of approximately 20 Å was predicted by extrapolation. However, a sample with a modulation wavelength of 15 Å showed a dramatic decrease of the modulation amplitude after the shortest practical annealing time (20 minutes at 256 °C), rather than the theoretically predicted increase. Measurements by Paulson and Hilliard [12] in the Au-Cu system showed that a plot of $\tilde{D}_B$ vs. B^2 deviated from linearity for short wavelengths, and did not become negative for any measured B. The authors suggest the possibility of Brownian motion or the effect of interactions beyond the nearest neighbors.

$\tilde{D}_B$ vs. B^2 is approximately linear in the present two-dimensional simulation, and it cannot explain the shape of the curve obtained by Paulson and Hilliard. However, the present study provides a qualitative indication that spinodal ordering in one dimension, i.e., a negative $\tilde{D}_B$, is not expected. Three dimensional simulations are presently being performed in an attempt to provide a better description of the observations.

The results obtained in this paper suggest that the assumption of linear diffusion behavior and therefore independence of orthogonal modulations is not justified for small modulation wavelengths even in the case of small amplitudes. The ordering process depends on the dimensionality of the system and the kinetics of one-dimensional long-range ordering cannot be treated without taking into account simultaneous ordering in orthogonal directions.

Conclusions

Interdiffusion in two-dimensional compositionally modulated materials has been studied by Monte-Carlo simulations. An initial non-linear decay of the logarithm of the amplitude has been shown to be caused by a change in diffusivity due to the simultaneous development of short-range order. This indicates that the regular solution model, in which a random chemical configuration is assumed, should be applied with caution. The present simulations also show that for a general modulation direction, earlier theoretical predictions of one-dimensional spinodal ordering below a critical wavelength are not fulfilled. It is suggested, that for modulation wavelengths of the order of several atomic diameters, the assumption of the independence of perpendicular modulation waves is not justified for any amplitude. A full theoretical description of spinodal ordering will therefore have to be three dimensional. In a future publication, the results of three-dimensional simulations and the coupling of perpendicular modulation waves will be discussed.

Acknowledgements

The author acknowledges useful discussions with Professor Frans Spaepen and Dr. John Cahn. This work has been supported by the Office of Naval Research under contract number N00014-85-1-C-0023. Additional support by an IBM post-doctoral fellowship is gratefully acknowledged.

References

1. M. Hillert, Sc.D. Thesis, Massachusetts Institute of Technology, 1956.
2. J. W. Cahn and J. E. Hilliard, J. Chem. Phys. 28, 258 (1981).
3. A. L. Greer and F. Spaepen, in Synthetic Modulated Structures, eds. L. L. Chang and B. C. Giessen (Academic Press, 1985), p. 419.
4. L. S. Darken, Am. Inst. Mining Met. Engrs. Inst. Met. Div. Metals Technol. 15, Techn. Publ. 2311 (1948), 2443 (1948).
5. H. E. Cook, D. de Fontaine, and J. E. Hilliard, Acta Met. 17, 765 (1969)
6. D. de Fontaine, in Solid-State Physics, eds. H. Ehrenreich, F. Seitz and D. Turnbull, Vol. 34 (Academic Press, 1979), pp. 74-172.
7. A. G. Khachaturian, Theory of Structural Transformations in Solids (Wiley, New York, 1983).
8. T. Tsakalakos, Thin Solid Films 86, 79 (1981).
9. W. T. Press, W. T. Vetterling, S. Teukolsky and B. P. Flannery, Numerical Recipes, (Cambridge University Press).
10. H. E. Cook, Acta Met. 18, 297 (1970).
11. M. Atzmon and F. Spaepen, Mat. Res. Soc. Symp. Proc. , Vol. 80 (MRS, Pittsburgh 1987).
12. W. M. Paulson and J. E. Hilliard, J. Appl. Phys. 48, 2117 (1977).

MoN-TiN SUPERCONDUCTING ARTIFICIAL SUPERSTRUCTURE FILM

KENJI KAWAGUCHI AND SHIGEMITSU SHIN
National Chemical Laboratory for Industry, Tsukuba, Ibaraki 305, Japan

ABSTRACT

We succeeded in synthesizing MoN-TiN superconducting artificial super-structure film (SASF). The structure of MoN layers is estimated to be γ-Mo$_2$N. The behavior of Tc in SASFs is recognized by a proximity effect using adjustable parameters of bulk Tc. Calculated parameters are, however, different from Tc of single layered films.

1. Introduction

Many combinations of SASFs have been investigated already. However, very few studies of SASFs composed of ceramic compounds are reported yet. It is important to develop the synthetic and analytic technique of ceramic SASF. Probably the present work is the first example of synthesizing of a ceramic SASF by an alternative reactive deposition method. The first reason why TiN-MoN combination was chosen is that stoichiometric B1-MoN is difficult to synthesize in spite of its predicted high-Tc. Multilayering is one of the effective methods to stabilize such unstable material as B1-MoN. Both MoN and TiN have the same B1-structure with a small lattice mismatch (TiN=0.423 nm, MoN=0.425 nm [1]). This condition is favorable for epitaxial growth. Secondly, B1-mono-nitrides and mono-carbides are among the simplest compounds and are suitable for an initial study. As many of them are superconductors with various lattice constants, we can modulate their superconducting properties and crystal structures by changing the constituent without destroying the epitaxial growing condition. In this paper, we report structural properties of MoN-TiN SASFs and preliminary analysis of their superconductivity.

2. Sample preparation and structural properties

Samples were prepared by an alternative reactive deposition method. The deposition chamber was first evacuated to 10^{-9} Torr. Several cleaved MgO single crystal plates were used for substrates and preheated at 500°C in ultrahigh vacuum for two hours. Typical size of substrates is 5x10 mm and the total film thickness is around 100 nm. Source metals, Ti and Mo, were evaporated in an ammonia atmosphere. The pressure of the atmosphere was about 1×10^{-5} Torr. Very slow deposition rate (typically 0.02 nm/s) was chosen to compensate the low atmospheric pressure to some degree. The substrate was maintained at 300°C during the deposition. The optimum substrate temperature to fully nitride the specimen and to minimize the interdiffusion at the interfaces has not been found yet. Further investigation is in progress.

Structural properties of MoN-TiN SASFs were studied mainly by X-ray diffraction using a standard diffractometer. Higher-order Bragg peaks at the low-angle and around the MgO(200) peak, as shown in Fig. 1, imply the formation of a well regulated artificial superstructure. Several peaks were also observed around the MgO(400) peak for some samples. Even in the case of the SASF composed of monatomic MoN layers (0.2 nm), higher order reflections at the low-angle and some satellite peaks at the middle-angle appear clearly in Fig. 1-(c). Fine waves modulating superstructure reflections at the low-angle seem to arise from the finite thickness of the coherent crystal structure. The estimated thickness is about 50 nm and half the total film thickness for Fig. 1-(c). Though direct measurement of (200) and (400) reflections from SASFs was difficult because of the intense substrate reflections, the satellite peaks

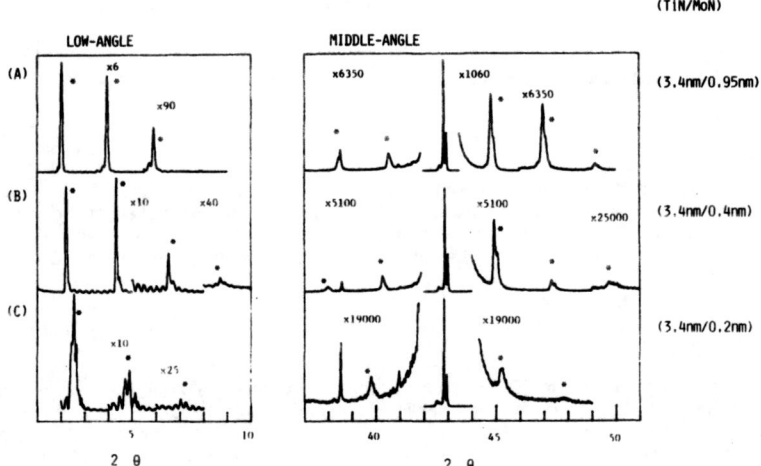

Fig. 1 X-ray diffraction patterns for three SASFs. (*) indicates the peak derived from the superstructure period.

imply the existence of those reflections. The peak position of (200) is midpoint between (-1) and (+1) satellite peaks in wave number. The position indicates the mean spacing of atomic planes in proportion to the composition. No other reflection without satellite peaks around the (200) and (400) reflections was detected. Some small pieces of the samples removed from the substrates were examined by transmitting electron diffraction. Spot pattern was observed distinctly and neither ring nor arc were found. Judging from these results, both TiN and MoN layers are epitaxially grown with the [100] direction perpendicular to the film plane, and the film must be a single-crystal. For the sample with extremely thin MoN layers in Fig. 1-(c), the peak positions of (200) and (400) indicate approximately the spacing of atomic planes in TiN layers and corresponds to the reported value of B1-TiN. The intensity of satellite peaks is evidently intense on higher angle side than lower angle side of the (200) reflection. In the case of stoichiometric B1-MoN, the intensity of satellite peaks on both sides is approximately equal because of the small lattice mismatch between B1-MoN and TiN. The diffraction profile is interpreted reasonably by assuming γ-Mo_2N. Those SASFs are considered to be composed of B1-TiN and γ-Mo_2N layers, and the interpretation is consistent with the results of MoN and TiN single layered films.

3. Superconductivity

Electrical resistance was measured by the conventional four-probe method. The measuring current density was less than 0.8 A/cm². We define Tc as a midpoint Tc. The temperature dependence of the resistivity for TiN and MoN single layered films is shown in Fig. 2. Observed Tc for MoN films is around 5 K and close to the reported value (5.6 K) [2]. Obvious transition is not found for TiN films above 1.8 K. For the stoichiometric B1-TiN, Tc is known to be 5.5 K [3]. There are two possibilities to cause the significant decrease of Tc. The first is the lack of nitrogen. The reduction of 10 % nitrogen composition

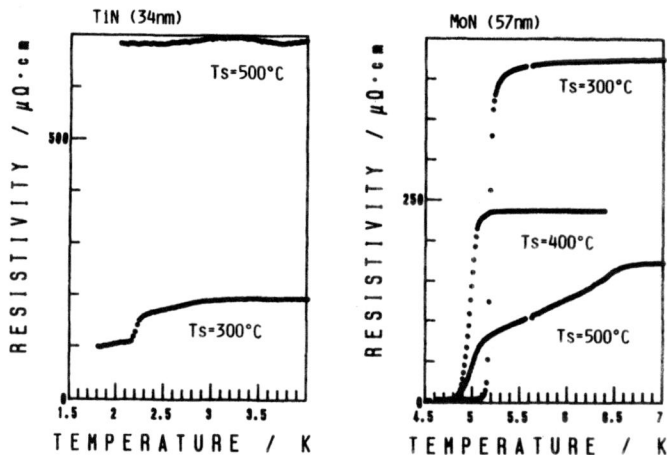

Fig. 2 The temperature dependence of the resistivity for the single layered films. Ts indicates the substrate temperature.

lowers the Tc of TiN to 1.7 K [3]. Secondly, the impurity of oxygen may suppress the superconductivity. A slight oxygen was detected by Auger electron spectroscopy.

A proximity effect and a localization effect are usually dominant in the superconducting behavior of SASFs composed of conductors. There is no special relation between Tc and sheet resistance. Consequently, only the proximity effect is considered in the present work. Experimental results of Tc for SASFs and the calculations on the basis of the trilayer model proposed by Triscone et al. [4] are plotted together as a function of MoN layer thickness in Fig. 3. Electronic specific-heat coefficient for γ-Mo$_2$N is estimated to be 80 % of the theoretical value for stoichiometric B1-MoN [1]. Interfacial layers are assumed to be a homogeneous compound $(Mo_2N)_{0.5}(TiN)_{0.5}$ in the region of thick MoN layers. In the case of MoN layers thinner than interface layers, a $(Mo_2N)_x(TiN)_{1-x}$ compound is assumed, where the composition x=0.5t(MoN)/t(i) and t(MoN),t(i) represent the thickness of MoN layers and TiN layers, respectively. All parameters for the interface used in calculation are the weighted mean values of Mo$_2$N and TiN in proportion to x. Bulk Tc was supposed to be 5 K for the MoN layer and 1.5 K for the TiN layer, at first. The values are derived from the results of the single layered films. Such calculations can not reproduce the experimental results in Fig. 3(a,b,c), even if interface layers with various thicknesses are taken into account. Next bulk Tc for each layer was taken to be an adjustable parameter. Those calculations exhibit a good agreement with experimental results in Fig. 3(A,B,C). Obtained bulk Tc are 3.7 K for MoN and 2.7 K for TiN layer, and are almost independent of the interfacial thickness. These bulk Tc are significantly different from those of single layered films. Probably, the inconsistency is due to the structural strain. Further detailed study is necessary to solve the problem. The calculation curve with the 0.8 nm interface (Fig. 3(c)) shows a little disagreement arising from its linear behavior in the range where the thickness of MoN layers is thinner than that of the interfacial layers. We can evaluate the interface to be thinner than 0.8 nm, which is only four atomic layers.

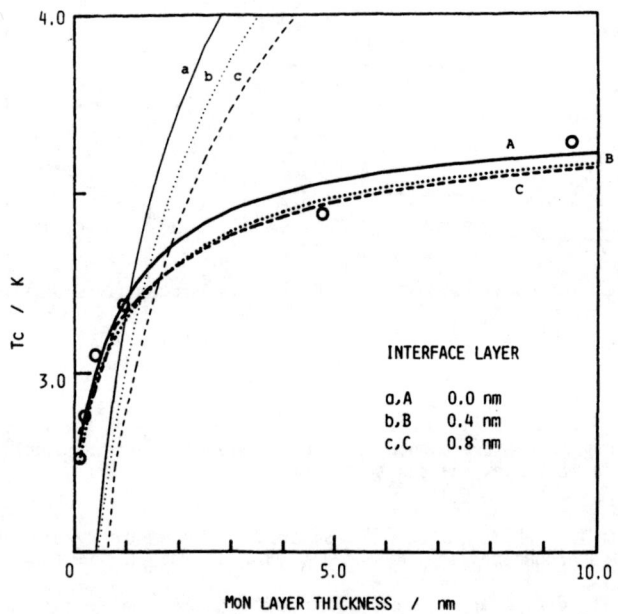

Fig. 3 Transition temperatures Tc of SASFs as a function of MoN layer thickness. The thickness of TiN layer is 3.4 nm. Circles represent the experimental results. (A,B,C) and (a,b,c) are calculated curves with and without adjustable parameters, respectively.

References

1. D.A.Papaconstantopoulos, W.E.Pickett, B.M.Klein and L.L.Boyer, Phys.Rev.B 31 (2), 752 (1985)
2. B.Cendrewska, A.Morawski and A.Misiuk, J.Phys.F 17, L71 (1987)
3. L.E.Toth, Transition Metal Carbides and Nitrides, edited by J.L.Margrave (Academic Press, New York and London, 1971), p.220
4. J.M.Triscane, D.Ariosa, M.G.Karkut and O.Fischer, Phys.Rev.B 35 (7), 3238 (1987)

PART V

Properties

SUPERCONDUCTING SUPERLATTICE STRUCTURES

JOHN L. MAKOUS,* JOHN A. LEAVITT,* LAURENCE C. McINTYRE,Jr.,* LUIGI
MARITATO,** RUGGERO VAGLIO,*** ANNAMARIA CUCOLO***, and CHARLES M. FALCO*
*Department of Physics and Optical Sciences Center, University of Arizona,
Tucson, Arizona 85721
**INFN-LNF, Via E. Fermi 40, 00044 Frascati, Italy
***Dipartimento di Fisica, Università di Salerno, 84100 Salerno, Italy

ABSTRACT

Superlattice superconductors exhibit behavior different from other
"homogeneous" superconductors because of their layered structures.
Sputtering has proved to be an excellent technique for producing such
multilayered systems. Also, systems naturally having layered structures,
such as the new high-T_c materials, can be fabricated by sputtering. Here we
describe the preparation of superconducting superlattices by dc triode
sputtering and techniques useful for characterizing them. In particular, we
discuss Mo/Ta superconducting superlattices and high T_c thin films of
$YBa_2Cu_3O_{7-x}$.

INTRODUCTION

The properties of elemental superconductors are insensitive to non-
magnetic impurities and to microstructural defects. This is as predicted by
the BCS theory. However, because the coherence length in many superconductors is long, typically of order 50 Å to 1 μm, it is possible to affect
the properties of a superconducting film by the proximity effect with a
normal metal, or by forming superlattices. These metallic superlattices are
fabricated by alternately layering thin layers 2 Å to 5000 Å of a superconductor with another material, which may or may not be a superconductor.
The superconducting properties of these systems then become structure
dependent, as exhibited, for instance, by dimensional crossovers [1,2],
proximity effects on the T_c's and the energy gaps [3,4], and by anisotropies
in the critical field [5].

The "new" high T_c superconductors by comparison are very structure
dependent, both in the bulk and as thin films. They also exhibit critical
current and critical field anisotropies, which might be expected from their
naturally layered or "superlattice" structure.

Sputtering techniques have been shown to be very useful for producing
metallic superlattices [6]. In addition, sputtering of metallic
superlattices of Y, Ba, and Cu in various layer combinations recently has
been successfully used as a method for producing thin films of the high T_c
oxide materials [7]. In this paper we briefly describe the preparation of
superlattices by sputtering as well as techniques used to characterize their
structure. These techniques include several x-ray analysis methods and
Rutherford backscattering spectroscopy (RBS). Specifically we discuss the
preparation and characterization of Mo/Ta superlattices and of high T_c thin
films of $YBa_2Cu_3O_{7-x}$. Physical properties of these Mo/Ta superconductors
will be described, including the resistivity and layer thickness dependence
of T_c, and tunneling results. Finally, we include a discussion of the
structure and resistive transitions of the high-T_c $YBa_2Cu_3O_{7-x}$ films prepared
by the sputtering of Y/Ba_2Cu_3 superlattices.

SUPERLATTICE PREPARATION BY SPUTTERING

We fabricate our metallic superlattices by alternately passing a rotating
substrate platform over targets which are sputtered using a magnetically

enhanced, dc triode sputtering system previously described [8]. Feedback control of the sputtering rates and microprocessor control of the substrate platform motion enable us to keep layer thickness deposition constant to ±0.3% [8,9]. Layer thickness calibration is carried out using a Mirau interferometer and RBS [9]. RBS is a useful tool for measuring film thickness as well as chemical stoichiometry and impurity content.

Accurate control of the deposition process allows us to produce superlattices with modulation wavelengths down to an atomic monolayer if so desired. We have produced superlattices in which the wavelengths are modulated by alternating integer atomic planes of Mo and Ta, including samples in which this integer has a value of one, resulting in a "monolayer" superlattice. This superlattice can be thought of as a new simple cubic material (in one dimension) with 2 atoms per unit cell.

In producing high T_c films of $YBa_2Cu_3O_{7-x}$ we deposit monolayers of Y alternating with the appropriate thickness of a Ba_2Cu_3 layer sputtered from a metallic Ba_2Cu_3 alloy target. This process is repeated to create an Y/Ba_2Cu_3 alloy superlattice of total thickness 5000 Å to 1 μm. Monolayers of Y separated by the proper amount of Ba_2Cu_3 are used to imitate the final structure of the high T_c material as closely as possible. Here RBS analysis is important for determining the sputtering conditions under which the 2:3 ratio of Ba to Cu in the target is maintained in the film. In addition, it is useful for calibrating layer thicknesses to give the required 1:2:3 stoichiometry in the films. Pure Ar gas, without O_2, is used during sputtering, and a post-deposition ex-situ anneal in O_2 is necessary to make these films superconducting.

CHARACTERIZATION

After preparing a superlattice it is necessary to determine the actual structure, composition, and impurity content of the film. As mentioned above, RBS analysis is very useful for measuring film thickness and stoichiometry, including impurity content in the film. With RBS the stoichiometry of the film can be determined with an accuracy of 0.1% under normal conditions [10]. However, this technique can be damaging to the film, and sometimes more benign analysis methods are desired as an alternative, or to give complementary information.

X-ray diffraction techniques are very useful for superlattice characterization because they are non-destructive and they can give information on both amorphous and crystalline multilayers. In Fig. 1 we show Cu-Kα x-ray diffraction spectra measured using a θ-2θ Bragg-Brentano diffractometer on a series of Mo/Ta superlattices. The wavelengths shown range from Λ = 82.2 Å, which corresponds to 18 atomic planes of Mo and Ta per superlattice wavelength, to Λ = 4.57 Å, which consists of alternating monolayers of Mo and Ta. These are the high-angle x-ray diffraction spectra in which the spread of the satellite peaks decreases as the modulation wavelength of the superlattice increases [11], as seen in Fig. 1. The value of the modulation wavelength can be obtained from the angular separation of the satellites [11]. For the Mo/Ta samples in Fig. 1 the values obtained from the satellites agree to within ±2% (the uncertainty of the measurements obtained from the spectra) with the values calculated from the sputtering parameters and rotation speed of the substrate table.

Additional information can be obtained from the linewidth of the Bragg peak (2θ ≈ 39.2° in Fig. 1). Using the Scherrer equation [12] the structural coherence length perpendicular to the substrate can be determined from the measured linewidth. In Fig. 1 it is seen that there is essentially no linewidth broadening of the Bragg peak as the wavelength goes to the monolayer limit. This indicates that the perpendicular structural coherence length in Mo/Ta superlattices remains unchanged as the wavelength is decreased to the monolayer limit. This behavior contrasts with behavior previously found in other metallic superlattice systems such as Nb/Cu [13] and Mo/Ni [14].

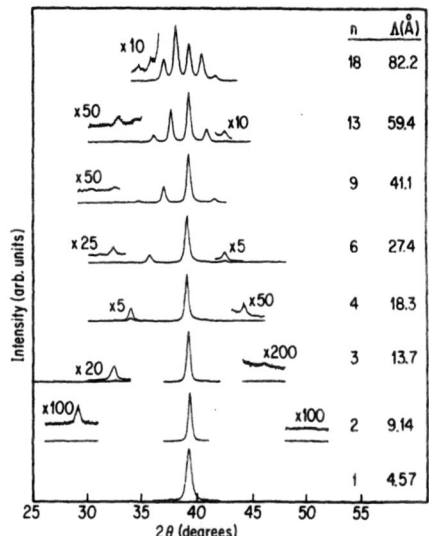

Figure 1. θ-2θ x-ray diffraction spectra of Mo/Ta for modulation
wavelengths from Λ = 4.57 Å to 82.2 Å.

X-ray diffraction using a θ-2θ diffractometer for high-Tc films is fairly
complicated due to a more complex unit cell [15,16]. Diffraction from the
films is similar to that from the bulk since there is no additional
artificial structure imposed when making the films.

The repeat distance in a multilayer gives rise to low-angle x-ray
diffraction peaks which can, therefore, be used to obtain the modulation
wavelength for both crystalline and amorphous multilayers. If the modulation
wavelength is an integer number of atomic planes, the low-angle superlattice
peaks will exactly coincide with high-angle satellite peaks because of the
structure of the unit cell.

Wide-film Debye-Scherrer x-ray diffraction was used to determine
crystallite orientation in the plane of the film. With the Mo/Ta superlattices we find there is a preferred orientation of the crystallites with
respect to one another. In contrast, diffraction from the high-Tc films
showed complete arcs, indicating random orientation of the crystallites in
the plane.

PHYSICAL PROPERTIES OF METALLIC SUPERLATTICE SUPERCONDUCTORS

Resistivity versus temperature measurements were made on the Mo/Ta
superlattices using a four-probe technique in a closed cycle refrigerator.
The temperature dependence of the superlattices remain positive over the
entire modulation wavelength scale from a "bulk" value of 700 Å to the atomic
monolayer limit of 4.57 Å. In addition, the absolute resistivity values
remain well inside the metallic regime, i.e. << ~150 $\mu\Omega$-cm. These results
contrast with behavior found in other metal-metal superlattices such as
Nb/Cu [13] and Mo/Ni [14], where a change from positive to negative
temperature dependence occurs as the wavelength is decreased below ~20 Å, and
where the absolute resistivities approach ~150 $\mu\Omega$-cm.

The superconducting transition temperatures of Mo/Ta samples with different layer thicknesses were measured resistively using the standard four-probe technique. Each sample showed a sharp transition with $\Delta T < 0.05$ K. The measured T_c's vs. superlattice wavelength, Λ, are shown in Fig. 2. The monotonic decrease of T_c with Λ is generally well described by the de Gennes-Werthamer proximity effect [17,18] in the region where the layer thickness d is larger or comparable to the coherence length ξ_s of the superconductors. For $d < \xi_s$ the lowering of T_c can be attributed to a smearing of the density of states $N(E)$ and a lowering of $N(E_F)$. This is related to the reduction of the mean free path due to finite-size effects.

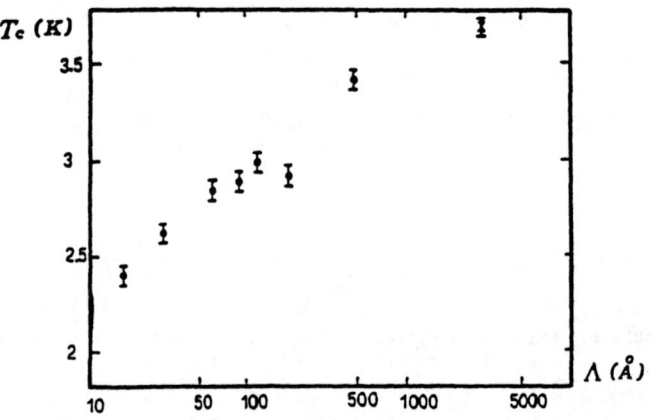

Figure 2. T_c vs. Superlattice Wavelength Λ for Mo/Ta

Some of the Mo/Ta superlattices prepared under inferior sputtering conditions exhibited higher normal state resistivity values and lower T_c's compared to other samples of the same wavelengths. On a plot of T_c vs. ρ_{300} the data for both the "good" and the "bad" samples lay on the same curve. This result is consistent with the hypothesis that resistivity is the single "universal" measure relevant for determining T_c. Similar T_c vs. ρ behavior has been found in other superlattice superconductors such as Nb/Cu [3]. This kind of behavior of T_c with resistivity has been previously found in A15 compounds [19], transition metals, and alloys [20] where the behavior was due to various kinds of disorder, defects or impurities in these materials. It seems that this type of behavior is independent of whether the effects are caused by intrinsic disorder or by layering in the films.

Tunneling is an ideal probe to determine the energy gap $\Delta(0)$, and to investigate the possible existence of quasi-particle states in the gap. All samples used for tunneling measurements were made with a topmost Ta layer of 200 A. Tunnel junctions were made by forming an oxide barrier on this topmost layer in air over several days. A Pb counter-electrode of 5000 A was deposited on top of the oxide layer and etched by photolithography to complete the junction. This procedure resulted in extremely good quality barriers.

Standard tunneling techniques were used to measure dV/dI. For all the superlattices measured, the value of $2\Delta(o)/kT_c$ was consistent with the BCS weak-coupling prediction of 3.52.

SPUTTERED HIGH-T_c THIN FILMS

As mentioned above, the as-deposited Y/Ba_2Cu_3 superlattices were not superconducting, and a post-deposition anneal in flowing O_2 was necessary to make the superconducting oxide. The films are reactive with most substrates during annealing, and optimization of the annealing procedure is still under way to maximize the structural formation and minimize substrate interaction. The best annealing procedure found by us to date consists of two heat treatments: one at 850 °C for 5 min followed by a slow cool, and the second at 920 °C for 1 min followed by a slow cool to room temperature. Both heat treatments are conducted in flowing O_2.

We have deposited high T_c films on single-crystal sapphire, MgO (100) single crystals, and random cut Y stabilized zirconia, all with and without metallic buffer layers to limit substrate interaction. The buffer layers are pre-deposited in a separate sputtering process.

Because of substrate diffusion during annealing, the structure and physical properties of these high T_c films are dependent on the substrate/buffer layer combination, annealing procedure, and film thickness. We find that the properties of the films after annealing are not extremely sensitive to the Y:Ba:Cu stoichiometric ratio as deposited in the Y/Ba_2Cu_3 alloy superlattice. The oxygen content seems to be a more important factor as indicated by x-ray diffraction data and resistivity measurements.

Resistivity measurements were made using a four-probe technique in a closed cycle refrigerator able to reach 7.5 K from room temperature. X-ray diffraction measurements were made on a θ-2θ Bragg-Brentano diffractometer as described above.

Films deposited directly on uncoated sapphire and MgO are found to be non-superconducting above 7.5 K. X-ray diffraction spectra indicate the presence of mostly the semiconductive $YBa_2Cu_3O_{6+x}$ phase [15] with minor indications of the (006) and (005) diffraction peaks of the superconducting phase, $YBa_2Cu_3O_{7-x}$ [15]. These results are consistent with resistivity measurements which find high values at 300 K and semiconductive behavior with temperature. This is shown in Fig 3 by the "o's", which are the resistance values of a film deposited on uncoated sapphire. These measurements were scaled down by a factor of 14 for comparison with the other curve. In almost all of the films deposited on substrates without a buffer layer there is a change in this semiconductive behavior at $T \approx 60$ K, at which point the resistivity drops sharply as T decreases toward zero, indicating the presence of the superconducting phase. This behavior is clearly seen in Fig. 3.

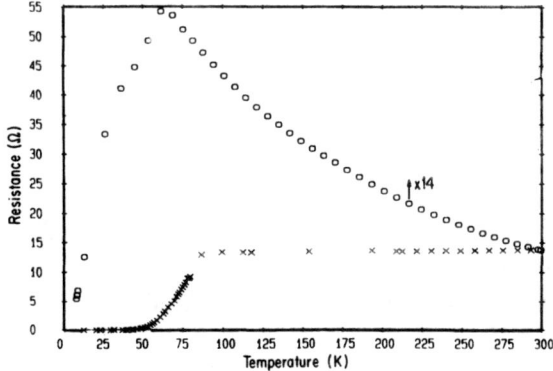

Figure 3. Resistance vs. temperature for $YBa_2Cu_3O_x$ deposited on uncoated sapphire (o's) and sapphire with a Ag buffer layer (x's). The curve for the uncoated sapphire is scaled down by a factor of 14 for comparison.

Our best results to date have been obtained with films deposited on substrates having metallic buffer layers separating the high T_c film from the substrate. The other curve in Fig. 3, represented by the "x's", is the temperature dependence of resistance of a film deposited on sapphire with a Ag buffer layer of approximately 1500 Å. This sample exhibits a T_c onset at ~85 K and zero resistance by ~45 K. X-ray diffraction spectra of this film showed the presence of mostly the superconducting phase with large relative intensities of the (002), (003), (005), and (006) peaks. The high relative intensities of the (00n) peaks indicate a preferred grain growth of the superconducting phase with the c axis perpendicular to the substrate plane.

The highest transition temperature we have obtained so far is with a film deposited on the MgO/Ag combination. It exhibited an onset transition at $T \sim$ 89 K with a T_c midpoint of 75 K and the full transition by 65 K.

SUMMARY

We have shown that dc triode sputtering can be very successfully used in preparing metallic superlattices. X-ray diffraction and RBS are particularly important characterization tools for superlattices. Metallic superconducting superlattices exhibit interesting properties due to the layered structure. Those discussed here were a layer thickness dependence of T_c and a resistivity behavior of T_c resulting from layering effects in Mo/Ta superlattices. In addition, tunneling measurements indicate that these superlattices are BCS weak-coupled superconductors.

Also we have shown that sputtering is useful for producing materials which have multilayering inherent in their structures. We demonstrated this by describing the preparation of $YBa_2Cu_3O_{7-x}$ high T_c films using the superlattice technique. We have found that this method enables the accurate calibration and deposition of the 1:2:3 stoichiometric ratio of the Y, Ba, and Cu. Post-deposition annealing in O_2 produces superconducting thin films with onset transitions as high as 89 K.

ACKNOWLEDGEMENTS

This work was funded by the U.S. DOE under Contract No. DE-FG02-87ER45297. We are grateful to John Cronin, G. P. Rajendran, and Donald Uhlmann for useful discussions and collaborations on related work and to P. Stoss, M. D. Ashbaugh, B. Desfouly-Arjomandy, J. Oder, Z.- M. Yang, and G. van Zijll for assistance with the RBS data acquisition and analysis. The Ion Beam Analysis Facility at the University of Arizona is partially supported by the Air Force Office of Scientific Research through the University Research Initiative Program.

REFERENCES

1. Ruggiero, T. W. Barbee, and M. R. Beasley, Phys. Rev. Lett. <u>45</u>, 1299 (1980).
2. R. Vaglio, A. M. Cucolo, and C. M. Falco, Phys. Lett. A <u>118</u>, 89 (1986).
3. I. Banerjee, Q. S. Yang, C. M. Falco, and I. K. Schuller, Solid State Commun. <u>41</u>, 805 (1982).
4. C. M. Falco and I. K. Schuller, in <u>Superconductivity in d- and f- Band Metals</u>, edited by W. Buckel and W. Weber (Kernforschungszentrum, Karlsruhe, 1982), p. 283.
5. I. Banerjee and I. K. Schuller, J. Low Temp. Phys. <u>54</u>, 501 (1981).
6. See, for example, C. M. Falco and I. K. Schuller, in <u>Synthetic Modulated Structures</u>, edited by L. L. Chang and B. C. Giessen (Academic, New York, 1985), Chap. 9.

7. See, for example, M. Gurvitch and A. T. Fiory, Appl. Phys. Lett. 51, 1027 (1987), or M. Hong, S. H. Liou, J. Kwo, and B. A. Davidson, Appl. Phys. Lett. 51, 694 (1987).
8. C. M. Falco, J. Phys. Colloq. 45, C5-499 (1984).
9. W. R. Bennett, PhD dissertation, University of Arizona, 1985.
10. L. C. McIntyre, Jr., M. D. Ashbaugh, and J. A. Leavitt, to appear in MRS Symposia Proceedings – Materials Modification and Growth Using Ion Beams – Symposium C, 1987 Spring Meeting.
11. I. K. Schuller, Phys. Rev. Lett. 44, 1597 (1980).
12. H. P. Klug and L. E. Alexander, in X-Ray Diffraction Procedures, 2nd ed. (Wiley, New York, 1974), chap. 9.
13. T. R. Werner, I. Banerjee, Q. S. Yang, C. M. Falco, and I. K. Schuller, Phys. Rev. B 26, 2224 (1982).
14. M. R. Khan, C. S. L. Chun, G. P. Felcher, M. Grimsditch, A. Kueny, C. M. Falco, and I. K. Schuller, Phys. Rev. B 27, 7186 (1983).
15. P. K. Gallagher, H. M. O'Bryan, S. A. Sunshine, and D. W. Murphy, Mater. Res. Bull. 22, 995 (1987).
16. J. D. Jorgensen, M. A. Beno, D. G. Hinks, L. Soderholm, K. J. Volin, R. L. Hitterman, J. D. Grace, I. K. Schuller, C. U. Segre, K. Zhang, and M. S. Kleefisch, Phys. Rev. B 36, 3608 (1987).
17. P. G. de Gennes and E. Guyon, Phys. Lett. 3, 168 (1963).
18. J. J. Hauser, H. T. Theuerer, and N. R. Werthamer, Phys. Rev. 136, A637(1964).
19. P. W. Anderson, K. A. Muttalib, and T. V. Ramakrishnan, Phys. Rev. B 28, 117 (1983).
20. C. Camerlingo, P. Scardi, C. Tosello, and R. Vaglio, Phys. Rev. 31, 3121 (1984).

CRITICAL FIELD MEASUREMENTS ON SUPERCONDUCTING GRAPHITE–KHG MULTILAYERS[a]

A. CHAIKEN*, P.M. TEDROW**, and G. DRESSELHAUS**
*Massachusetts Institute of Technology, Cambridge, MA 02139
**Francis Bitter National Magnet Laboratory[b], Cambridge, MA 02139
[a]Supported by AFOSR Contract #F49620–83–C–0011.
[b]Supported by NSF.

ABSTRACT
Upper critical fields of graphite–KHg multilayers with 10Å periodicity were measured as a function of angle and temperature. The $H_{c2}(\theta, T)$ data were compared to the anisotropic Ginzburg–Landau model and were found to be in qualitative agreement, except at the lowest reduced temperatures, where significant deviations are found. The primary deviations from the anisotropic GL model are first, that the values of the critical field at the lowest temperatures are found to be higher than that predicted by the model, and secondly, the critical field anistropy ratio is found to be temperature–dependent. These deviations are discussed in light of more detailed models of anisotropic superconductivity.

INTRODUCTION
The class of layered superconducting materials has assumed increased importance as a result of the recent discovery of the high-T_c perovskite superconductors, which possess highly anisotropic properties.[1] High-T_c perovskites belong to the group of chemically grown multilayers, which includes graphite intercalation compounds (GIC's), and transition metal dichalcogenides and their intercalation compounds. Another new group of layered superconductors could be termed synthetically structured, since they are prepared using sputtering and evaporation techniques.[2] There are several striking similarities in the superconducting properties of these various layered materials, such as T_c–dependence on layer thickness[3] and unusual temperature variation of the upper critical field H_{c2}.[4] An understanding of the nature of these and other anomalies in the superconductivity of one type of layered material, the superconducting GIC's, may give some insight into the properties of the entire class.

The graphite intercalation compounds are particularly interesting to study because of their unique structure, which in the stage n compound consists of n graphite layers in alternation with an intercalant metal unit (here, a trilayer of K and Hg). The structure[5] of C_4KHg, a typical superconducting GIC, is shown in Figure 1.

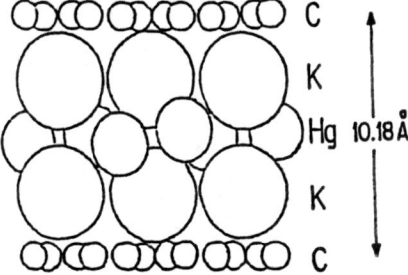

Figure 1: Structure of the stage I KHg–GIC C_4KHg.

The upper critical fields of the synthetically layered superconductors have been among their most widely studied properties, since H_{c2} parallel to the layer planes is strongly affected by the layering.[2] A thorough experimental study of H_{c2} in KHg– and RbHg–GIC's was performed by Iye and Tanuma,[6] who interpreted their data in terms of an anisotropic version of the standard Ginzburg–Landau theory.[2] This model describes the critical field anisotropy in terms of a single parameter

$$\epsilon \equiv \frac{H_{c2,\|\hat{c}}}{H_{c2,\perp\hat{c}}} = \left[\frac{m_a}{m_c}\right]^{1/2} \tag{1}$$

which, since ϵ is related to the effective masses, should be temperature–independent. The fundamental assumption of this model is that the Cooper pairs have different coherence lengths parallel and perpendicular to the layer planes, and that these coherence lengths have the temperature–independent value $\epsilon = \xi_c/\xi_a$. In the GL model, $H_{c2}(T)$ is linear near T_c and should show critical field saturation at low reduced temperatures, as described by the Werthamer, Helfand, and Hohenberg (WHH) model that is appropriate for isotropic type II superconductors.[7] Iye and Tanuma found a linear $H_{c2}(T)$ for the stage I compounds C_4KHg and C_4RbHg, and for the stage II compound (two layers of carbon per intercalant trilayer) C_8KHg, but saw no sign of the low–temperature saturation expected in the WHH model. They furthermore found strong evidence for a positive curvature ($d^2H_{c2}/dT^2 > 0$) of the critical fields in the stage II compound C_8RbHg. Iye and Tanuma found $T_c \approx 0.7$ K for C_4KHg$_x$.[6] Here the H_{c2} measurements are extended to C_4KHg samples with higher T_c's (≈ 1.5 K) and thus to lower reduced temperatures.

EXPERIMENTAL DETAILS

The samples used in these studies were prepared with standard techniques.[5,8] They were characterized with x-ray diffraction, which showed single–phased super-lattice peaks and a layering periodicity of about 10.2Å, in agreement with earlier work.[6] Chemical analysis showed that a typical $T_c = 1.5$ K sample had the composition C_4K$_{0.94}$Hg$_{0.98}$.

The superconducting transitions were measured with an ac inductance bridge using a frequency about 500 Hz and a modulation field amplitude of about 0.03 gauss (applied in the layer planes). The temperature was monitored using the ^{3}He vapor pressure and a Ge resistance thermometer. T_c was defined as the temperature where the inductance had dropped to 90% of its value in the normal state. Zero-field transition widths were on the order of 50 mK. H_{c2} was defined as the intersection of a line drawn tangent to the transition and the level upper portion of the sweep, just as in Ref. [6]. Other definitions of H_{c2} were tried in analyzing the data; while they slightly changed the results quantitatively, they had no effect on the *shape* of any of the curves described below.

RESULTS

Figure 2 shows typical H_{c2} versus reduced temperature (T/T_c) results for a C_4KHg sample with $T_c \approx 1.5$ K. The two sets of data correspond to the critical fields with $\vec{H}$ applied perpendicular to the layer planes ($\| \hat{c}$) (Fig. 2a and 2b) and parallel to them ($\perp \hat{c}$) (Fig. 2a). Displayed with each data set are two fits, one the best line fit, and the other corresponding to the best fit using the WHH theory. The linear fit is clearly better than the WHH fit for both field orientations, an effect not predicted by the anisotropic Ginzburg–Landau model. The lack of saturation in $H_{c2}(T)$ is qualitatively similar to that seen in the other layered superconductors, such as the transition metal dichalcogenides[4], and also seen in all the other superconducting GIC's[6].

Another way to compare the data for different materials is to look at the values of

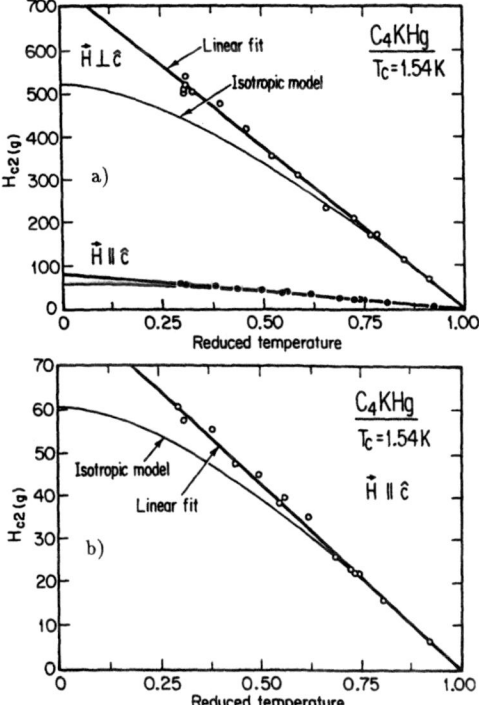

Figure 2(a) Upper critical fields for C$_4$KHg with the applied field parallel and perpendicular to the graphite c-axis. For $\vec{H} \perp \hat{c}$, two fits are shown: a linear fit, with $\chi^2 = 3.3e - 3$, and a WHH theory (isotropic model) fit, with $\chi^2 = 1.6e - 2$. (b) More detailed look at the $\vec{H} \parallel \hat{c}$ data shown in a). For the linear fit, $\chi^2 = 1.3e - 3$, and for the WHH theory fit, $\chi^2 = 1.2e - 2$.

the reduced critical field, h*, which is defined by:

$$h^*(T) \equiv \frac{H_{c2}(T)}{\left(\frac{dH_{c2}}{dT}\big|_{T_c}\right) * T_c} \tag{2}$$

The WHH theory[7] predicts that h*(0) should be about 0.69 for an isotropic type II superconductor, compared to the h*(0) = 1 that extrapolation of the simple linear temperature dependence would give. A plot of h* versus t for four different samples for both orientations of the applied field, shows that at the lowest reduced temperature obtained (t ≈ 0.3), all samples have already reached h* ≈ 0.7, the usual zero-temperature value.

So far the discussion has centered on the temperature dependence of the critical field at constant angle. For the anisotropic superconductors, it is also important to consider the angular dependence of the critical field at constant temperature. For the case of an anisotropic Ginzburg–Landau superconductor with uniaxial symmetry, this dependence is of the form:

$$H_{c2}(\theta) = \frac{H_{c2}(0)}{\sqrt{(\epsilon^2 \sin^2 \theta + \cos^2 \theta)}} \tag{3}$$

Iye and Tanuma[6] reported good fits to this equation for their $T_c \approx 0.7$ K samples. The new data for the $T_c = 1.5$ K samples also fit this equation well, although the calculated curves tend to underestimate the value of H_{c2} near $\theta = 0°$slightly.

The most interesting aspect of $H_{c2}(\theta)$ is its change as the temperature is varied. In the context of the anisotropic GL model, the quantity $H_{c2}(\theta)/H_{c2}(0)$ should be dependent only on θ and ϵ. If the parameter ϵ of the model is truly temperature–independent, then plots of $H_{c2}(\theta)/H_{c2}(0)$ at different reduced temperatures should all lie on top of one another. Examination of Figure 3 shows that this is indeed the case for the two higher reduced temperature curves, in good agreement with the theory. However, the peak of the curve for the lowest reduced temperature of 0.30 ($1/\epsilon = (9.5 \pm 0.5)$) clearly lies above the other two traces ($1/\epsilon = (7.6 \pm 0.5)$ at t = 0.57 and 0.76). The temperature dependence of ϵ has been observed in H_{c2} data of several C_4KHg samples, and in addition was seen in C_8KHg data by Iye and Tanuma[6] ($1/\epsilon = 17.6$ at t = 0.81 and $1/\epsilon = 21.6$ at t = 0.23). Temperature dependence of the anisotropy is also a common feature in the other layered superconductors.[3,9]

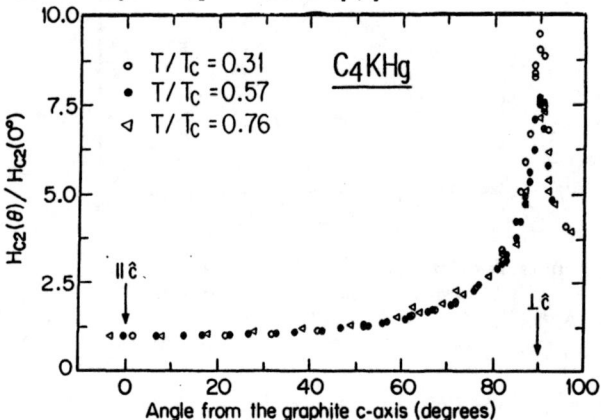

Figure 3: $H_{c2}(\theta)/H_{c2}(0)$ for a $T_c = 1.5$ K sample at reduced temperatures of 0.31, 0.56, and 0.67. The two higher temperature curves give $1/\epsilon = (7.6 \pm 0.5)$, and the lower temperature curve gives $1/\epsilon = (9.5 \pm 0.5)$.

The results that the anisotropy parameter is temperature–dependent and that $H_{c2}(T)$ is linear for both orientations of the field seem inconsistent, since linear temperature dependences in both directions should give a temperature–independent ϵ. However, the angular dependence of H_{c2} is a much more accurate way of determining the anisotropy than the $H_{c2}(T)$ experiment. In addition, the amount of deviation from linearity in $H_{c2}(T)$ that would be expected from the temperature–dependent anisotropy numbers is smaller than the size of the error bars in the field measurement. Therefore it appears most likely that there is a low-temperature upturn in the $H_{c2,\perp\hat{c}}$ data or downturn in the $H_{c2,\|\hat{c}}$ data which is obscured by random errors in the $H_{c2}(T)$ measurement. This interpretation seems consistent with the C_8KHg results of Iye and Tanuma[6], who reported linear $H_{c2}(T)$ for both field orientations, just as is seen here for C_4KHg, and yet also saw a temperature–dependent anisotropy ratio, as was discussed above.

DISCUSSION AND CONCLUSIONS

The experimental evidence described above suggests that the anisotropic GL model is in overall good agreement with the data. Yet it is inadequate to describe the detailed low temperature behavior of C_4KHg. The observation of small deviations from this model should not be surprising given the previous observation of positive curvature in $H_{c2}(T)$ for C_8RbHg by Iye and Tanuma, and the report of positive curvature and temperature–dependent anisotropy in C_8KHg by Pendrys. et al. [13] The failure of the anisotropic GL model to predict all details of the observed phenomena is not a reflection on its appropriateness for the system at hand, but rather a consequence of the fact that one parameter is not sufficient to describe all the effects of the anisotropic crystal and band structure of the layered superconductors.

Therefore the observations warrant consideration of more sophisticated microscopic models of $H_{c2}(\theta, T)$ in layered superconductors. There are quite a number of detailed calculations available, including some that have produced excellent fits to data for the synthetic superlattices.[14,15,16] One model that predicts linear or superlinear temperature dependence of H_{c2} is the dimensional crossover theory of Klemm, Luther, and Beasley[16]. This theory predicts anomalies in $H_{c2}(T)$ only when the c–axis coherence length ξ_c is on the order of the layer spacing, though. In C_4KHg ξ_c is about 200 Å, twenty times the layer spacing. Even in the stage II GIC's, ξ_c is about 4 times the layer spacing, so this theory would not predict any deviations from the anisotropic GL model for GIC's.

Other authors[14,15] have taken a different approach to fit superlinear $H_{c2}(T)$ data for multilayers in the three-dimensional regime. They developed a model which conceptually divides the superconductor into superconducting and normal layers, and assigns different microscopic parameters (T_c, Fermi velocity, and mean free path) to them. While this model seems very successful for proximity-coupled superlattices with layer thicknesses on the order of 100 Å, one hesitates to apply it to a system like the GIC's which have a multilayer repeat distance on the order of 10 Å (see Fig. 1), since the justification for the assignment of different microscopic parameters to such thin layers would seem to be tenuous, especially when the mean free path along the $\hat{c}$ direction is greater than or comparable to the layer thickness, as is the case for C_4KHg. In addition, models developed for the synthetic superlattices predict that a superconductor/normal metal superlattice must have a lower T_c than the bulk superconductor, while C_4KHg and C_8KHg actually has a higher transition temperature than the KHg alloy ($T_c = 0.9K$).

In order to determine which model of multilayer superconductivity is most appropriate for the case at hand, one needs to consider what is already known about the properties of GIC's. Extensive studies of stage I alkali metal GIC's[17] closely related to C_4KHg have shown that their Fermi surface is well described by a zone-center piece of 3-D character, and a zone-boundary piece of graphite π–character. This information, if extended to the similar alkali metal–mercury GIC's, suggests consideration of models for superconductivity which emphasize Fermi–surface anisotropy and/or multiple–band effects. These models can be fit to data from the transition metal dichalcogenides, which have $H_{c2}(T, \theta)$ behavior similar to the GIC superconductors.[9]

One such model is that of Entel and Peter[18], who have explicitly derived an $H_{c2}(T)$ that is linear to low reduced temperatures by assigning different diffusivities and electron–phonon couplings to different bands. This parameterization seems more appropriate for the thin–layer limit than that used in the synthetic superlattice models.[14,15] Youngner and Klemm[19] considered both Fermi–surface anisotropy and energy gap anisotropy in a calculation of anomalous $H_{c2}(T)$ behavior. Toyota *et al.* [9] found that by including both types of anisotropy they could fit both their $H_{c2}(\theta)$ and temperature-dependent anisotropy data on $NbSe_2$. It is not yet possible to decide which of these two microscopic models would be more appropriate for the graphite/KHg multilayers. The similarities between the anisotropic–gap/multiband models and the C_4KHg data are suggestive and strongly motivate more experiments.

Finally, it should be noted that the ideas discussed above in relation to GIC's, with their small layer thicknesses, may well be applicable to superconducting multilayers with 10 Å-scale layers. To date, most multilayers that have been synthesized with such small layer thicknesses have either shown a dimensionality crossover or been interdiffused.[2] Therefore, the unusual behavior that has been observed so far in the critical fields of the synthetic multilayers has a different cause than the anomalies described here for the superconducting GIC's. However, should it be possible in the future to synthesize superconducting superlattices with small layer thicknesses and no dimensional crossover, it would seem that they too should be describable by the more microscopic models of $H_{c2}(T, \theta)$ discussed above.

Acknowledgements

The authors would like to thank M.S. Dresselhaus and T.P. Orlando for useful advice, and J.E. Tkaczyk and J.A.X. Alexander for providing the algorithms on which the WHH fitting program was based. The authors A.C. and G.D. gratefully acknowledge support from AFOSR grant F49620-83-C-0011.

1. A. Ourmazd, J.A. Rentshler, J.C.H. Spence, M. O'Keeffe, R.J. Graham, D.W. Johnson Jr., and W.W. Rhodes, Nature 327, 308 (1987).

2. S.T. Ruggiero and M.R. Beasley, in Synthetic Modulated Structures, edited by L.L. Chang and B.C. Giessen (Academic Press, New York, 1985), p. 365.

3. I. Banerjee and I.K. Schuller, J. Low Temp. Phys. 54, 501 (1984).

4. J.A. Woollam, R.B. Somoano, and P. O'Connor, Phys. Rev. Lett. 32, 712 (1974).

5. P. Lagrange, M. El-Makrini, D. Guérard, and A. Hérold, Synthetic Met. 2, 191 (1980).

6. Y. Iye and S. Tanuma, Phys. Rev. B25, 4583 (1982).

7. T.P. Orlando, E.J. McNiff, S. Foner, and M.R. Beasley, Phys. Rev. B19, 4545 (1979).

8. G. Roth, A. Chaiken, T. Enoki, N.C. Yeh, G. Dresselhaus and P. M. Tedrow, Phys. Rev. B32, 533 (1985).

9. N. Toyota, H. Nakatsuji, A. Hoshi, N. Kobayashi, Y. Muto, and Y. Onodera, J. Low Temp. Phys. 25, 485 (1976).

10. L.A. Pendrys, R.A. Wachnik, F.L. Vogel, and P. Lagrange, Synth. Met. 5, 277 (1980).

11. K.R. Biagi, V.G. Kogan, and J.R. Clem, Phys. Rev. B32, 7165 (1985).

12. S. Takahashi and M. Tachiki, Phys. Rev. B33, 4620 (1986).

13. R. A. Klemm, A. Luther, and M. R. Beasley, Phys. Rev. B12, 877 (1975).

14. H. Kamimura, Ann. de Phys. 11, Suppl. 2, 39 (1986).

15. P. Entel and M. Peter, J. Low Temp. Phys. 22, 613 (1976).

16. D.W. Youngner and R.A. Klemm, Phys. Rev. B21, 3890 (1980).

CHARACTERIZATION OF MULTILAYERS AS X-RAY DISPERSION DEVICES

JOHN V. GILFRICH*, DENNIS B. BROWN, DAVID L. ROSEN*# AND RALPH K. FREITAG
Condensed Matter Physics Branch, Naval Research Laboratory, Washington, DC
20375-5000

ABSTRACT

The appropriate application of multilayers as x-ray dispersion devices requires that their diffraction characteristics be understood. Conventional models, based on perfect-crystal and mosaic-crystal theories, predict diffraction efficiencies (integral reflection coefficients) significantly larger than values measured experimentally. It has been shown that introduction of surface roughness effects into the model can promote agreement between experimental and theoretical values, while the presence of other types of defects produce changes too small in magnitude to explain the discrepancy. Because it is reasonably well agreed that the resolving power of multilayers is only moderate, compared to the more conventional "crystal" dispersing devices, it is important to be able to predict or measure that parameter in order to assess the usefulness for a particular application. Experimental measurements and theoretical calculations have been carried out on multilayers (almost exclusively tungsten/carbon) prepared to have 2d-spacings from 50 to 140Å. The experimental work used both single-crystal and double-crystal spectrometers; the calculations used the crystal diffraction model, as modified to include surface roughness.

INTRODUCTION

The goal of an effort to study the behavior of multilayers as x-ray dispersion devices is to understand the dependence of their performance on wavelength, number of layers and their spacing, substrate and interface roughness, and the order of diffraction. The effects of these parameters are being investigated both experimentally and theoretically. The experimental effort consists of measuring integral reflection coefficients (R-values) using both single- and double-crystal spectrometers, and the rocking curve widths using the double-crystal instrument. The theoretical work involves the calculation of these same values using conventionally accepted models of crystal diffraction theory. Previous work [1] has demonstrated that a model for the influence of substrate roughness on the integral reflection coefficient can produce acceptable agreement between measurement and calculation for higher orders of diffraction as well as first order. It was further shown [1,2] that other forms of defects in multilayers (e.g. thickness variability) produced effects on the integral reflection coefficient which were either zero or were too small to explain the experimental results which were measured. Rocking curve widths, measured on those same samples [2], showed marked differences between the two samples studied. Extension of the work to a second set of samples was inaugurated to shed further light on the problem.

* Also at Sachs/Freeman Associates, 1401 McCormick Drive, Landover, MD 20785.
# Present Address: New Mexico State University, Las Cruces, NM 88001.

THEORETICAL

Previous work [1,2] has developed both a simulation method and an analytical solution for treating the effect of substrate surface roughness on the integral reflection coefficient. The results for the analytical solution will be summarized briefly. It will be assumed, for the relatively long wavelengths used in the present paper, that the dependence of the integral reflection coefficient on the structure factor can be obtained from kinematical theory as

$$R = c \ |F|^2 \tag{1}$$

where R is the integral reflection coefficient, F is the structure factor, and c is a constant of proportionality. Further, for (00ℓ) reflections

$$F = \sum_m e^{i \ 2\pi \ \ell \ w_m} \sum_n f_n \ e^{i \ 2\pi \ \ell \ z_n} \tag{2}$$

where z_n is the normalized position of the nth atom within the unit cell, f_n is the atomic scattering factor for the nth atom, and w_m is the position representing the displacement of the mth unit cell due to surface roughness. Both position components are in the direction normal to the surface plane. The first summation in Eqn. 2 represents a correction for surface roughness, while the second summation is the standard structure factor for a lattice without defects.

It is possible to replace the first summation in Eqn. 2 by an integration using a probability distribution function $P(w')$, where $w' = 2\pi w$. That is, $P(w')$ is the probability of finding unit cells with a displacement due to surface roughness given by the value w'. It has been shown by Rosen et al. [1] and by Gilfrich et al. [2] that a Gaussian probability distribution for surface displacements is not in agreement with their experimental data. Further, it has been shown by those authors that a good first approximation to their experimental data can be obtained using a rectangular probability distribution, viz.

$$P(w') = \frac{1}{A} \frac{1}{2 \ \pi}, \text{ for } -\pi A < w' < +\pi A \tag{3a}$$

$$P(w') = 0, \text{ for } w' < -\pi A \text{ and } w' > +\pi A \tag{3b}$$

where A is the parameter which specifies the width of the distribution. Using this distribution, the first summation of Eqn. 2 can be rewritten as

$$G_\ell = \frac{1}{\pi A \ell} \sin(\pi A \ell) \tag{4}$$

where G_ℓ is a correction for substrate surface roughness [1,2].

EXPERIMENTAL

The primary experimental effort was conducted using a single-crystal spectrometer to measure the R-value of the multilayers over a range of wavelengths, using techniques described previously [3]. As described before [1], the roughness correction can accommodate the difference between experimental first order R-values and a "perfect-crystal" calculation for two multilayers, W-C with a 2d-value of 76Å and 150 layers, and V-C with a 2d-value of 80Å and 200 layers, over the wavelength range of 8 to 15Å. More importantly, it shows fair agreement for five and four orders of diffraction, respectively, at Al K_α (8.34Å). Figure 1 is an example of the manner by which the correction improves the agreement between the measurements and the calculations in

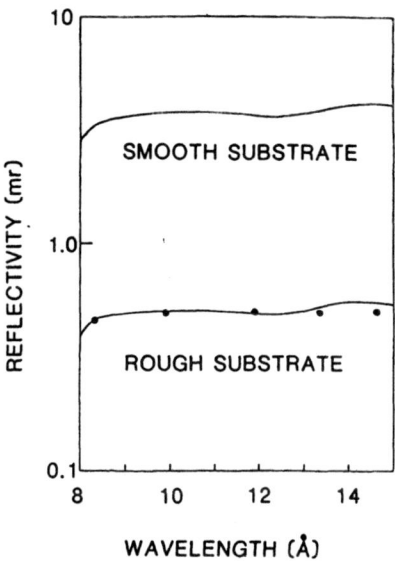

Fig. 1. Experimental data for first
order R-values of the W-C multilayer
2d = 76Å) compared to theory, assuming
a smooth substrate (RMS roughness = 0)
and a rough substrate (RMS roughness =
7.4Å). Solid lines represent the
calculations for the kinematic theory;
the circles represent experimental
points. (From Reference 1)
Only the kinematic calculation is shown
here; in this wavelength range it and
the perfect crystal (with absorption)
theory are approximately the same.

first order for the W-C multilayer. Double-crystal measurements of the two
multilayers, illustrated in Figure 2, showed a remarkable difference in the
rocking curve widths for the two samples, explaining, at least in part, the
similarity in the magnitude of the R-values, contrary to the expected
difference due to the difference in atomic number between vanadium and
tungsten. The FWHM of the rocking curve, after correcting for the effects of
the first crystal, for the W-C was 17 minutes of arc, while for the V-C it
was 138 minutes of arc.

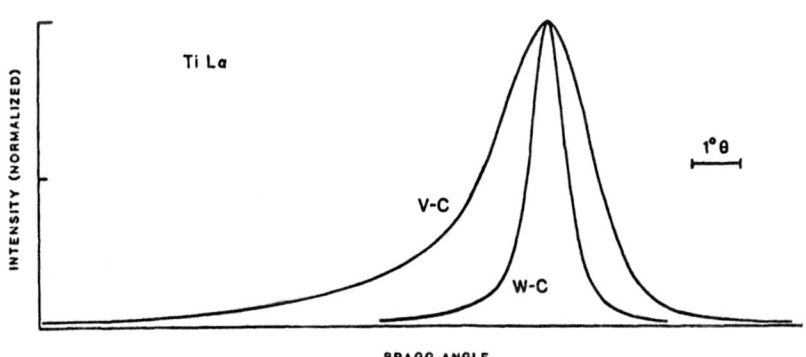

Figure 2. Double-crystal rocking curves for two multilayers.

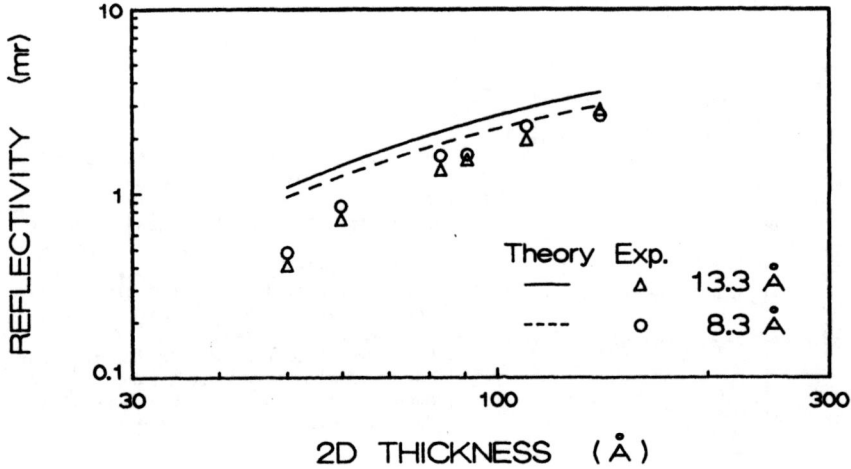

Fig. 3. Measured and calculated R-values for six W-C multilayers.

The second set of multilayers studied consisted of six samples of W-C, having 2d-values of 50, 60, 83, 91, 110 and 140Å, with 75, 40, 125, 100, 75 and 75 layers, respectively. Single-crystal R-value measurements were made at Al K_α (8.34Å) and Cu L_α (13.3Å), and initially compared to Henke's calculations [4]. Those calculations, however, were based on a tungsten layer to total bilayer thickness of 0.4. Since the thickness ratio of these samples was somewhat lower than that, new calculations were made using better values [1]. Figure 3 shows that comparison; the theoretical curve is almost straight while the experimental data is quite curved. The curvature is to be expected if the multilayer is other than perfectly smooth because any partic-

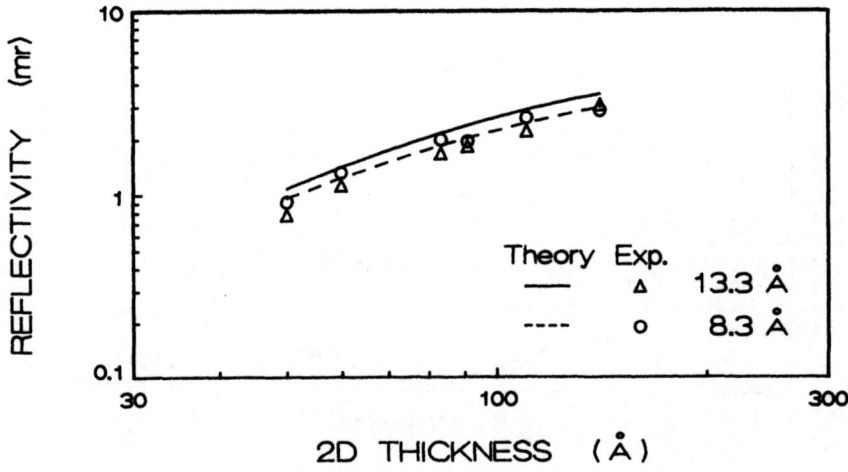

Figure 4. Calculated R-values compared to roughness-corrected measurements.

ular value of roughness is a larger fraction of the lower 2d-values than the higher ones. By applying a correction assuming an RMS roughness of 3.1Å, the comparison becomes as is shown in Figure 4. The agreement is excellent for the 8.3Å (AlK_α) data; for Cu L_α the agreement is somewhat poorer, the experimental points being about 20% below the calculation. However, the roughness correction does make the relationship match the slight curvature of the theoretical curve for both wavelengths.

Spiller and Rosenbluth [5] have presented calculations (see especially their Fig. 4) which show that a roughness correction factor may be smaller for long wavelengths (they used 66.3Å) as compared to short wavelengths (they used 1.54Å). Although the Spiller and Rosenbluth calculations are not directly comparable with the present work (they treated peak diffracted amplitude while the present work treats integral diffracted intensity), nevertheless a brief comment is appropriate. The Spiller and Rosenbluth result is analogous to a standard result from crystal diffraction theory [6]. This theory states that a correction for the thermal motion of atoms to the formulae for the integral reflection coefficient takes a different form in the kinematical limit (absorption dominates with negligible extinction) and the perfect crystal limit (extinction dominates with negligible absorption). Specifically, the thermal correction takes the form exp(-2M) in the kinematical limit and the form exp(-M) in the perfect crystal limit. It will be observed that the data in Fig. 4 of Spiller and Rosenbluth are approximately related as a power of two. In the present work, for comparison, both of the wavelengths used are sufficiently long that the kinematical approximation is approximately correct. As a result, the calculated integral reflection coefficient is reduced by multiplying by $G_\ell{}^2$.

CONCLUSION

It has been demonstrated that a roughness correction can accommodate the difference observed between theoretical calculations and measured integral reflection coefficients for multilayers being used as x-ray dispersion devices. Double crystal measurements, in order to determine the rocking-curve width, are underway to shed further light on the diffraction process by comparison to the theoretical resolving power.

ACKNOWLEDGMENT

The set of six W-C multilayers of varying d-spacing were graciously provided by Ovonics, Energy Conversion Devices, Troy, MI. We thank them for their collaboration.

REFERENCES

1. D.L. Rosen, D. Brown, J. Gilfrich and P. Burkhalter, "Multilayer Roughness evaluated by X-Ray Reflectivity," accepted for publication by the Journal of Applied Crystallography.
2. J.V. Gilfrich, D.B. Brown and D.L. Rosen, SPIE Proceedings 688, 115 (1986).
3. J.V. Gilfrich, D.J. Nagel and T.W. Barbee, Jr., Appl. Spectrosc. 36, 58 (1982).
4. B.L. Henke, P. Lee, T.J. Tanaka, R.L. Shimabukuro and B.K. Fujikawa, Atomic Data and Nucl. Data Tables 27, 1-144 (1982).
5. E. Spiller and A.E. Rosenbluth, SPIE Proceedings 563, 221 (1985).
6. See, for example, R. W. James, "The Optical Principals of the Diffraction of X-Rays" (G. Bell and Sons, London, 1958), pp. 50-51, 59-60, and 227.

COMBINED MICROSTRUCTURE X-RAY OPTICS; MULTILAYER DIFFRACTION GRATINGS

Troy W. Barbee, Jr.
Lawrence Livermore National Laboratory, Livermore, CA 94550

ABSTRACT

Multilayers are man-made microstructures engineered to vary in depth that are now of sufficient quality to be used as x-ray, soft x-ray and extreme ultraviolet optics. Gratings are in-plane man-made microstructures which have been used as optic elements for most of this century. Joining of these two optical elements to form combined microstructure optics has the potential for greatly enhancing both the resolution and the throughput attainable in these spectral ranges. Experimental results for multilayer gratings are presented and discussed. It will be demonstrated that multilayer diffraction gratings act as x-ray prisms and are high efficiency dispersion elements.

INTRODUCTION

Multilayer structures, such as interference filters and reflection or antireflection coatings were developed [1] more than 50 years ago for the visible, infrared and ultraviolet portions of the light spectrum. However, multilayer structures with sufficient quality for advanced optical research in the x-ray region [2,3,4] where wavelengths are 10 to 10000 times shorter than those in the visible have only been fabricated in the last decade. This capability now provides us with the challenge to design new reflective optical elements that can focus, disperse or otherwise manipulate x-rays, soft (lower energy x-rays) and extreme ultraviolet light.

In this paper the properties of an advanced multilayer optical element, a multilayer diffraction, grating - are presented and discussed. Diffraction gratings, which are in plane microstructures, were developed in the early 1900's and are a standard dispersive element used in moderate to high resolution applications. I note here that the in-plane scale of diffraction gratings ranges from approximately 300 nm to 5000 nm. This is to be contrasted with the in-depth micro-structural scale typical of multilayers which ranges from 1.0 nm to 40 nm. The convolution of the properties of these two periodic structures and the properties which result are both unique and technologically useful. Of particular interest is the potential for very high resolution - high efficiency optics operating at near normal angles of incidence in the soft x-ray and extreme ultraviolet spectral ranges.

Multilayer Diffraction Gratings

Multilayer diffraction gratings are combined microstructure devices as shown in the schematic cross-section in Fig. 1. The in-plane or grating structure has a period d_g^o and consists of flat topped bars of width $d_g^o/2$. This structure is a simple laminar amplitude grating as light is only diffracted from the tops of the grating bars by the multilayers. The multilayer structure is periodic in depth with a period d_o as shown. Light incident at grazing angle θ and is diffracted by the multilayer at an exit angle θ in zeroeth order. Interference of light diffracted by the multilayer on each grating bar results in intensity maxima (grating dispersed multilayer diffracted light) at angles $m\phi$ relative to zeroeth order where m (=±1, ±2, ±3 ----) are the grating orders. In the following the relationship between the dispersion angle ϕ, the structural parameters d_g and d_o, the angle of incidence θ, and the composition of the multilayer is developed. A short discussion of Bragg diffraction by the multilayer is presented first.

Multilayers reflect light by a diffraction process that can be exactly modeled using optical multilayer codes [1,3,5] and can also be described by Bragg's equation [6] as normally applied to x-ray diffraction by crystalline materials. Simply stated Bragg's equation analytically relates the wavelength of incident light λ and the angles θ at which constructive interference of light scattered by the periodic multilayer structure of period d_o occurs. In its simplest form it is

$$n \lambda = 2d_o \sin \theta \tag{1}$$

where n is the order of Bragg reflection. For a multilayer the period d_o is the sum of the thicknesses t_A and t_B of layers A and B respectively in the multilayer. It is necessary to modify Eq. 1 to account for refraction resulting from the interaction of light of wavelength λ and the multilayer component elements (i.e. A, B).

The x-ray refractive [6], η, is given as

$$\eta = 1 - \delta - i \beta \tag{2}$$

where δ is a term related to scattering and β a term related to absorption. n is always less than unity except in the vicinity of the low energy absorption edges of a number of materials. Both δ and β are wavelength dependent and for an element A are given by

$$\delta_A = \frac{\lambda^2}{2\pi} N_A r_e f_{A1} \tag{3}$$

and

$$\beta_A = \frac{\lambda^2}{2\pi} N_A r_e f_{A2} = \frac{\lambda}{4\pi} \mu_A \tag{4}$$

where μ_A is the linear absorption coefficient of A, N_A is the atom density of A per unit volume, r_e the classical radius of an electron and f_{A1} and f_{A2} the number of electrons per atom [7] active in scattering or absorption respectively. For the case of Bragg diffraction it is generally only necessary to consider the effects of the scattering term δ. By use of Snells Law it can be shown that the refraction corrected form of Braggs equation is

$$n \lambda = 2d_o \sin\theta \left[1 - \frac{2\delta - \delta^2}{\sin^2 \theta} \right]^{1/2} \tag{5}$$

Since multilayers consist of layers of materials A and B of thicknesses t_A and t_B respectively we calculate a value of δ for a multilayer period to be

$$\bar{\delta} = \frac{t_A}{d_o} \delta_A + \frac{t_B}{d_o} \delta_B \tag{6}$$

which is used in Eq. 5 when applied to multilayers. This refraction corrected form of Bragg's equation is that used in the following analysis of the dispersion of multilayer gratings.

The response of gratings is described in scalar theory [8] by the generalized grating equation:

$$m \lambda = d_g^0 (Sin \alpha + Sin \beta) \qquad (7)$$

where α and β are angles relative to the normal to the grating surface as defined in Fig. 1 and m the grating order of the dispersed light. Noting that $Sin\beta = Sin[-(\alpha-\phi)]$ for $m = \pm 1$ Eq. 7 may be rearranged to give

$$Sin \phi = \frac{m\lambda}{d_g^0 Sin\theta} \qquad (8)$$

This equation may also be derived by analogy of the reflecting multilayer grating to a transmission grating by assuming the period of the transmission grating is $d_g^0 Sin\theta$.

The properties of the multilayer are now included by noting that the refraction corrected form of Braggs equation (Eq. 5) can be rearranged to give

$$\frac{\lambda}{Sin\theta} = \frac{2d_0}{n} \left[1 - \frac{2\bar{\delta} - \bar{\delta}^2}{Sin^2\theta} \right]^{1/2} \qquad (9)$$

which when substituted in Eq. 8 gives

$$Sin \phi = \frac{2md_0}{nd_g^0} \left[1 - \frac{2\bar{\delta} - \bar{\delta}^2}{Sin^2\theta} \right]^{1/2} \qquad (10)$$

This relationship between ϕ, d_0, d_g^0, δ and θ has several unique characteristics.

First, the dispersion angle ϕ is essentially constant; the only angular or wavelength dependence being contained in the refraction correction term. The refraction correction term is typically less than 0.1 and is often nearly constant over broad spectral ranges. This property allows these multilayer gratings to be applied in essentially the same manner as simple multilayers. Second, the multilayer grating separates the Bragg orders diffracted by the multilayer, the dispersion angle, ϕ, being inversely proportional to the Bragg order n. Third, since the dispersion angle ϕ is independent of wavelength multilayer diffraction gratings are constant resolution dispersion elements for a given instrument geometry. Fourth, ϕ (independent of refraction) is determined for m=n=1 by the ratio of the multilayer period (d_0) to the grating period (d_g). All these effects have been experimentally observed and are discussed in the following.

Experimental Results

These multilayer diffraction grating structures have been experimentally studied at the Stanford Synchrotron Radiation Laboratory on Beamline III-4, [9] a differentially pumped line gathering 0.6 mr of synchrotron light from a bending magnet. The beamline consists of the bending magnet source, a gold coated collimating mirror giving a high energy cutoff of approximately 3.5 keV, a differential pumping system allowing the experimental chamber to be at pressures up to 10^{-5} torr, a single or double multilayer dispersion element monochromator, and an adjustable slit scanning detector. Facilities for introduction of incident spectrum limiting filters, transmission absorp-

310

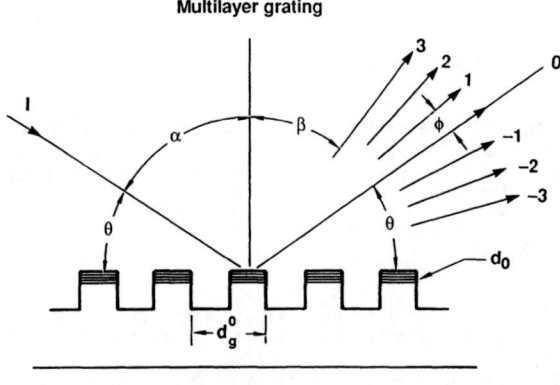

Multilayer grating

θ — Bragg angle
0 — Bragg diffracted beam
±1, ±2, ±3 — Grating dispersed
Bragg light

Variables — θ, d_0, d_g^0, α, β, ϕ

Fig. 1   A schematic cross-section of a multilayer
diffraction grating is shown.

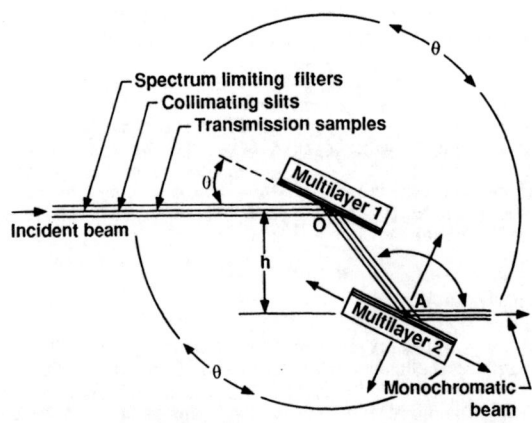

Fig. 2   A schematic of the two element monochromator
used in this work is shown.

tion cross-section samples, and transmission gratings for independent spectral calibration of the light transmitted by the multilayer optics are also available.

The monochromator is shown schematically in Fig. 2. Multilayer 1 is mounted on a large externally driven rotary table with its front surface on the center of rotation at point 0. In the two dispersion element mode multilayer 2 is mounted on the same rotating table and has three additional degrees of motion. First, it is mounted on an x-y stage giving linear motions normal and parallel to its front surface. These motions allow the diffracted beam from multilayer 1 to be followed as θ is varied from 2° to 80° while maintaining, h, the spacing of the incident and exit beams constant. The third motion is rocking of multilayer 2 about axis A for alignment and rocking curve measurements. A second rotating table allows a detector to be mounted in the monochromator chamber. When multilayer 2 is moved out of the beam diffracted by multilayer 1 this detector is used to scan the dispersed beam or to measure the rocking curve of multilayer 1 as θ is varied. Another detector is mounted in a chamber down beam of the monochromator. This detector may be scanned both vertically and horizontally and is typically slitted to a 100 μm aperture.

Three types of experiments have been performed with multilayer gratings. First, the grating is mounted in the multilayer 1 position. Data is taken by scanning the grating in θ (grating scan) with the detector fixed or by scanning the detector while holding θ constant (detector scan). Second, a two dispersion element configuration is used with the grating mounted in the multilayer 2 position. Again, either the grating is rocked with a fixed detector or the detector is scanned with a fixed grating. Third, an absorption sample is introduced into the incident beam, characteristic absorption edges providing energy calibration.

The dispersed light distribution observed in the two dispersion element configuration in both the detector scan and grating scan modes is shown in Fig. 3. The multilayer grating consisted of a 8.0 nm period rhodium/carbon multilayer containing 40 layer pairs deposited onto a 2000 nm period simple amplitude grating. The angle of incidence, θ, on multilayer 1 was 50 deg; and the energy of the transmitted light 108 eV. Only a limited number of orders (m $\leq$ ± 3) are shown in the detector scan due to its limited range of motion. The grating scan data presented shows m=±7 dispersed orders. At higher detector sensitivities more than m=±15 orders were observed.

The detector scan data is shown in more detail in Fig. 4 with estimates of absolute intensity in the m=o, ±1 orders. Note that the bandpass of this multilayer is ~5% so that the intensities are in a ~ 6 eV energy band. The intensities shown were measured using a gallium arsenide-phosphide Schottky diode. This detector was illuminated with a 2mm wide beam slitted to a 0.1 mm height. The intensities shown should be multiplied by 500 to give an intensity normalized to cm^2 area. Note that the available beam area was 16 mm^2 so that the available photon flux on this beamline was approximately 80 times that shown the Fig. 4.

Also shown in Fig. 4 is a comparison of the uncorrected dispersion angle, ϕ_0, and the observed dispersion angle ϕ. At this energy (~108 eV) the refraction effect is substantial being approximately 9%. Note that it is possible to calculate the average refractive index, δ, for the rhodium/carbon multilayer for 108 eV light from the experimental values of ϕ and θ given d_0, d_g^o, m and n are known. This is the subject of another paper.

Another important characteristic of multilayer diffraction gratings is demonstrated by measurement of the aluminum K edge at 1558 eV using a single multilayer grating mounted in the multilayer 1 position and scanning the dispersed spectrum with a detector θ being fixed (i.e. detector scan). This experiment has been performed using four separate multilayers deposited onto simple laminar gratings (d_g = 2000 nm). Th multilayers used were: Molybdenum/Silicon, d=3.0 nm; Rhodium/carbon, d=3.0 nm; Tungsten/carbon, d=4.4 nm; Rhodium/carbon, d=4.0 nm. In this experiment the angular position

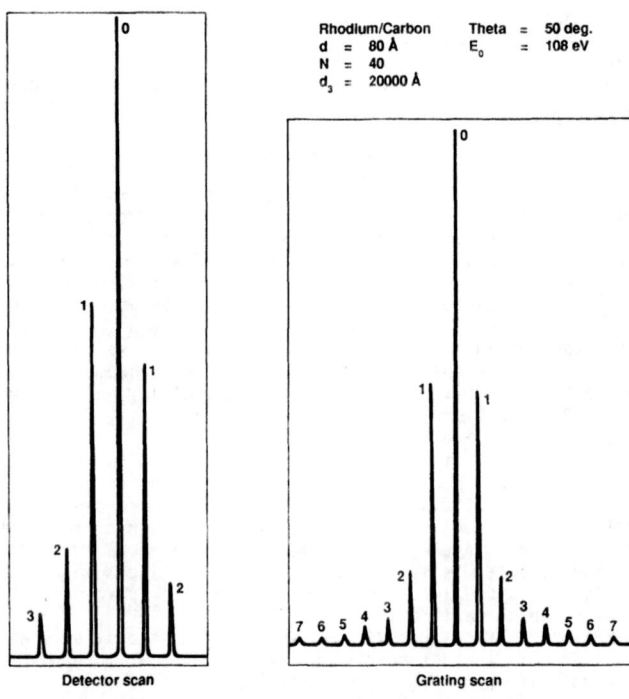

Fig. 3 Experimentally observed dispersion patterns for a multilayer grating obtained by both rocking the grating and scanning the detector are shown.

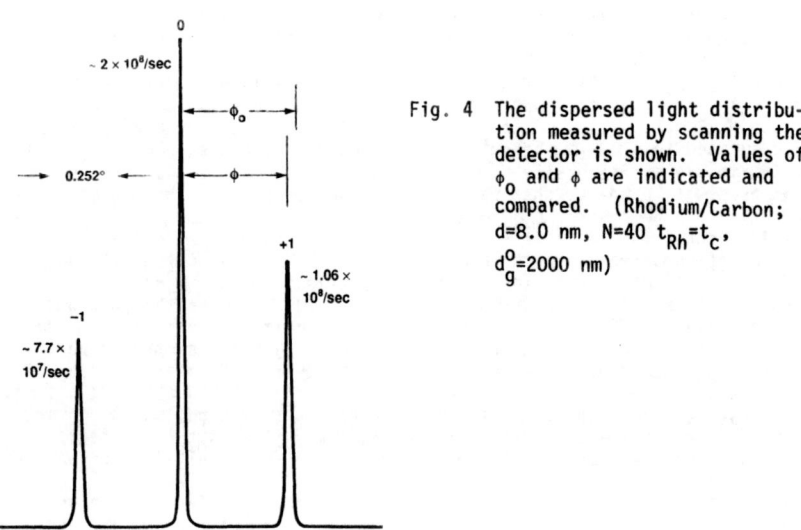

Fig. 4 The dispersed light distribution measured by scanning the detector is shown. Values of ϕ_0 and ϕ are indicated and compared. (Rhodium/Carbon; d=8.0 nm, N=40 $t_{Rh}=t_C$, d_g^0=2000 nm)

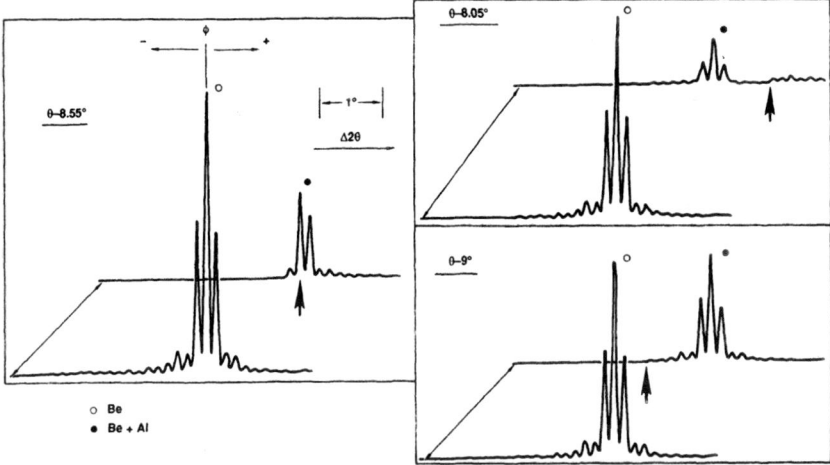

Fig. 5 Grating dispersed light observed as a function of detector position is shown for fixed angles of incidence θ (8.05°, 8.55°, 9.0°) of a white soft x-ray beam (800 to 3000 eV) onto a multilayer grating (d_0=3.0 nm, d_g=2000 nm) with and without a 4.25 μm thick aluminum absorber in the x-ray beam. Note the position of the Al K edge (1550 eV) in these scans as denoted by the vertical arrows.

that light of approximately 1558 eV was dispersed was determined by following the position of the Al K absorption edge introduced into the dispersed spectrum using a 4.25 μm thick aluminum absorber. This corresponds to approximately 5 eV intervals for the 3.0 nm period multilayers. Representative data for the 3.0 nm period molybdenum/silicon multilayer coated grating at θ values of 8.05 deg (~1655 eV), 8.55 deg (~1558 eV) and 9.0 deg (~1470 eV) are shown in Fig. 5. Scans both with and without the aluminum absorber are shown. The positions of the aluminum K edge is denoted by the vertical arrows on the aluminum absorber scans. Note a 25 μm Be filter absorber was also used to give a low energy cutoff of ~800 eV for the incident beam.

The important result demonstrated in Fig. 5 is that the position of the aluminum edge is to the low angle side of the Bragg peak (higher energy, m negative) of the bragg peak for θ = 9 deg (~1470 eV), directly on the Bragg peak for θ=8.55 deg (~1558 eV) and to the high angle side (lower energy, m positive) of the Bragg peak for θ=8.05 deg (~1655 eV). Analysis of all data for this multilayer grating (30 detector scans) demonstrated that the angular position of the aluminum K edge was fixed relative to the incident beam at an angle of 162.9° corresponding to a Bragg angle of 8.55 deg. This is just the Bragg angle expected for 1558 eV light for this multilayer. This result demonstrates that these multilayer gratings disperse light in a continuous manner with high energies appearing at larger angles relative to the incident beam (small Bragg angles) and lower energies at lower angles relative to the incident beam (larger Bragg angles). Multilayer gratings therefore act as X-RAY PRISMS, as a consequence of the convolution of the effects of the two periodic structures - i.e. the multilayer and the grating. This is a unique and unexpected result that allows fixed angle of incidence spectrographs to be fabricated with resolutions and spectral ranges defined by the structural parameters of the multilayer and grating.

314

CONCLUSIONS

The data presented have demonstrated that multilayer gratings disperse wavelengths Bragg diffracted by the multilayer so that all constructive interference occurs at essentially constant angles relative to the zeroeth order Bragg diffracted beam. Additionally, the energy of the dispersed light varies monotonically from high energy (m negative) to low energy (m positive) demonstrating that these gratings act as x-ray prisms. I note that it is possible to characterize the absorption edge of a material with the resolution of the multilayer by simply ratioing the intensities of the grating orders with and without the absorber in the beam incident to the grating. Also, these results are general in nature and not limited to the soft x-ray or extreme ultraviolet spectral domains. Additionally, the effects should be observable with neutrons when multilayer grating structures specifically designed for neutrons are used.

ACKNOWLEDGMENTS

Work performed under the auspices of the U.S. Department of Energy by Lawrence Livermore National Laboratory under contract #W-7405-Eng-48.

REFERENCES

1. Max Born and Emil Wolf, Principles of Optics, Pergamon Press, New York (1983).
2. T.W. Barbee, Jr., "Multilayers for X-ray Optics," Opt. Eng. 25, 989 (1986).
3. T.W. Barbee, Jr., "Sputtered layered synthetic microstructure (LSM) dispersion elements; layered synthetic microstructures (LSM)," in Low Energy X-ray Diagnostics - 1981, D.T. Attwood and B.L. Henke, eds., AIP Conf. Proc. No. 75, p. 131, AIP New York (21981); "Reflecting media for x-ray optic elements and diffracting structures for the study of condensed matter," Superlattices and Microstructures 1, 311 (1985).
4. E. Spiller, "Evaporated multilayer dispersion elements for soft x-rays," in Low Energy X-Ray Diagnostics - 1981, D.T. Attwood and B.L. Henke, eds., AIP Conf. Proc. No. 75, p. 124, IP, New York (1981).
5. J.H. Underwood, T.W. Barbee, Jr., and D.L. Keith, "Layered synthetic microstructures: properties and applications in x-ray astronomy," in Space Optics: Imaging X-Ray Optics Workshop, M. Weisskopf, ed., Proc. SPIE 184, 123 (1979); J.H. Underwood and T.W. Barbee, Jr., "Synthetic multilayers as Bragg diffractor for x-rays and extreme ultraviolet: calculations and performance," in Low Energy X-Ray Diagnostics - 1981, D.T. Attwood and B.L. Henke, eds., AIP Conf. Proc. No. 75, p. 170, AIP, New York (1981).
6. R.W. James, The Optical Principles of the Diffraction of X-rays, Oxbow Press, Woodbridge, Conn. (1982).
7. B.L. Henke, P. Lee, T.J. Tanaka, R.L. Shimabukuro, and B.K. Fujikawa, "Low energy x-ray interaction coefficients: photoabsorption, scattering and reflection. E = 100-2000 eV, Z = 1-94," in Atomic and Nuclear Data Table 27, Academic Press, New York (1982).
8. M. C. Hutley, Diffraction Gratings (Academic, New York, 1982).
9. W.K. Warburton, "Beamline III-4 at SSRL" Nucl. Instrum. Methods 172, 387 (1980).

ELASTIC PROPERTIES OF ARTIFICIALLY LAYERED THIN FILMS

ROBERT C. CAMMARATA
Department of Materials Science and Engineering, The Johns Hopkins University, Baltimore, MD 21117

ABSTRACT

Enhancements in the elastic moduli by factors of two or more in compositionally modulated metallic thin films have been observed for a certain range of composition modulation wavelengths. The experimental and theoretical understanding of this phenomenon, known as the supermodulus effect, is reviewed. Also, the mechanical properties of other artificially layered and composite materials are discussed and compared with the behavior of metallic superlattice thin films.

INTRODUCTION

Perhaps the most remarkable property that has been observed in artificially layered thin films is the "supermodulus effect," reported by Hilliard and coworkers in 1977 [1]. The supermodulus effect is the anomalous enhancement, by factors of two or more, in the elastic moduli of highly textured compositionally modulated metallic thin films for a certain range of composition wavelengths (layer repeat lengths). The supermodulus effect has been the subject of several reviews [2-5].

Thus far, no completely satisfactory explanation of the supermodulus effect has been given. Theories based on singularities in the electronic energy due to interactions of the Fermi surface with "artificial" Brillouin zones, as well as models based on higher order elastic effects caused by the presence of large coherency strains, have been proposed. However, as will be discussed, there are serious deficiencies in both of these theories, and the origin of the supermodulus effect must still be considered a mystery.

While the mechanical properties of metallic superlattices have received the most attention, ceramic and polymeric multilayer thin films have been investigated as well, and many of their properties are very similar to those found in metallic films. Also, composites with small diameter filaments have displayed elastic modulus enhancements that are nearly identical to those of the supermodulus effect. These results suggest that many of the mechanical properties of artificially layered structures (as well as composites in general) may depend on certain universal features of interfaces, independent of the actual materials that make up the layers.

ELASTIC PROPERTIES OF ARTIFICIAL METAL SUPERLATTICES

The artificially layered metal system that has received the greatest experimental attention with regard to the supermodulus effect is Cu/Ni, in which monomodal enhancements in the biaxial [6] and flexural [7,8] moduli, and bimodal enhancements in Young's modulus and the shear modulus C_{44} [7,8] have been reported. Figure 1 and Figure 2 show schematically the flexural modulus and Young's modulus, respectively, as a function of composition modulation wavelength λ, measured in Cu/Ni layered films. Similar elastic modulus enhancements have been observed for a variety of other f.c.c. metal/f.c.c. metal systems, such as Cu/Pd, Au/Ni, and Ag/Pd [9]. In all of these systems, which were produced by multisource evaporation, the supermodulus effect has been found for a narrow range of composition modulation wavelengths centered at about 2 nm. Also, elastic anomalies were seen only in multilayered films that were highly textured and coherent; that is, there had to be a high degree of lattice matching between the layers of the films without any misfit dislocations at the interfaces. [All the f.c.c. metal/f.c.c. metal layered films that have displayed the supermodulus effect were produced with (111) texture.] Some attempts to duplicate the above results for Cu-Pd [10] and Cu-Ni [11] were unsuccessful, though in each case experimental difficulties could be cited as the reason [7]. One system for which no enhancements have been measured is Cu/Au.

The elastic properties of sputter deposited superlattice thin films composed of alternating layers of (111) textured f.c.c and (110) textured b.c.c. metals have been investigated by Grimsditch et al. [12-15]. The shear modulus C_{66} of Cu/Nb, Ni/Mo, Ni/V, and Au/Cr layered films were determined by sound velocity measurements. While the Au/Cr system displayed a modulus increase similar to that found in the f.c.c. metal/f.c.c. metal superlattices, decreases in the shear moduli of the Cu/Nb, Ni/Mo, and Ni/V films were observed. There are, however, serious questions concerning the significance of these latter softening results. X-ray diffraction characterization of the Ni/Mo films revealed evidence of a crystalline to amorphous phase transition as the composition modulation of the films was reduced. This transformation occurred in the same wavelength range where the shear modulus was observed to decrease [13]. Furthermore, the X-ray superlattice peaks associated with the composition modulation disappeared for both the Ni/Mo and Cu/Nb films at modulation wavelengths where the measured shear moduli were at a minimum [15]. (No extensive X-ray data for Ni/V films have been reported.) These results strongly suggest that major microstructural changes had occurred in the Ni/Mo and Cu/Nb systems as the composition wavelength was varied, and this could account for any changes in the elastic moduli.

It should be noted that the intentional production of small modulation wavelength amorphous multilayer films by sequential sputtering from pure elemental metal targets has been reported [16-18]. A similar process may be occurring during the sputter deposition of small wavelength Ni/Mo and Cu/Nb films, probably caused by the mechanism of solid state amorphization [19]. The reported decreases in the shear moduli are consistent with the expected softening caused by a crystalline to amorphous solid state transformation [3].

Vibrating reed measurements of Young's modulus in compositionally modulated amorphous Cu/Zr thin films revealed no supermodulus effect for the same range of modulation wavelengths for which enhancements were found in the

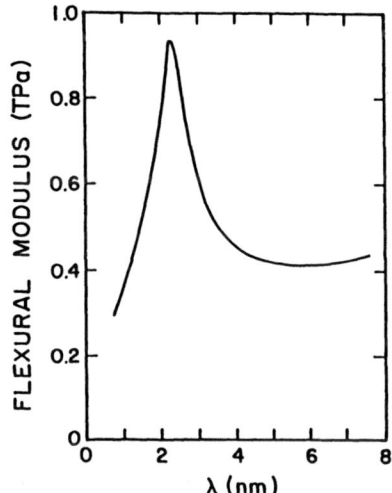

Figure 1.   The flexural modulus of
             Cu/Ni compositionally
             modulated thin films as
             a function of modulation
             wavelength λ [7,8].

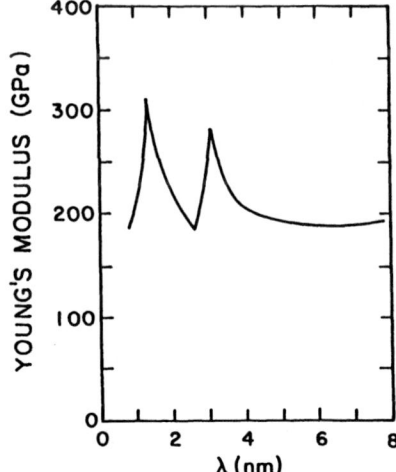

Figure 2.   Young's modulus of Cu/Ni
             compositonally modulated
             thin films as a function
             of modulation wavelength
             λ [7,8].

crystalline systems [20]. However, the experimental procedure used in that study has been called into question [21].

MECHANICAL PROPERTIES OF OTHER ARTIFICIALLY LAYERED MATERIALS

Though most of the work concerning the mechanical properties of artificial superlattices has concentrated on metallic thin films, other types of layered materials have also displayed interesting behavior similar to the supermodulus effect. Helmersson et al. [22] have recently reported on the preparation of compositionally modulated ceramic thin films. Strained-layer-superlattices of TiN/VN with (100) orientations were produced by reactive magnetron sputtering, and the microhardness of these films was measured using a diamond tip indenter. Figure 3 shows schematically the Vickers hardness of these films as a function of the composition modulation wavelength λ. As can be seen in the figure, an enhancement peak centered at a composition wavelength near 5 nm was obtained, very similar to the anomalous elastic enhancements found in metal superlattices. Though hardness is a plastic property, involving the flow of dislocations, it is nevertheless tempting to attribute the enhancement peak to the supermodulus effect. Indeed, it could be argued that since dislocation self-energies depend on the

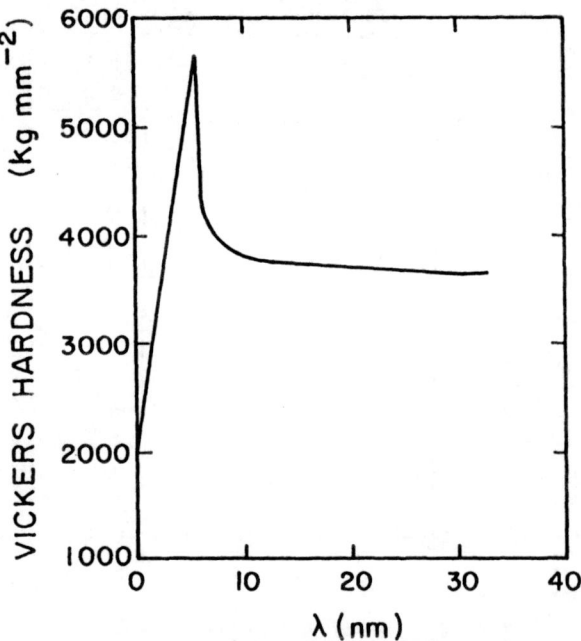

Figure 3. The Vickers hardness as a function of the repeat length λ in TiN/VN superlattices [22].

elastic constants, any anomalies in the elastic constants will also be manifested in the plastic behavior. It should be pointed out that similar enhancements in the plastic properties of Cu/Ni compositionally modulated thin films have also been reported [4,7,8].

Multilayer polymeric films with layer thickness as small as a few hundred angstroms have been produced by the coextrusion of two thermoplastic materials [23]. These types of films have displayed both enhanced and reduced plastic behavior when compared to the properties of the layered materials. Also, an extension of the maximum linearly elastic strain has been observed. Despite the fact that these results are in some ways similar to the mechanical properties of multilayer metal and ceramic films, it is possible to explain the polymer behavior without invoking the existence of enhanced elastic constants [24]. Nevertheless, a detailed study of the micro-mechanisms involved in both elastic and plastic deformation of multilayer polymer films may shed more light on the mechanical properties of metal and ceramic superlattices.

In many respects, artificial multilayer thin films can be considered to be ultrafine composite materials, and thus it is worth comparing their behavior with the elastic properties of other types of composite materials. The mechanical properties of composites made up of 5 to 300 nm diameter Nb filaments in a Cu matrix have been studied in detail by Bevk [24]. Enhancements in Young's modulus (as well as in plastic properties) by factors of two to three were observed. These results are quite intriguing, since they are essentially identical to the supermodulus effect found in Cu/Ni modulated films (and in marked contrast to the behavior reported for Cu/Nb films that displayed evidence of solid state amorphization).

It is quite evident that there are striking similarities in the mechanical properties of artificially layered metallic, ceramic, and polymeric thin films. Though any attempts at broad generalization must be viewed with suspicion, there may be certain universal structural features of artificial multilayers that manifest themselves in the mechanical behavior, independent of the actual layer materials.

THEORIES OF THE SUPERMODULUS EFFECT

Two rival theoretical approaches have been proposed to explain the large elastic modulus enhancements observed in metallic modulated films [2-5]. One of these approaches considers the changes in the electronic structure of the films due to Brillouin zone folding caused by the composition modulation. The second theoretical model is based on the existence of large coherency strains present in many of the films when the composition wavelength is not too large. Both of these theories contain serious deficiencies, and neither provides an adequate explanation of the supermodulus effect.

Fermi Surface-Brillouin Zone Interactions

The periodic composition modulation in a multilayered thin film creates artificial Bragg planes, which in turn give rise to "artificial" (or reduced) Brillouin zones. When the boundaries of these reduced zones interact with the Fermi surface, dramatic changes in the electronic structure may occur. Theories based on Fermi surface-Brillouin zone interactions have been used to explain the formation of superlattices in bulk Cu-Au alloys [25], the thermodynamically stabilizing effect of naturally produced composition modulations in Au-Ni alloys [26], and a variety of other solid state phenomena [3].

One feature of these Fermi surface-reduced Brillouin zone interactions is the existence of screening singularities in the electron dielectric function. It has been suggested that these singularities affect the electronic structure in such a way as to cause discontinuous changes in the elastic moduli near the composition modulation wavelength associated with the interacting reduced Brillouin zone. Screening singularities in the band structure energy [27] and the ion-ion interaction energy [28] have been proposed as causing the supermodulus effect. In fact, any screened interaction (e.g., spin wave interactions) could conceivably give rise to anomalous elastic properties.

Several shortcomings with models based on the above electronic effects have been pointed out [3]. One problem is they predict that both enhancements and dehancements in the moduli should be observed in the same system as the entire composition modulation wavelength range is scanned, a result that is not seen experimentally. Furthermore, it has yet to be shown that the effect is large enough to be significant. The singularities that are at the heart of these theories are based on calculations performed for a temperature of 0 K, and therefore it is assumed that there exists a perfectly sharp Fermi surface. It is this sharpness that causes the electronic singularities. There is some question as to whether or not these large enhancements will persist when the calculations are made for a finite temperature.

Perhaps the most troubling difficulty with these Fermi surface-Brillouin zone interaction theories is that, as was stated above, the energy changes associated with such interactions have been used to explain the thermal stability of periodic structures. However, modulated films that have displayed a supermodulus effect have also homogenized when subjected to relatively low temperature anneals [2,3]. Hence, the composition modulation produces little, if any stabilizing effect, which strongly suggests that any electronic energy changes associated with the modulation are small, and unlikely to affect the elastic moduli.

Coherency Strain Effects

It has been proposed that the existence of large coherency strains, that can be created in superlattice thin films when there is a high degree of lattice matching between the layers, is the basis of the supermodulus effect [2-5]. Since these coherency strains can be quite large, as high as a few percent, they can result in atomic displacements out of the Hookean region of the interatomic potential. The result is that higher (third) order elastic constants are manifested as enhancements in the (second order) elastic moduli. Jankowski and Tsakalakos [29] have performed calculations employing Born-Mayer potentials to show that within each layer these higher order effects may become significant for some systems.

However, since these changes are due to an odd (third) order effect, they can cause either increases or decreases in the elastic constants, depending on whether the layer is under compression or under tension. For example, if a modulated film with large coherency strains is considered (e.g. Au/Ni), the layer containing the material with the larger equilibrium lattice parameter (Au) will be under a compressive stress, while the layer with the smaller equilibrium lattice parameter material (Ni) will experience a tensile stress. Though the compressed layer may exhibit an increased stiffness, the layer under tension will display a nearly equivalent decrease. The net result is that these effects will for the most part cancel, and no major change is expected to be displayed by the film as a whole. This was in fact the result obtained from a continuum calculation using theoretically determined higher order elastic constants [30], and is also the result of calculations using Born-Mayer potentials [31].

There have been some molecular dynamics calculations that have suggested that the presence of internal strains in layered films may produce significant changes in the elastic constants [32,33]. In particular, Imafuku et al. [33] calculated huge modulus enhancements for Cu/Ni and similar systems. However, they stated that the supermodulus effect will be most pronounced in "incoherent structures with periodic strains." It is not completely clear what they mean by this, and it is possible that they may have been considering strain states other than those actually found in coherently strained films. In any event, it is difficult to understand why it is necessary to employ a "dynamics" calculation, or why it would produce results that differ from a "static" potential calculation. Since a great deal of the physics in molecular dynamics simulations is hidden in the many computations, it still remains to be seen whether the results of such computations for the supermodulus effect have any physical meaning.

Perhaps the most serious objection to the coherency strain model is that it does not seem to correlate well with experimental results. For example, Cu/Au modulated films, which would be expected to have large coherency strains since there is a large difference between the equilibrium lattice parameters of the two components, should display greater elastic modulus enhancements than Cu/Ni films, where the equilibrium lattice parameters are much closer. However, as was discussed earlier, there is no supermodulus effect found in Cu/Au films, while there are large modulus increases measured in Cu/Ni films.

Empirical Correlations

If neither the Fermi surface-Brillouin zone interaction theory nor the coherency strain model provides the complete answer, what could be the other possibilities? No satisfactory alternative theories exist at present; one can only speculate based on trends found in the experimental results. One interesting correlation [2,20] is that all metal superlattices, familiar to the author, which have displayed a supermodulus effect have contained either a ferromagnetic (Ni, Fe), antiferromagnetic (Cr), or strongly paramagnetic (Pd) element. Furthermore, the one system that has shown no enhancement (Cu/Au) contains only paramagnetic components. This strongly suggests that there may be magnetoelastic effects, similar to the ΔE effect (changes in elastic moduli of ferromagnetic materials by several percent due to magnetomechanical hysteresis phenomena [34]) that could be causing, or at least contributing to, the supermodulus effect. Calculations which considered the effects of coherency strains on the energy of magnetization in films with ferromagnetic layers produced no significant enhancement [31]. However, some films that have displayed the supermodulus effect, such as Cu/Ni, have also shown anomalous magnetic properties [35].

It has been proposed that the supermodulus effect is strictly an interface effect. For example, a phase different from those found in the bulk of the layers might form at the interfaces, and give the layered structure greater stiffness. This possibility has been previously rejected because work on Ni/Mo superlattices, where the ratio of Ni to Mo was varied in order to keep the interface density constant, showed that the observed decreases in the shear modulus were not correlated with the number of interfaces [13]. However, as was discussed earlier, any results from the Ni/Mo system must be viewed with suspicion due to probable microstructural variations in films with different modulation wavelengths. The fact that similar mechanical properties have been found in metal, ceramic, and polymer layered thin films, as well as in multifilamentary composites, suggests that some universal interfacial phenomenon may be present. The modulus enhancements in the Cu-Nb filamentary composites have been attributed to thermally induced strains at the filament/matrix interface [24]. Enhancements in the plastic properties of multilayer polymeric films have been found to depend in part on interfacial adhesion [23].

FUTURE DIRECTIONS

It is clear that there is still much more work that needs to be done before the supermodulus effect can be said to be well understood theoretically. Any model of the supermodulus effect will have to explain several experimental observations. One question that would have to be addressed is why the effect occurs for only a limited range of composition modulation wavelength, or, equivalently, why the effect is lost at small and long wavelengths. At long wavelengths, multilayered metal films have been observed to become incoherent, and this loss of coherency could conceivably cause the loss of the enhanced modulus. On the other hand, the ceramic superlattices displaying the behavior found in Figure 3 showed no loss of coherency even at the longest wavelengths.

The lack of a supermodulus effect at short modulation wavelengths may be qualitatively attributable to various factors [3]. It is not surprising that films with layer thicknesses of the order of a few atomic distances display different properties from those containing more bulk-like layers. Nevertheless, no completely convincing discussion has been put forth to explain the small wavelength behavior; specifically, why does the supermodulus effect always

disappear and the elastic modulus return to the long
modulation wavelength value as the composition modulation
wavelength goes to zero?

Another experimental feature that any successful theory
of the supermodulus effect must explain is the curious bimodal
enhancements found for Young's modulus and the shear modulus
C_{44} in Cu/Ni films. Previous attempts [29] to explain this
double-peak behavior have not been convincing [3].

Certain experimental issues with regard to the
measurement of the supermodulus effect (as well as the
measurement of elastic moduli in general) need to be
addressed. One question concerns what modulus is actually
measured by the vibrating reed test. Baral et al. [7,8] have
identified it as the flexural modulus. However, it may be
that when the reed is longer than it is wide, as was the case
in the experiments that produced Figure 1, the elastic
constant measured is Young's modulus [3,20]. This is an
important issue, since Young's modulus measured by a tensile
test produced a bimodal enhancement, while the modulus
measured in the vibrating reed test produced only a monomodal
enhancement (see Figure 1 and Figure 2).

Another aspect that deserves further experimental
examination concerns the possibility that major
microstructural changes occur in compositionally modulated
films as the modulation wavelength is varied. There are still
questions as to what effect loss of coherency has on the
physical properties in general, and the mechanical properties
in particular, of modulated films. Furthermore, the
possibility that many apparent modulus anomalies in f.c.c
metal/b.c.c. metal films are due to solid state amorphization
should be investigated further. It would appear that
transmission electron microscopy, especially on cross-
sectionally prepared specimens, would be an effective method
for studying many of these issues.

Other types of superlattice films should be studied to
see if they possess the supermodulus effect. One class of
multilayered films for which little detailed study of the
elastic and plastic properties has been carried out, in so far
as the author is aware, is semiconductor superlattices. It
would appear that a great deal of information could be gleaned
from such experiments, especially since the electronic
properties of these films have been extensively studied. It
may be possible to correlate effects of the composition
modulation and the coherency strains on the elastic and
plastic properties of semiconductor strained-layer-
superlattices with their well characterized effects on the
band structure. Perhaps in this way the validity of the
theories of the supermodulus effect that are based on
electronic energy considerations may be tested. Finally,
interesting mechanical properties may be observed if the
artificially layered thin film is composed of different types
of materials. For example, a film composed of alternating
layers of ceramic and metal materials would be expected to
display enhanced plastic properties similar to those found in
bulk composites made up of brittle and ductile materials. It
is also possible that unusual elastic properties similar to
the supermodulus effect may be displayed by these "hybrid"
films.

In conclusion, a great deal of experimental and
theoretical work has yet to be done before even a partial

understanding of the elastic properties of artificially
layered thin films has been achieved.

ACKNOWLEDGEMENTS

The author wishes to thank T. Tsakalakos and F. Spaepen
for useful discussions, and S.M. Prokes for a critical reading
of the manuscript.

REFERENCES

1. W.M.C. Yang, T. Tsakalakos, and J.E. Hilliard, J. Appl.
    Phys. 48, 876 (1977).
2. P.C. Clapp, in Modulated Structure Materials, NATO ASI
    Series, Applied Sciences, ed. T. Tsakalakos, Martinus
    Nijhoff, Dordrecht, p. 465 (1985).
3. R.C. Cammarata, Scripta Metall. 20, 479 (1986).
4. T. Tsakalakos and A.F. Jankowski, Ann. Rev. Mater. Sci.
    16, 293 (1986).
5. R.W. Cahn, Nature 324, 108 (1986).
6. T. Tsakalakos and J.E. Hilliard, J. Appl. Phys. 54, 734
    (1983).
7. D. Baral, Ph.D. Thesis, Northwestern University (1983).
8. D. Baral, J.B. Ketterson, and J.E. Hilliard, J. Appl.
    Phys. 57, 1076 (1985).
9. A summary table of most of work on metal superlattices
    can be found in the article by I.K. Schuller, IEEE 1985
    Ultrasonics Symposium, edited by B.R. McAvoy (IEEE, New
    York, 1985) p. 1093.
10. B.S. Berry and W.C. Pritchett, Thin Solid Films, 33, 19
    (1976).
11. H. Itozaki, Ph.D. Thesis, Northwestern University (1983).
12. A. Keuny, M. Grimsditch, K. Miyano, I. Banerjee, C.M.
    Falco, and I.K. Schuller, Phys. Rev. Lett. 48, 166
    (1982).
13. M.R. Khan, C.S.L. Chun, G.P. Felcher, M. Grimsditch, A.
    Keuny, C.M. Falco, and I.K. Schuller, Phys. Rev. B 27,
    7186 (1982).
14. R. Danner, R.P. Huebener, C.S.L. Chun, M. Grimsditch, and
    I.K. Schuller, Phys. Rev. B 33, 3696 (1986).
15. I.K. Schuller and M. Grimsditch, J. Vac. Sci.
    Technol. B 4, 1444 (1986).
16. C.J. Lin, F. Spaepen, and D. Turnbull, J. Non-Cryst. Sol.
    61/62, 767 (1984).
17. B.M. Clemens and J.C. Buchholz, Mat. Res. Soc. Symp.
    Proc. 37, 559 (1985).
18. E. Chason, H. Kondo, T. Mizoguchi, R.C. Cammarata, F.
    Spaepen, B. Window, J.B. Dunlop, and R.K. Day, Mat. Res.
    Soc. Symp. Proc. 58, 72 (1986).
19. R.B. Schwarz and W.L. Johnson, Phys. Rev. Lett. 51, 415
    (1983).
20. R.C. Cammarata, Ph.D. Thesis, Harvard University (1985).
21. K. Rouillard and F. Spaepen (unpublished).
22. U. Helmersson, S. Todorova, S.A. Barnett, J.E. Sundgren,
    L.C. Markert, and J.E. Greene, J. Appl. Phys. 62, 481
    (1987).
23. W.J. Schrenck and T. Alfrey, Jr., in Polymer Blends,
    edited by D.R. Paul and S. Newman (Academic Press, New

York, 1978), vol. 2, p. 129.
24. J. Bevk, Ann. Rev. Mater. Sci. 13, 319 (1983).
25. H. Sato and R.S. Toth, Phys. Rev. 124, 1833 (1961); 127, 469 (1962).
26. N. M. Dunaev and M. S. Zakharov, JETP Lett. 20, 336 (1974).
27. G. Henein, Ph. D. Thesis, Northwestern University (1979).
28. T.-B. Wu, J. Appl. Phys. 53, 5265 (1982).
29. A.F. Jankowski and T. Tsakalakos, J. Phys. F: Met. Phys. 15, 1279 (1985).
30. Amitava Banerjesa and John R. Smith, Phys. Rev. B 35, 5413 (1987).
31. R.C. Cammarata, to be published.
32. I.K. Schuller and A. Rahman, Phys. Rev. Lett. 50, 1377 (1983).
33. M. Imafuku, Y. Sasajima, R. Yamamoto, and M. Doyama, J. Phys. F: Met. Phys. 16, 823 (1986).
34. Magnetic Properties of Metals and Alloys, American Society for Metals (American Society for Metals, Cleveland, 1959) p. 258.
35. Report on Artificially Structured Materials, National Research Council (National Academy Press, Washington, 1985) p. 54.

EXPLORING THE RELATION BETWEEN INTERFACE STRUCTURE
& MECHANICAL PROPERTIES IN MULTILAYER MATERIALS

A.F. JANKOWSKI*, D.M. MAKOWIECKI*, M.A. MCKERNAN*,
*Lawrence Livermore National Laboratory, Chemistry & Materials
Science, P.O. Box 808, Livermore, California 94550
S.R. NUTT** AND K. GREEN**
**Brown University, Division of Engineering, Box D, Providence,
Rhode Island 02912

ABSTRACT

The relationship between microstructure and physical behavior is especially pronounced in synthetic multilayer materials.
Insight to the mechanisms responsible for changes in the mechanical properties can be investigated through a careful examination of the multilayer microstructure. A dominant feature of the
multilayer structure is the interface. The population of interlayer boundaries, that is interfaces, is directly proportional
to the multilayer period for any given film thickness. In this
paper, we will evaluate "TEM" images of multilayer systems. The
interface structure will be viewed in cross-section and a range
of layer thicknesses will be considered. Variation in the elastic modulus, yield stress, and microhardness have been observed
for noble-transition metal systems over a wide range of multilayer periods, from less than 1 nm to greater than 1000 nm. In
epitaxial systems, the extent of superlattice perfection (coherency effects) is closely tied with changes in physical behavior.
Emphasis will thus be placed on the structure and strain distribution from the interface, and its role in determining the mechanical properties of multilayers.

INTRODUCTION

The basis of developing an understanding of the mechanical
properties rests in the microstructure. Epitaxially grown metallic multilayers fall within the general category of strained
layered superlattices. The atomic structure by which strain is
accommodated from interface through interlayer regions is not
precisely known, however. Alternating layers of compression and
tension repeat with the multilayer period. In the simplest case,
an A/B binary system, the composition modulation (a product of
physical vapor deposition, for example) consists of a periodic
repetition of A and B atoms. The atoms have different radii,
thus lattice displacements are introduced through the epitaxial
layers to accommodate the atomic size difference. The biaxial
alternation between compressive and tensile stresses, creates a
"strain wave" that is coincident with the composition modulation. A "strain wave" description, serving as a phenomenological
basis, of the supermodulus effect in metallic mulitlayers has
been proposed [1]. Commensurately strained layers in a superlattice where stress and strain are relaxed from coherent interfaces is postulated as a plausible structure. The need to determine the exact nature of the "strain wave", from samples where
the mechanical behavior is known, is the objective of this
investigation. Transmission electron microscopy serves as an
excellent analytical method for this purpose, in which samples
are examined in cross-section to the direction of modulation,
that is, deposition.

Mat. Res. Soc. Symp. Proc. Vol. 103. ©1988 Materials Research Society

The nature of misfit dislocations in the microstucture of GaAs/GaAs$_5$P$_5$ multilayers was examined by Matthews and Blakeslee [2]. The interfaces between layers in these "CVD" epitaxially grown multilayers consisted of large coherent areas separated by long straight misfit dislocations, as interpreted from electron microscopic observations. The misfit dislocations are produced in the process of slight atomic adjustments in the interface which lower the energy of a system containing an interphase boundary [3-5] (in this case the interfaces between the alternating layers under compresion and tension). The interface will exhibit regions of good atomic fit separated by regularly spaced constricted regions of severe mismatch, following the process of atomic relaxation at the interface where atoms seek their lowest energy position. These regions of mismatch constitute the cores of interfacial misfit dislocations. Subsequent motion of dislocations through the multilayer interfaces is restricted, as coherency strains give rise to stresses that aid the motion of dislocations through one of the materials present but oppose its motion through the other [6].

The use of lattice fringe imaging has provided a detailed view of the local atomic displacements about interlayer boundaries. Interplanar spacing variations (modeled to either a sinusoidal or linear curve) have been determined from interfringe spacing measurements [7]. For the modulated structure of a spinoidally decomposed Au-77at.pct.Ni alloy, a continuous composition variation was clearly illustrated by the interplanar spacing modulation as determined from the microdensitometer analyisi of the lattice image [8,9]. The spacing modulation periodicity was in excellent agreement with electron diffraction measurements. In some regions however, the modulation amplitude was greater than the maximum possible from the equilibrium phase compositions. This may be due to localized concentrations of Au or Ni, or perhaps regions under compression or tension. Similarly, an elastic relaxation of shear stress accompanying a quasiperiodic lattice modulation was observed for a spinodally decomposed alloy of In$_x$Ga$_{1-x}$As$_y$P$_{1-y}$[10]. The lattice modulation examples discussed thus far are of "layered" structures that formed through phase transformation processes.

Cu/Ni, produced by "PVD", and the GaAs/GaAs$_5$P$_5$ multilayers are examples of systems in which the lattice modulation has been "artifically" introduced. The successive deposition of Cu and Ni allows the retention of a layered structure (a Frank-van der Merwe growth mode [4,5]) by preventing a rearrangement of the layers. Thus, the formation of a three dimensional island structure (Stranski-Krastanov [11] or Volmer-Weber [12] growth modes) as a result of surface diffusion effects is avoided. (A review of these growth modes may be referenced elsewhere [13-15]). Amplitude contrast, observed for Cu/Ni multilayers with periods as low as 1.6 nm (viewed in cross-section using "TEM"), may be attributed to a variation in coherency strains extending from the layer interfaces [16]. A detailed look at the atomic displacements coinciding with coherency strains from the multilayer interfaces should therefore be of practical interest.

FABRICATION AND PREPARATION FOR ANALYSIS

The multilayer structures used in this study were fabricated by a magnetron sputtering process. A 5 mtorr flow of Ar gas was used to sputter the target materials, where the cryogenically pumped, system base pressure was 0.01 µtorr. The basic sys-

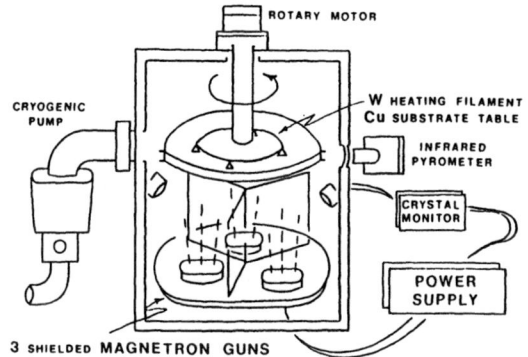

Fig 1. A schematic of the three source magnetron sputtering gun assembly used in the fabrication of the metallic multilayers. A feedback loop from the crystal monitors to the power supplies controls the deposition rates of the three materials.

tem features are schematically illustrated in Figure 1. A three gun circular array with individual source shielding rests 20 cm below an oxygen-free, Cu substrate table. The table, resistantly heated by an 80 mil W filament wire, rotates above the magnetron guns. Quartz crystal monitors rest alongside the substrate materials which are clamped to the table bottom. The rate measured by the crystals is fed back into the power supplies to maintain a constant deposition flux, typically 0.1-1.0 nm/s using a 50-400 watt power setting. The layer thicknesses are therefore controlled with respect to the frequency of the rotating substrate table. An infrared pyrometer controls the substrate temperature within 1 °C. Epitaxial Si wafers, heated to 350°C, were used as substrate materials for the deposited noble-transition metal superlattices.

The structure of the multilayers is best studied in cross-section (i.e. interface-parallel). In this way, the individual layers can be viewed discretely, avoiding the problems of superposition encountered when viewed parallel to the surface normal. The preparation of cross-section specimens for transmission electron microscopy has been procedurally outlined in detail by Bravman and Sinclair [17]. In brief, the procedure may be described as follows. Epoxy is applied to the surface of individually coated substrate wafers, which are then stacked under an applied pressure to produce a sandwiched composite structure. A diamond saw is used to cut 0.5 mm sections from the stack. A series of grinding papers and polishing cloths are then used to thin the composite, with motion always parallel to the sandwiched wafer interfaces. A slurry saw then cuts a 3 mm disc from the thinned cross-sectioned stack. A dimpling machine then produces a concave impression in the sample which is subsequently ion-milled before viewing.

CHARACTERIZATION AND STRUCTURE

The multilayer repeat period is most easily measured using x-ray diffraction. As an example, consider the $\theta/2\theta$ scan for the Au/Ni film shown in Figure 2. The reflections at low (glancing) angle are representative of the layer thicknesses, i.e. repeat period. Information about the crystallographic orientation is provided at the higher 2θ values. For this epitaxial sample, only the (111) Bragg reflection is present. The satellites ob-

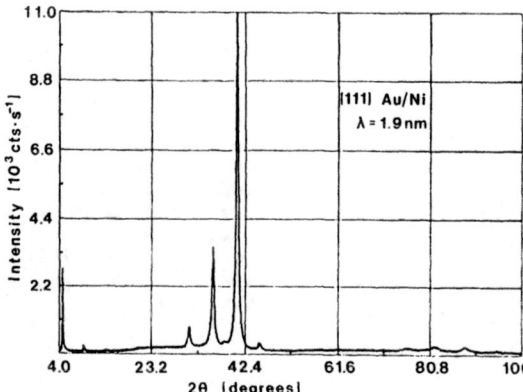

Fig 2. An x-ray θ/2θ scan of a Au/Ni multilayer. The 0.2 um thick film was examined using a Cuk source. The Bragg reflection is flanked by equally spaced (+) and (-) satellites, from which the multilayer period may be computed as 1.9 nm.

served above and below the Bragg angle are attributable to the artifical composition modulation. Calculations for the composition modulation and amplitude are detailed elsewhere [18]. The relatively intense (-) vs (+) satellites are a result of the larger scattering factor for Au as compared to that for Ni [19].

Transmission electron micrographs of cross-sectional specimens support the x-ray diffraction analysis and provide additional microstructural detail. Au/Ni multilayers of 1.9 and 2.8

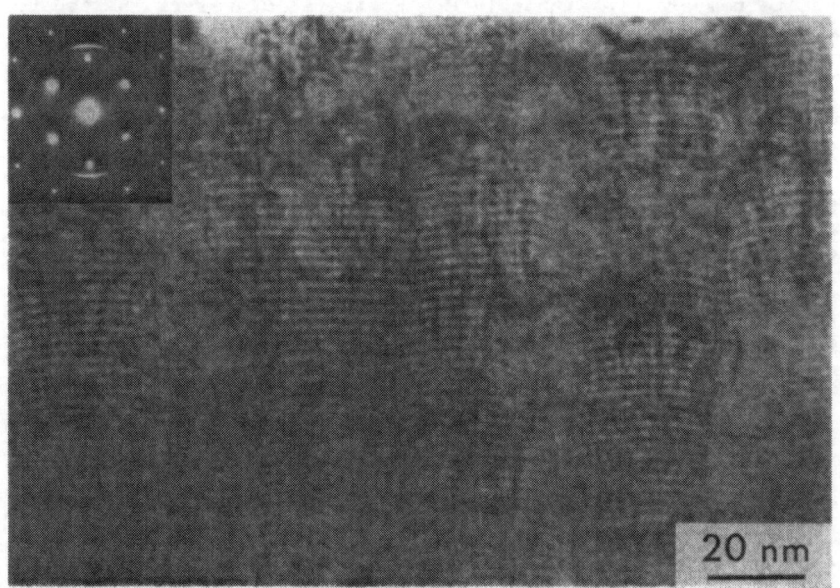

Fig. 3 A "TEM" cross-sectional view of the Au/Ni multilayer corresponding to Fig 2. Bulge test measurements of samples with this orientation and period have shown modulus enhancement [20].

nm periods are shown in Figures 3 and 4, respectively. The pronounced alternating light-dark contrast of the layers is caused
by the substantial structure factor difference between gold and
nickel. Hence, the structure shows a distinct composition modulation with sharp interfaces between the layers. The apparent
curvature of the layers in some regions is a consequence of the
cone-shaped growth structure which develops during the deposition process. Streaks of dark contrast extend normal to the
layers and appear to be associated with boundaries between the
conical growth features. The dark contrast of these streaks can
be explained by a Moire effect, attributable to the partial overlap of the cone-shaped crystallites, although lattice strain
and dislocations may also be involved. The inset diffraction
pattern reveals strong texturing of the multilayer film. The
silicon substrate spot pattern is superimposed over the two
strong arcs which arise from the (111) multilayer planes. These
arcs correspond to an interplanar spacing of 0.22 nm and a lattice constant of 0.379 nm, roughly midway between the lattice
constants of Au and Ni. The arcs are caused by relative misor-
ientations that exist between the columnar grains in the film.
The weaker subsidiary arcs inside and outside the strong arcs
are generated by the multilayer periodicity. The spacing of
these satellite arcs is in good agreement with the x-ray data.
Despite the pronounced compositional modulation, the structure
has a unique lattice parameter. A value which is intermediate
to that of gold and nickel. Thus, the large atomic misfit in
this epitaxial (111) fcc system appears to be accomodated relatively uniformly within the layers and not solely at the inter-

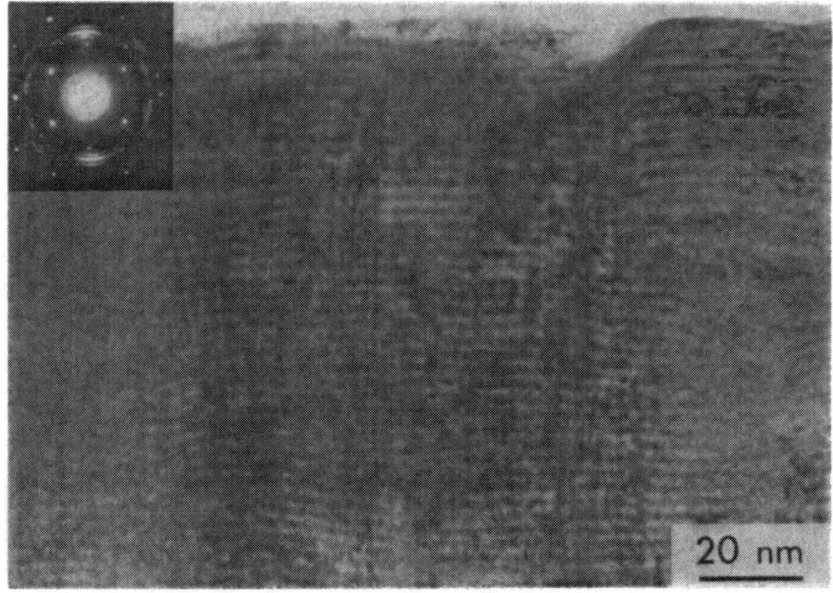

Fig 4. A "TEM" cross-sectional view of a 2.8 nm period Au/Ni
multilayer structure.

faces between layers. (Similar observations have been suggested in the Cu/Ni case [16].) An appropiate model for such a metallic multilayer structure, with a square composition profile, would include the use of atomic displacements that periodically repeat with the layer sequence and form a "strain wave" [1].

DISCUSSION

The mode of stress relaxation from internal phase boundaries regulates the material behavior of an alloy. This is especially pronounced in the enhancement of elastic moduli in strained layered metallic superlattices. Size (atomic lattice misfit) and periodicity (repeat layer thickness) effects are observed to play a major role. For the fcc noble-transition metal multilayers, in which modulus enhancement is known to occur, the effect is clearly observable for a periodicity range corresponding with a commensurate structure composed of 4-6 atomic planes per elemental layer (e.g. in a (111) Cu/Ni multilayer with periods ranging from 1.6 to 2.4 nm). The dimensional constraints on material selection required for coherent interfaces in superlattices have been recently outlined by Kobayashi and Das Sarma [21]. Using a rigid-lattice approximation and considering the lattice constant ratio, relative orientation and displacement, and adatom cluster size, static energy calculations for an interface composed of symmetrically dissimilar structures are performed. It's recommended that for an A/B superlattice, where a_A is less than a_B, the layer B thickness should be made less than that of layer A. Milchev and Markov [22] have similarly investigated the effect of anharmonicity in epitaxial interfaces with respect to substrate-induced dissociation of finite epitaxial islands, equilibrium structure of thin epitaxial films and mean dislocation density.

To produce multilayer superlattices composed of 4 to 6 commensurate atomic planes per layer with elements having an atomic misfit greater than 2-4 % does not ensure the occurence of modulus enhancement, however. Si/Ge superlattices, produced by an ultra-high vacuum, molecular beam epitaxy technique [23], evidenced no modulus anomalies in multilayers with the prescribed dimensional features [24]. An important difference between the Cu/Ni and Si/Ge systems may be not only the precise manner in which lattice strain relaxes away from the interfaces, i.e. the "strain wave", but the constituent material properties as well. "HREM" images and "RBS" data of the Si/Ge interfaces describe a structural configuration in which the full strain misfit was accomodated within a few atomic planes of the interface. This range of relaxation is the minimal extreme as that postulated for the supermodulus materials [1]. Differences between the "strain wave" profiles of the multilayer superlattices as well as which material layers (those elastically compliant or stiff) are under compression or tension [1] are crucial for investigating the phenomena of modulus enhancement.

ACKNOWLEDGEMENTS

Work performed, in part, under the auspices of the United States Department of Energy by Lawrence Livermore National Laboratory under Contract W-7405-Eng-48.

REFERENCES

1. A.F. Jankowski, J.Phys.F:Met.Phys. $\underline{18}$, in press (1988).
2. J.W. Matthews and A.E. Blakeslee, J.Cryst.Growth $\underline{27}$, 118 (1974); $\underline{29}$, 273 (1975); $\underline{32}$, 265 (1976).
3. G.J. Shiflet, Matls.Sci.Eng'g. $\underline{81}$, 61 (1986).
4. F.C. Frank and J.H. van der Merwe, Proc.Roy.Soc.(London) A $\underline{198}$, 205,216 (1949); $\underline{200}$, 125 (1949).
5. J.H. van der Merwe, Proc.Phys.Soc.(London) A $\underline{63}$, 616 (1950); J.Appl.Phys. $\underline{34}$, 117,123 (1963); $\underline{41}$, 4725 (1970); Treatise Mater.Sci.Technol. $\underline{2}$, 1 (1973).
6. J.W. Cahn, Acta Metall. $\underline{11}$, 1275 (1963).
7. D.J.H. Cockayne and R. Gronsky, Phil.Mag. A $\underline{44}$, 159 (1981).
8. R. Sinclair, R. Gronsky and G. Thomas, Acta Metall. $\underline{24}$, 789 (1976).
9. R. Gronsky and G. Thomas, in Modulated Structures (AIP Conf. Proc. $\underline{53}$, 1979) 266.
10. M.J. Treacy, J.M. Gibson and A. Howie, Phil.Mag. A $\underline{51}$, 389 (1985).
11. I.N. Stranski and L. Krastanov, Sitzungsber.Akad.Wiss.Wien, Math-Naturwiss,Kl,IIb $\underline{146}$, 797 (1938).
12. M. Volmer and A. Weber, Z.Phys.Chem. $\underline{119}$, 277 (1926).
13. E. Bauer and H. Poppa, Thin Solid Films $\underline{12}$, 167 (1972).
14. E. Bauer, Z.Kristallogr. $\underline{110}$, 372 (1958).
15. R.W. Vook, Inter.Met.Rev. $\underline{27}$, 209 (1982).
16. S. Nakahara, R.J. Schultz and L.R. Testardi, Thin Solid Films $\underline{72}$, 277 (1980).
17. J. Bravman and R. Sinclair, J.Elec.Micr.Tech. $\underline{1}$, 53 (1984).
18. G.E. Henein and J.E. Hilliard, J.Appl.Phys. $\underline{54}$, 728 (1983).
19. S.C. Moss, in Local Atomic Arrangements Studied By X-Ray Diffraction (Metallurgical Soc. Conf. $\underline{36}$, 1965) 114.
20. W.M.C. Yang, T. Tsakalakos and J.E. Hilliard, J.Appl.Phys. $\underline{48}$, 876 (1977).
21. A Kobayashi and S. Das Sarma, Phys.Rev. B $\underline{35}$, 8042 (1987).
22. A. Milchev and I. Markov, Surf.Sci. $\underline{136}$, 503,519 (1984); $\underline{145}$, 313 (1984).
23. J.C. Bean, L.C. Feldman, A.T. Fiory, S. Nakahara and I.K. Robinson, J.Vac.Sci.Technol. A $\underline{2}$, 436 (1984); J. Bevk, J.P. Mannaerts, L.C. Feldman, B.A. Davidson and A. Ourmazd, Appl.Phys.Lett. $\underline{49}$, 286 (1986); T.P. Pearsall, J. Bevk, L.C. Feldman, J.M. Bonar, J.P. Mannaerst and A. Ourmazd, Phys.Rev.Lett. $\underline{58}$, 729 (1987).
24. J. Bevk, "Elastic Properties of One-Two Dimensional Micro-composites" (ASM Materials Week, Cincinn., October 13, 1987) - comments that in the Si/Ge superlattice, the displacements of the tetragonally distorted Ge lattice were modeled successively using bulk elastic constants.

MAGNETIC SUPERLATTICES

IVAN K. SCHULLER
Physics Department B-019, University of California - San Diego
La Jolla, California 92093

ABSTRACT

Magnetic superlattices serve as model systems for the study of thin film, interfacial, proximity, coupling and superlattice phenomena. Due to these phenomena, the physical properties of magnetic superlattices can be tuned in a reproducible fashion by proper control of the preparation process.

Magnetic measurements in conjunction with detailed structural characterization provide a fruitful area of research, especially in understanding basic phenomena in magnetism. We describe here briefly a few experimental examples from our work which illustrate the possibilities magnetic superlattices offer for the study of basic phenomena in magnetism.

INTRODUCTION

The recent interest in the study of magnetic superlattices was motivated by the report of enhanced magnetization in Ni/Cu superlattices above the magnetization of pure Ni [1]. This report was coincidental with the development of novel preparation and characterization techniques for thin films, especially geared towards metallic systems [2-4]. As a consequence, great interest was devoted to the exploration of new magnetic superlattices, the study of magnetic phenomena at different length scales and the engineering of new magnetic properties into materials by careful control of preparation conditions. Among the large number of causes that could be invoked for the property modification of superlattices, changes in the electronic structure, proximity effects, and variations of thickness compared to a characteristic magnetic length (RKKY, dipolar and exchange) have received particular attention. In general, the properties of magnetic superlattices can be conveniently categorized according to the physics that gives rise to these properties. In increasing order of complexity (as far as number of layers required) these are: thin film, two dimensional, interfacial, proximity, coupling and superlattice effects. The recent literature in this field is quite extensive and beyond the scope of this brief review. For a comprehensive review, the reader is referred to several recent books on the subject [2-4]. To illustrate the type of effects present we will use examples from our own work.

PHYSICAL PHENOMENA

a)   Thin Film and Two Dimensional Effects.

Thin film effects are due to the fact that a superlattice is made out of a collection of thin films. Although in principle, the observation of thin film effects do not require the incorporation into a superlattice, in practice, these present considerable technical advantages. Because a superlattice is made of a large collection of single layers, the total volume available for study is quite considerable so a number of studies can be performed which otherwise are not possible. In addition, in many cases single films require *in-situ* studies since surface oxidation presents a major problem when the sample is removed from the vacuum system. Since in superlattices only a small fraction of the sample is oxidized (typically 100 Å out of 10,000 Å) surface oxidation does not pose a problem.

Mat. Res. Soc. Symp. Proc. Vol. 103. ©1988 Materials Research Society

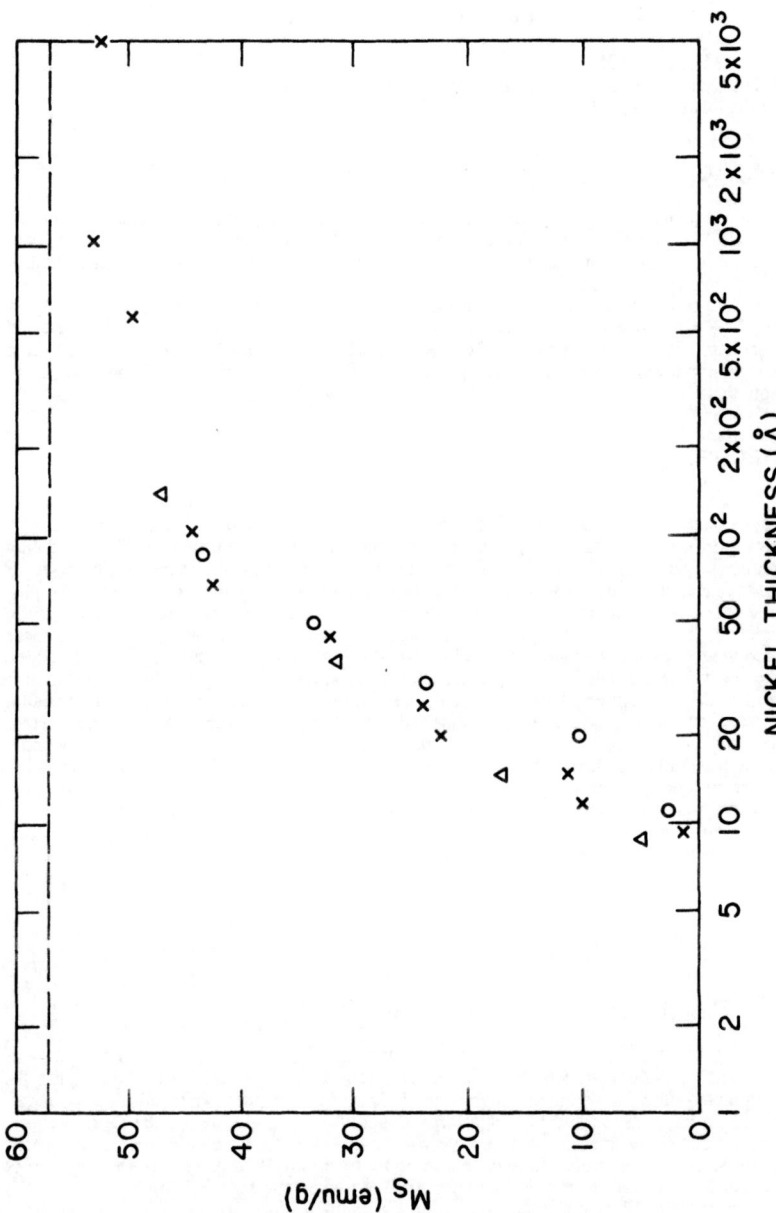

Fig. 1 Saturation magnetization (M_S) versus thickness in Mo/Ni
superlattices. The dashed line indicates the saturation
magnetization of bulk Ni.

An example of this type of effect is the decrease in magnetization and Curie temperature of Ni in Mo/Ni superlattices [5]. Fig. 1 shows the saturation magnetization (M_s) of Mo/Ni superlattices as a function of Ni thickness (d_{Ni}) at 5K. Close to $d_{Ni} = 10$ Å the saturation magnetization of the sample is below the detection limit, indicating that the sample is paramagnetic. A model assuming one to two dead layers of Ni at the interface between Mo and Ni explains quite well this curve. In addition, the Curie temperature (Fig. 2) shows a similar behavior. Note that these measurements were performed in multilayered samples after removal from the vacuum system, and the measurements were performed quite easily using a SQUID magnetometer.

The observation of dimensional effects in magnetic superlattices have been claimed by a number of groups [1,6,7]. In all these cases a remarkable linear temperature dependence of the saturation magnetization was observed close to monolayer superlattices. In order to uniquely identify whether this is a two-dimensional effect it is important to ascertain whatever roughness, pinholes, islands and other defects could give origin to this remarkable behavior.

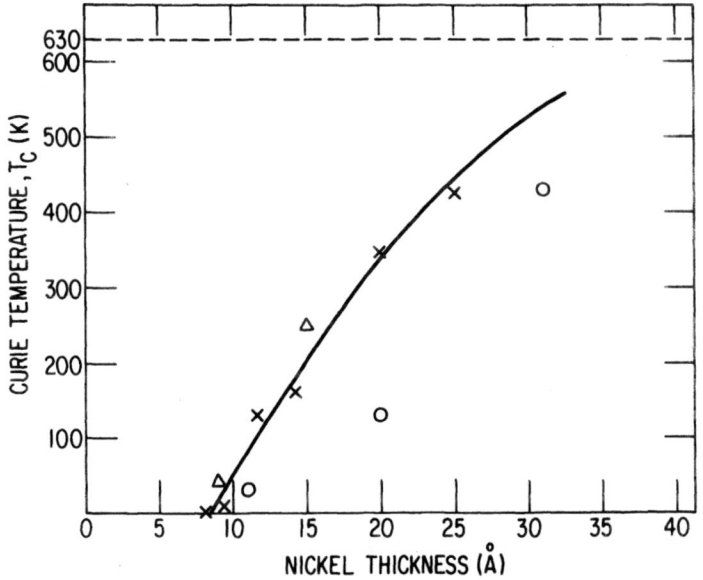

Fig. 2 Curie temperature extracted from Arrot plots
        for Mo/Ni superlattices. Note that Ni becomes
        non-magnetic close to a thickness of 10Å which
        corresponds to two dead layers at each interface.

b)   Interfacial and Proximity Effects

The presence of an interface can cause a number of interesting effects either due to the presence of the interface (through the development of interfacial electronic states for instance) or because of proximity effects such as electron transfer.

One property which has received considerable attention in thin films is the behavior of the surface anisotropy. Very early [8] it was noticed that the anisotropy extracted from ferro magnetic resonance (FMR) and DC magnetization disagreed considerably. In addition, the perpendicular field dependence of the magnetization was not linear as expected in a thin film. In order to address some of these questions, we have performed an extensive study comparing FMR and DC magnetization in Mo/Ni superlattices [9]. The curvature in the DC magnetization (shown in the insert of Fig. 3) can be very well fitted for all fields assuming that second order anisotropy is important. Of course, this is the first and most natural explanation for the curvature. To compare the anisotropy from the parallel FMR and the parallel and perpendicular DC magnetization only first order anisotropies should be taken into account. The physical reason for this is that parallel FMR senses the anisotropy through small precesions of the magnetic moment from the parallel direction. The anisotropy measured using DC magnetization on the other hand, requires tipping the moment 90 degrees from the easy axis (parallel to the film) into the perpendicular direction to the film. This type of analysis brings into agreement the first order anisotropy $H_a^{(1)}$ for thin films as shown in Fig. 3. In equal layered superlattices, an increasing discrepancy with the number of interfaces is observed. Although the origin of this discrepancy has not been uniquely identified at the present time it is believed that this discrepancy arises from a surface anisotropy which is sensed by the high frequency FMR measurements [9].

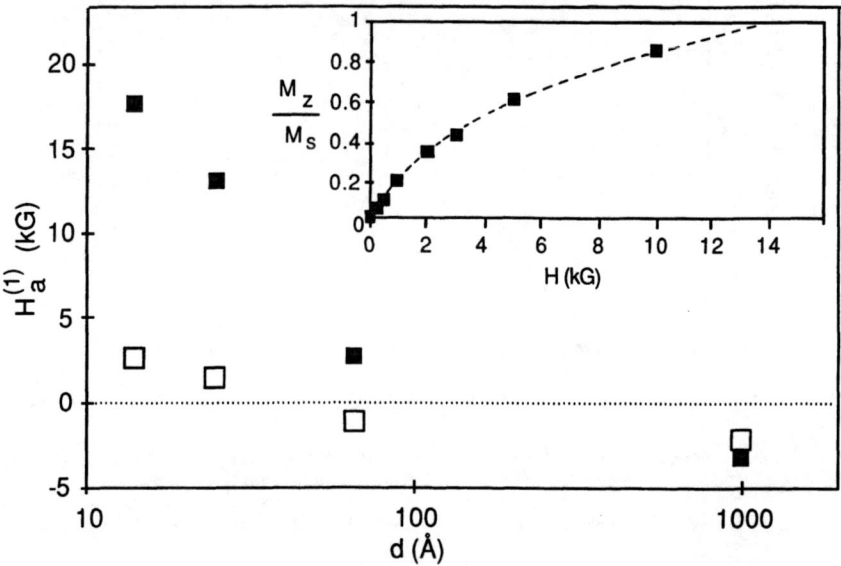

Fig. 3   The insert shows the magnetization in the perpendicular direction as a function of field. Dotted line is a fit assuming first and second order anisotropies. First order anisotropies obtained from DC magnetization (open squares) and FMR (closed square) measurements.

c)   Coupling and Superlattice Effects

The coupling of magnetic layers across a normal metal has received attention for some time [10]. Recently, RKKY coupling in Gd/Y [11] and propagation of spiral magnetism in Dy/Y [12] superlattices have been claimed.

Since the coupling mechanisms investigated (RKKY and spiral magnetism) in these studies have all a decay length of the order of 10 Å, extreme control (at the atomic level) over the layer thickness, roughness, inter-diffusion, etc., has to be invoked. In addition, it is important to rule out the possibility of slight interdiffusion, pinholes, roughness, or other defects explaining the effects.

A coupling mechanism which occurs at relatively long length scales (larger than 100Å) is the magnetic dipolar coupling. In this case, the requirements on structural integrity is not that stringent. This coupling mechanism has allowed the observation of superlattice effects, i.e., effects which not only depend on coupling across non-magnetic layers, but also rely on the periodic nature of the superlattice.

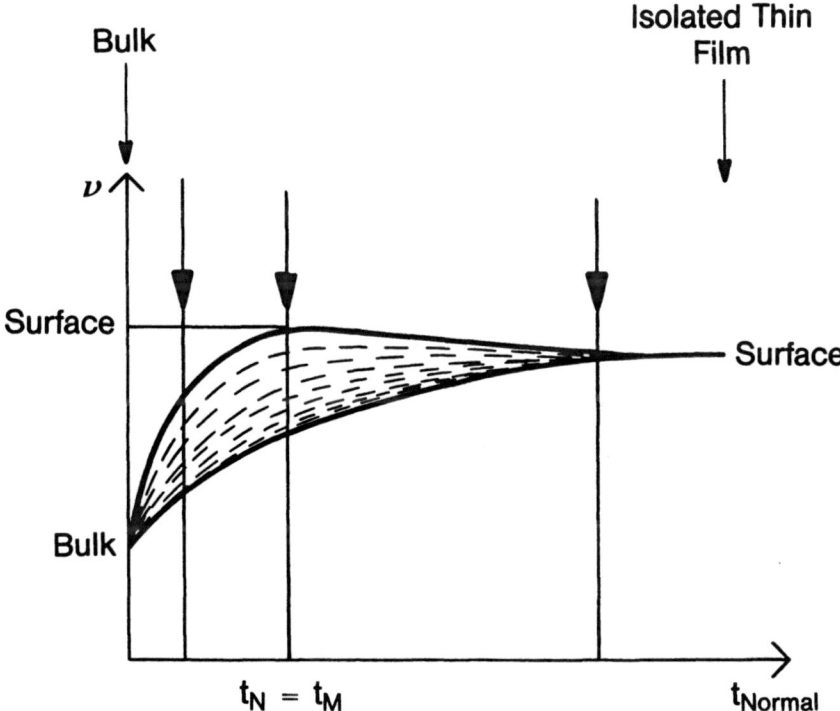

Fig. 4   Expected dependence of frequency versus thickness of normal metal.

Fig. 4 shows qualitatively the theoretical prediction for the frequency of magnons in a magnetic/normal superlattice as a function of normal metal thickness. For thick normal metal separator (i.e. isolated magnetic films) a single mode should be observed. When the normal layer thickness (t_N) becomes comparable to the magnetic layer thickness (t_M) one or two modes should appear depending on experimental broadening of the magnon lines. When the normal metal thickness becomes small, two distinct modes should be observed because the band of modes shown in the figure have a higher density of states at the bottom of the band. These effects have been calculated in detail [13,14] as a function of all parameters in the problem; t_N, t_M, magnetic field (H), scattering vector and saturation magnetization.

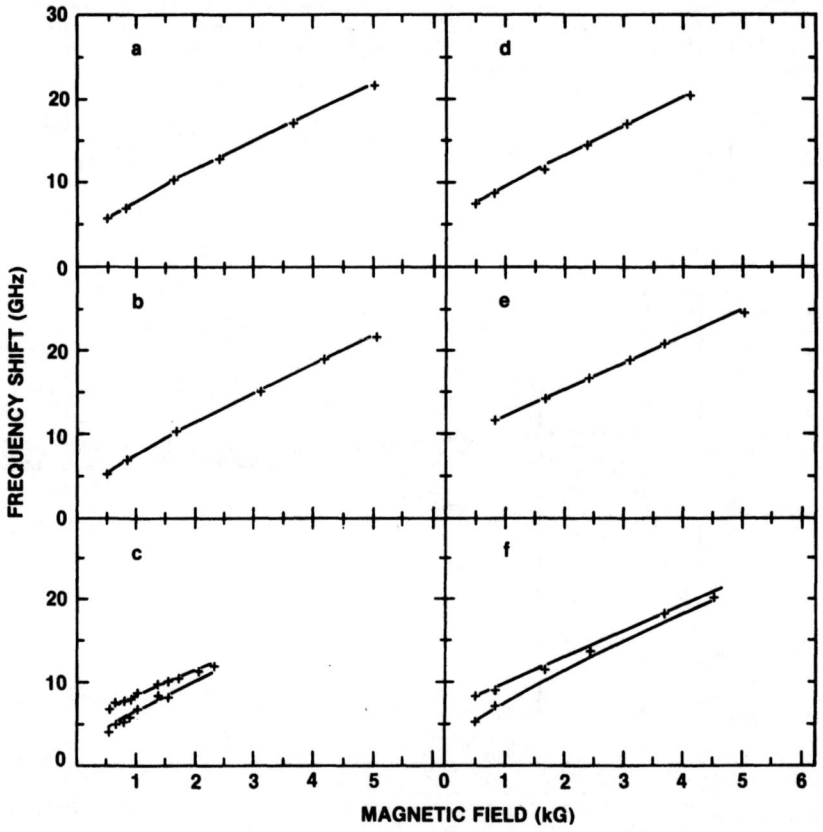

Fig. 5 Magnon frequency versus magnetic field for a series of Mo/Ni samples. The solid lines are theoretical fits as explained in the text.

The experiments [15,16] are in excellent, quantitative agreement with these theories without any adjustable parameters. For instance, the frequency shift as a function of H is shown in Figure 5, together with the theoretical prediction indicated by the solid lines. We should stress at this point, that the thickness at which superlattice effects due to dipolar coupling are observed are much larger than any imperfections possibly present in these materials.

CONCLUSIONS

In conclusion, magnetic superlattices have provided a very useful area of research. The phenomena that have been studied to date depend critically on the length scales which determine the physics. It is important to emphasize that the structural perfection needed for the observation of a particular effect varies considerably and therefore detailed structural characterization is imperative.

ACKNOWLEDGEMENTS

I thank my collaborators, M.R. Khan, M.J. Pechan, M. Grimsditch and A. Kueny for long years of interaction. I also thank my colleagues A.S.Arrot, M. Brodsky, G.P. Felcher, A.J. Freeman, E.M. Gyorgy, C.F. Majkrzak, D.B. McWhan, and G.P. Prinz for useful conversations. Work supported by DOE Grant #DE-FG03-87ER45332.

REFERENCES

1. B.J. Thaller, J.B. Ketterson and J.E. Hilliard, Phys. Rev. Lett. 41, 336 (1978).
2. "Synthetic Modulated Structures", L.L. Chang and B.C. Giessen, eds., Academic Press, Inc., Orlando (1985).
3. "Interfaces, Superlattices and Thin Films", J.D. Dow and I.K. Schuller, eds., Material Research Society Publishers, Pittsburgh, (1987).
4. "Physics, Fabrication and Applications of Multilayered Structures", P. Dhez and C. Weisbuch, eds., (in press).
5. M.R. Khan, P. Roach and I.K. Schuller, Thin Solid Films, 122, 183 (1986).
6. G. Xiao and C.L. Chien, J. Appl. Phys. 61 ,4061 (1987).
7. H.K. Wong et al., J. Appl. Phys. 55, 2494 (1984).
8. J.F. Dillon Jr., E.M. Gyorgy, L.W. Rupp Jr., Y. Yafet and L..R. Testardi, J. Appl. Phys. 52, 2256 (1981).
9. M.J. Pechan and I.K. Schuller, Phys. Rev. Lett. 59, 132 (1987).
10. W.S. Zhou, H.K. Wong, J.R. Owers-Bradley and W.P. Halperin Physica 108B, 953 (1981).
11. J. Kwo et al., Phys. Rev. Lett. 55, 1402 (1985).
12. S. Sinha et al., J. Mag. Magn. Mat. 54, 773 (1986).
13. P. Grünberg and K. Mika, Phys. Rev. B27, 2955 (1983).
14. R.E. Camley, T.S. Rahman and D. Mills, Phys. Rev. B27,261 (1983).
15. M. Grimsditch, M. Khan, A. Kueny and I.K. Schuller, J. Appl. Phys. 51, 498 (1983).
16. A.Kueny, M. Khan, I.K. Schuller and M. Grimsditch, Phys. Rev. B29, 2879 (1984).

Multilayered Interconnections for VLSI

Donald S. Gardner and Krishna Saraswat

Integrated Circuits Laboratory

Stanford University, Stanford, Calif. 94305

Abstract

Interconnections are predicted to become the limit in performance and reliability at submicron dimensions. Layered structures are one possible solution to the problems of electromigration and hillocks. Aluminum alloys can improve the properties of pure aluminum but with a consistent increase in resistivity due to lattice distortions introduced by the alloy in solid solution. The rise in resistivity is minimal in layered films because aluminum layers act as parallel conductors. Titanium is the most effective refractory metal for eliminating hillocks in layered films; however vanadium, tantalum, and tungsten also reduce hillocks to a manageable level although high resistivity, the inability to dry etch, or lateral hillocks can become a problem. Layered films of aluminum with $TiSi_2$ encounter problems with hillocks. These differences are partially explained by the increased strength of the aluminide $TiAl_3$ that forms compared to the other compounds. Barrier metals are needed for aluminum layered with refractory metals because of the formation of ternary compounds, and several metals have proven to be satisfactory. Such barriers, however, are required for even aluminum at submicron dimensions for stable low-resistivity contacts.

1 Introduction

In future VLSI technology, several materials and alloys will be combined to fabricate multilevel interconnections because performance will be so limited by the interconnections that different materials will be chosen for their various properties. Aluminum has long been used to form the metal interconnections; however, as device dimensions are scaled, the resulting increase in current density will lower reliability [1]. With multiple levels of aluminum, failures between levels also become a problem. The major problems with pure aluminum are electromigration, hillock formation causing electrical shorts between successive levels of aluminum, and the high solubility (0.5 atm% at 450°C) and diffusivity of silicon in aluminum (diffusion length is 1.4 μm for 1 sec at 450°C) leading to poor contacts to shallow junctions.

Aluminum-copper can reduce the problems characteristic of pure aluminum [2], but these films are difficult to dry etch, easily corrode after etching [3], are susceptible to long-term corrosion [4] and have bonding problems [5] In addition, hillocks are not completely eliminated. In this work, the term level will be used to describe conductors which are separated by an insulator whereas the term layer will be used to describe different conductors tiered together in one level of interconnections

Hillocks are partly the result of the large difference between the thermal expansion coefficients of the metal and substrate [6] and have been observed in films of lead, tin, and

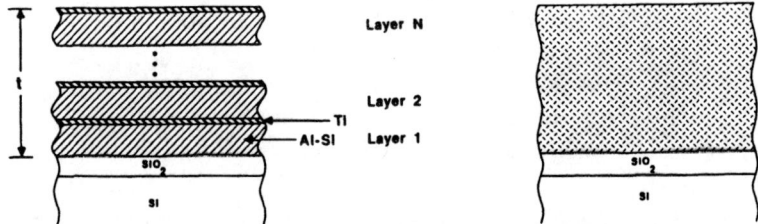

Figure 1. Cross section schematic of homogeneous and multilayered structures.

indium [7]. d'Heurle [8] demonstrated that surface reconstruction results in thin aluminum films on silicon and on SiO$_2$ after thermal cycling with hillocks and whiskers being observed.

The use of additional layers to reduce hillock density began in the early 1970's. Anodization of aluminum forming Al$_2$O$_3$ around the interconnection was found to lower hillock growth [9]. Faith [10] noted that, with both layered and unlayered Al-O films, hillock height drops to less than 0.5 μm with a 15 to 45 percent increase in resistivity; however, problems of manufacturability and etchability are encountered. Schreiber reported a surface roughness of 120 nm when a layer of titanium nitride in aluminum [11] was incorporated. He also noted that the surface roughness is 120 nm when using a layer of titanium within aluminum however, based on other studies [12,13], is converted into a homogeneous alloy after the thermal cycle and should be considered as such. Two other brief studies demonstrated that the deposition of TiWN on top of aluminum can lower surface roughness to below 200 nm [14]; the deposition of aluminum on WSi$_2$ also reduces aluminum surface roughness, but the reason is erroneously attributed to a gradual change in the thermal-expansion coefficient [15]. A study of a Ti-W alloy applied as the bottom layer revealed no noticeable drop in hillock density [16] although an increase in electromigration lifetime was observed [17].

A study of electromigration lifetimes by Howard [18] demonstrated improvements in lifetimes of a factor of 5 to 100 through an evaporation process of depositing refractory metals such as titanium, chromium, hafnium, or tantalum between two layers of Al-Cu. These metals were deposited separately because of the difficulties in evaporating aluminum and other materials simultaneously to form alloys [19]. When the copper is removed, however, the interdiffusion rate of the aluminum and titanium becomes high resulting in unacceptably high resistivities. Hinode [20] analyzed a layered Al/Ta/Al film evaporated without copper and observed two to three orders of magnitude improvement in electromigration lifetime.

None of these studies used more than one layer of a refractory metal and hillocks were largely ignored. Faith later stated that hillocks formed during furnace annealing could not be eliminated by any current method [21]. Shortly after, however, hillocks were eliminated using more than one layer of titanium with Al-Si (see Fig. 1) and without the added resistivity normally associated with homogeneous alloys and silicon was found to play a crucial role in keeping the films layered [12,13,22,23,24].

Layered films of Al/Ti without silicon were observed to be as smooth as films of layered Al-Si/Ti, but resistivity is somewhat higher in those without silicon. Further analysis with Rutherford backscattering showed that titanium diffuses through the aluminum and becomes uniformly distributed after annealing; however, films of Al-Si/Ti remain distinctly layered after thermal cycling. This was corroborated by resistivity measurements and cross-section TEM. X-ray diffraction and microdiffraction identified the aluminide TiAl₃ that forms a globular but localized layer between the Al-Si layers. After thermal cycling at 450°C, the resistivity of the layered films with pure aluminum increased because of inter-diffusion of titanium into the aluminum whereas the films with Al-1%Si were more stable. Also, the silicon lowers the maximum solid solubility limit of titanium in aluminum [25] which also reduces the resistivity [26,23].

Since then, further studies have been done on layered films of aluminum [26]–[45] including additional studies of the problem of silicon diffusion into aluminum [27,46,47]. It was established that multilayered Al-Si with titanium has lifetimes in excess of 1000 hours at 250°C using a current density of 1.5×10^6 amp/cm² [27]. Also, it was reported that a single layer of TiSi₂ or MoSi₂ on top of aluminum [48] can reduce hillocks and (50 nm) of TaSi$_x$ can almost eliminate hillocks on a 1 μm aluminum film. Having one layer only on top, however, does not provide the redundancy which leads to the high electromigration lifetimes.

In this paper, an investigation of hillocks, resistivity, and silicon diffusion into Al-Si films multilayered with either titanium, tantalum, tungsten, vanadium, or TiSi₂ is presented.

2 Experimental Procedure

The equipment used was a Balzers BAS 450 magnetron sputtering system. This system has four target positions which allows for sequential deposition of up to four vacuum compatible materials without breaking vacuum. The wafers are vertically mounted on a rotating drum which passes in front of all four target positions. Each target has an electrically activated shutter making it easy to deposit multilayer films when thick layers are used (> 100 nm). The rotational speed of the drum can be set up to 24 rpm which allows for the formation of a finely layered film. The films are then annealed at 450°C in H₂ + N₂ gas for 30 min. The base pressure was below 1.5×10^{-7} mbar in all cases and the argon gas pressure during sputtering was 2.0×10^{-3} mbar. The films were deposited at room temperature.

3 Hillock Measurements

When a film of pure aluminum is annealed, the surface becomes rough (see Fig. 2). Al-Cu films are smoother, but there are still some large hillocks. Surface profiles and SEM photomicrographs of layered films of Al-Si with titanium show significant improvements over Al and Al-Cu alloys.

Structures with three layers of tungsten resulted in a very smooth film with occasional large hillock-like pillars 2 to 3 μm high and 1 μm wide. CVD SiO₂ will conformally cover these in a test structure which makes them difficult to detect and creates many problems during lithography.

The surface profiles of layered films with tantalum, vanadium, and titanium silicide are compared in Fig. 3. As can be seen, the surface of the two-layered tantalum film exhibited hillocks 40 to 50 nm high after three thermal cycles. The surface of films with vanadium was good after three thermal cycles, with a final roughness of less than 10 nm; however, the resistivity became high. The roughness of a film layered with TiSi₂ degraded into hillocks in excess of 100 nm after a single thermal cycle.

Hillock sizes and density are reduced by the thin layer of metal on top of the aluminum film because this layer constrains the hillocks. This constraining effect is not one in which a

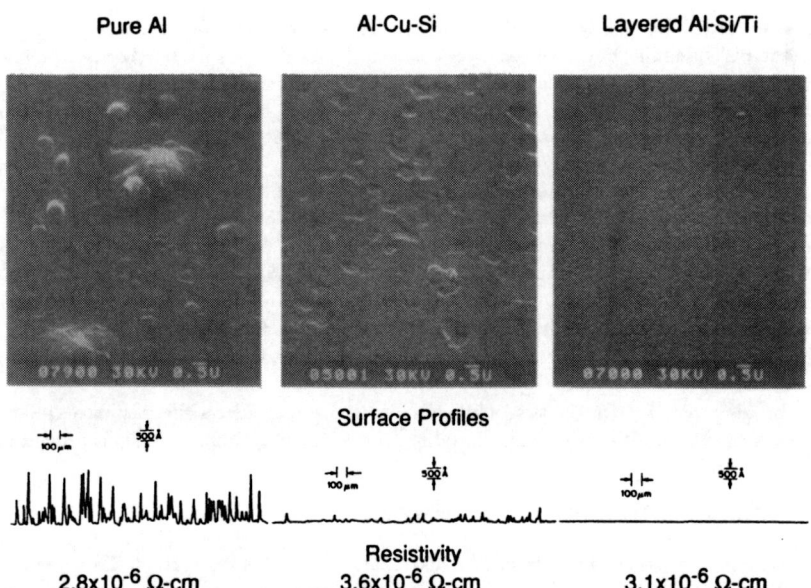

Figure 2. SEM photomicrographs, surface profiles and resistivities of pure aluminum, Al-Cu, and layered Al-Si/Ti films. The samples were thermally cycled to 450°C.

force perpendicular to the surface occurs; instead, any force that attempts to form a hillock is counteracted by adhesive forces to the aluminum at other points along the refractory metal layer. The complete film will plastically deform when the yield strength is reached, and this plastic deformation will appear in the direction of the free surface. As a result, the complete film as one unit will tend to expand upward. If there are any weak points in the top layer or if the layer cannot withstand the localized forces from the movement of material, a roughness can develop. With titanium, however, the resulting $TiAl_3$ is nearly two orders of magnitude stronger than aluminum. This model of adhesive forces can also be applied to layered films using a thick Al_2O_3 and to films with a damaged semi-amorphous top layer at the surface created by ion implantation of As^+ [49]. This reduces surface roughness; however after an initial thermal cycle, the surface damage recrystallizes and hillocks appear. With these structures, the improvement in electromigration reliability is small and resistivity is a concern. A multilayer film is preferable for these reasons.

4 Resistivity

Layered films can be modeled as parallel resistors where each resistor represents one layer of a film. If titanium or $TiAl_3$ exists, a lower resistivity film of aluminum will support most of the current.

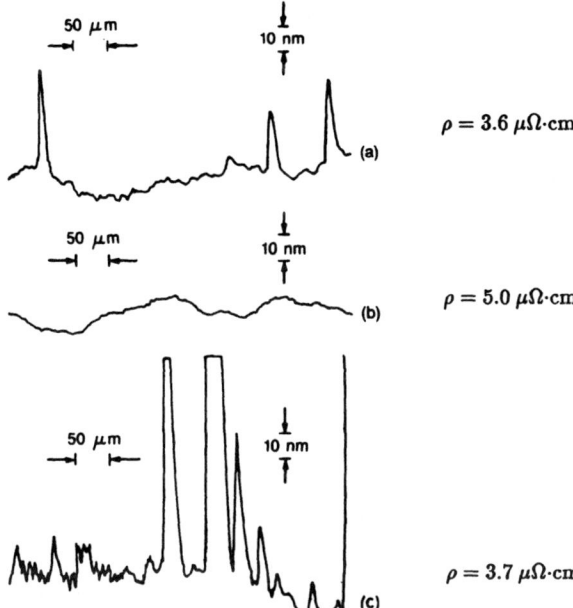

Figure 3. Surface profiles of multilayered aluminum films with tantalum, vanadium, and titanium silicide. (a) Two layers of 500 nm Al-1%Si with two layers of 50 nm tantalum after three thermal cycles. (b) Two layers of 500 nm Al-1%Si with two layers of 50 nm vanadium after three thermal cycles. (c) Two layers of 500 nm Al-1%Si with two layers of 50 nm $TiSi_2$ after one thermal cycle.

This resistivity before and after thermal cycling is a good indicator of the morphology of the film. For example, when Al-Si was layered with tungsten, the resistivity before and after cycling was similar to that of layered films of Al-Si/Ti which implies that the interdiffusion of tungsten into aluminum was minimal, but problems with large hillock-like pillars were encountered.

Films layered with tantalum, vanadium, and titanium silicide were also analyzed. The high resistance after thermal cycling of those layered with vanadium indicated that interdiffusion similar to that of Al/Ti without silicon had resulted [12], forming a homogeneous alloy. With $TiSi_2$, resistivity dropped after thermal cycling and the problems of contact spiking were eliminated without the need for a barrier metal, but some problems were encountered with hillocks forming. With tantalum, resistivity after thermal cycling actually dropped but some hillocks developed. This drop may be the combined result of two mechanisms; first, the film remained layered and second, the silicon in the aluminum layers interdiffused into the tantalum, thereby achieving purer aluminum layers. In addition, second-order effects that can contribute to a change include the removal of damage from the deposition process and the formation of $TaSi_2$ from the tantalum layer.

5 X-Ray Diffraction

Samples of layered Al/Ti and Al-Si/Ti were analyzed by X-ray diffraction before and after annealing [13]. This analysis revealed the formation of the $TiAl_3$ compound after annealing.

The X-ray diffraction measurements of layered films of aluminum with tantalum on SiO_2 could identify $TaSi_2$, but no tantalum aluminides were observed after thermal cycling. Similar measurements of aluminum layered with vanadium resulted in the identification of VAl_3 with no evidence of any compounds involving silicon after thermal cycling. VAl_3 was the compound previously reported to form at 450°C when vanadium was used as a barrier metal between aluminum and silicon [50,51]. These measurements were obtained from films on SiO_2 and, based on the various phase diagrams, the results are expected to vary when the films come in contact with silicon.

6 Contacts to Silicon

One problem with the use of refractory metals in aluminum is the reactions that may occur if the metallization comes in contact with silicon. Possible reactions include the formation of silicides or ternary compounds involving silicon. The result would be contact spiking that leads to junction failure [13]. Examination of the ternary phase diagram revealed an increase in the solid-solubility limit of silicon in the aluminide $(Ti(Al,Si)_3)$ and the presence of the ternary compound $Ti_7Al_5Si_{12}$. When the aluminum or the aluminide comes in contact with silicon through a contact hole, there is a large amount of silicon which would result in the phase $Ti_7Al_5Si_{12}$. This is the reason for the large pitting problem reported in the literature when titanium was used to lower contact resistance and to serve as a barrier metal between aluminum and silicon [52,50].

The results obtained from a layered structure based on titanium silicide are shown in Fig. 4a where enough silicon is supplied such that none is taken up from the substrate. The ternary compound $Ti_7Al_5Si_{12}$ is expected to form and, to satisfy the need for silicon, 12 atoms of silicon are required for every 7 atoms of titanium; with $TiSi_2$, 14 atoms are available for every 7 atoms of titanium. As illustrated in Fig. 4, the result is no contact spiking; however, because of the excess silicon, increased contact resistivity caused by the epitaxial regrowth of silicon is expected to be a problem with this metal system, especially in submicron contact holes. This is similar to the problems associated with using Al-1%Si with small contact holes (< 1 μm) because the solid solubility of silicon in aluminum is only 0.5 at% at 450°C. In addition, hillocks are not eliminated when titanium is replaced by $TiSi_2$.

Other refractory metals were also analyzed for the possibilities of contact spiking. With tantalum and vanadium, contact spiking became just as severe as with titanium. Figure 4b and 4c shows cross sections of contacts fabricated with these layered films in which deep contact spiking greater than 1 μm was observed.

A solution to this problem is to select a barrier metal such as TiN, TiW, or CVD W. An added advantage of a barrier metal is that contact resistance will be lower than that obtained with Al-Si. For submicron contacts, Al-Si will not make reliably good contact because of epitaxial regrowth and, as a result, a barrier metal is required because pure aluminum will generate contact spiking. Initial studies of a layered barrier consisting of 5 nm of titanium and 100 nm tungsten demonstrated that for Al-Si/Ti, this is an acceptable barrier at 450°C.

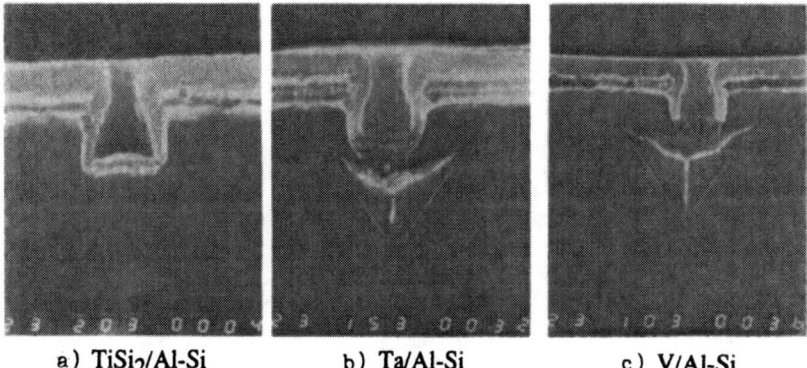

a) TiSi₂/Al-Si   b) Ta/Al-Si   c) V/Al-Si

Figure 4. Cross section of contacts fabricated with aluminum layered with titanium silicide, tantalum and vanadium.

7 Mechanical Stress versus Temperature

Stress measurements were made on all the films as a function of temperature using a laser to measure in situ the change in radius of curvature [6]. It was found that films of Al-Si would first exhibit elastic behavior with a slope of -2 MPa/°C. This can be theoretically calculated using Youngs Modulus E, and Poisson's ratio ν for aluminum [26,23] with a theoretical slope of -2.3 MPa/°C, which correlates well with the measured value.

These films were all deposited at room temperature, while the temperature at which the stress in the aluminum films changes from compressive to tensile is between 80°C and 100°C. This suggests that the aluminum has obtained some energy when sputtered from the target by the Ar⁺ ions.

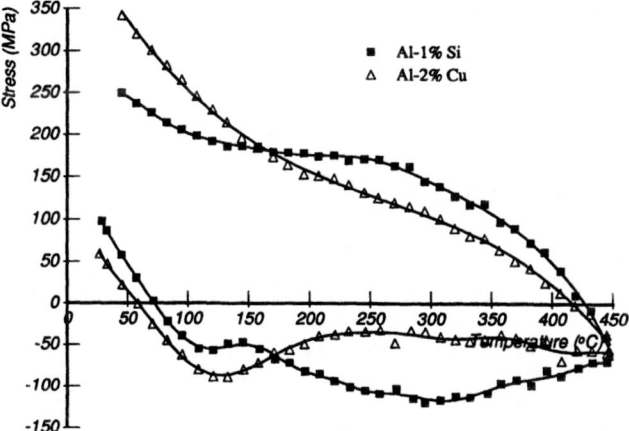

Figure 5. Stress versus temperature of aluminum and aluminum copper. Here, Al-Cu consists of 2% Cu, and the films were sputter deposited at room temperature.

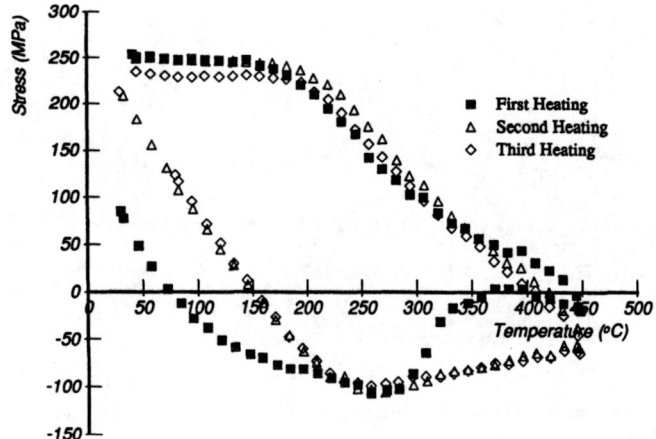

Figure 6. Temperature versus stress in a layered film of aluminum-silicon with titanium. The first, second, and third thermal cycles are shown. The film thickness was 0.96um thick and consisted of three 300 nm layers of Al-Si with three 20 nm layers of titanium.

The aluminum is seen to start to plastically deform at temperatures as low as 125°C (see Fig. 5) which is a typical postbake temperature for photoresist. At temperatures of 150°C, there is a dramatic decrease in stress which is due to recrystallization and grain growth. At 300°C the stress can be seen to gradually decrease which is due to the fact the the yield strength of aluminum decreases with temperature. When the films were cooled, the aluminum switched from being in a state of compression (negative) to one of tension (positive) and at some point began to plastically deform again.

Aluminum films layered with Ti, Ta, V, and TiSi₂ were tested [23] and exhibit behavior very similar to pure aluminum (see Fig. 6 for Al-Si/Ti). The films were first elastic in behavior, but then began to plastically deform. Finally, recrystallization and compound formation occurred. Upon further temperature cycling, it was found that the stress curves did not change. The samples were heated at 5 °C/min.

8   Stress in Finely Layered Films

The stress and surface roughness in finely layered Al-Si films with titanium were also measured. A significant difference was observed between a sample consisting of many layers of each material and one with fewer layers. Stress as a function of temperature in a film consisting of 150 layers of each material (Fig. 7) was comparable to the measurements of stress in films containing 60 layers [26]. In addition, surface roughness was also similar in that there were very few hillocks as can be seen in the insert in Fig. 7. When a film was fabricated with 20 initial layers, the major features of the temperature vs stress curve were the same, but the curve differed in that it was smooth. The deposition conditions were identical in both samples, including fabrication within the same day and obtaining the measurements one immediately after the other to minimize equipment variations. Surface roughness was substantial in the sample with fewer layers as illustrated in the insert in Fig. 7.

A probable explanation for the differences between the two samples is that, in the 150-layer film, the titanium layers are approximately 0.13 nm thick which is thin enough for this layer to actually be individual islands of material. In the other sample, the layers of titanium are approximately 1.0 nm thick which can be a continuous sheet. The first film will form numerous precipitates of aluminide whereas the second will form less frequent larger precipitates. These precipitates will make plastic deformation more difficult as is reflected in the measurement which reveals a more gradual change in stress during recrystallization and compound formation in addition to noisier behavior indicative of slower deformation. This difference between the two samples is significant because it demonstrates that the fabrication conditions must be carefully selected. If a homogeneous alloy is fabricated with too few layers, the film will have drastically different morphologies that will affect both the mechanical stress and electromigration lifetimes. If a film is fabricated with a single composite target, the morphology is more difficult to control without this added freedom to adjust the layering and control the precipitates. Since the individual elements sputter from the target at different rates, the composition of the deposited film will also change with time through the life of the target. Such parameters as temperature will have an impact on this because if the precipitates can be formed early during deposition as opposed to afterwards, the size will be different due to differences in surface mobility. It is perhaps better for controllability in manufacturing to deposit from more than one target.

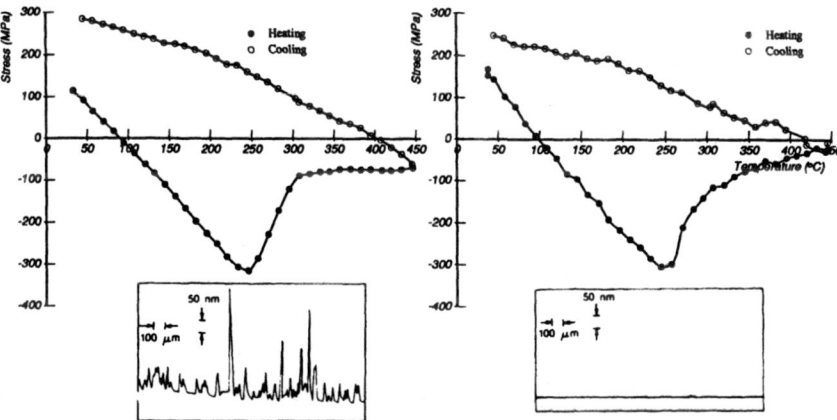

Figure 7. Stress behavior of coarsely and finely layered films of Al-Si with titanium. Left: 20 layers. Right: 150 layers. Thermal cycling produces a homogeneous alloy. The thickness is approximately 1 μm, in both films and Al-Si was the first layer deposited. The insert shows the surface roughness after thermal cycling.

9 Conclusion

Titanium, tantalum, tungsten, vanadium, and titanium silicide were investigated for layered films with Al-Si, and titanium was the only metal that resulted in smooth low-resistivity films. Contact spiking was predicted and later observed in titanium. Further experiments with tantalum, tungsten, and vanadium also revealed contact spiking in these metals; however, barrier metals such as TiW and TiN provide adequate protection from junction spiking and reduced contact resistivity. The use of $TiSi_2$ also eliminates contact spiking, but many hillocks formed and resistivity may be high in submicron contacts because of epitaxial regrowth of the excess silicon. Other metals, however, control surface roughness, and resistance is low when titanium, tantalum, titanium silicide, or tungsten are used with silicon doped aluminum, but not vanadium. Large hillock-like pillars appear in films with tungsten and possibly vanadium, and an optical microscope is invaluable in distinguishing these pillars. The impact of this research is that low resistivity, hillock free, dry etchable metal films can be fabricated and used in VLSI multilevel interconnects.

10 Acknowledgments

The authors wish to acknowledge B. W. Shen for providing the SEM cross sections. This work was jointly supported by SRC Contract No. 83-01-006 and DARPA Contract No. MDA 903-80-C-0432.

References

[1] D. S. Gardner, J. D. Meindl, and K. C. Saraswat. *IEEE Trans. on Electron Devices*, ED-34(3), pp. 633–643, 1987.

[2] F. M. d'Heurle. *Metallurgical Trans.*, 2, pp. 683–689, 1971.

[3] W. Lee and J. M. Eldridge. *J. Appl. Phys.*, 52(4), p. 2994, 1981.

[4] R. Rosenberg, M. J. Sullivan, and J. K. Howard. In *Thin Films. Interdiffusion and Reactions.*, pp. 48–54, 1978.

[5] S. Thomas and H. M. Berg. In *Proc. Intl. Reliability Phys. Symp.*, IEEE Electron Devices and Reliability Societies, pp. 153–158, 1985.

[6] P. A. Flinn, D. S. Gardner, and W. D. Nix. *IEEE Trans. on Electron Devices*, ED-34(3), pp. 689–699, 1987.

[7] H. L. Caswell, J. R. Priest, and Y. Budo. *J. Appl. Phys.*, 34(11), pp. 3261–3266, 1963.

[8] F. d'Heurle, L. Berenbaum, and R. Rosenberg. *Trans. AIME*, 242, pp. 502–511, 1968.

[9] C. J. Dell'oca and A. J. Learn. *Thin Solid Films*, 8, pp. R47–R50, 1971.

[10] T. J. Faith. *J. Appl. Phys.*, 52(7), pp. 4630–4639, 1981.

[11] B. Grabe and H. U. Schreiber. *Solid-State Electronics*, 26, p. 1023, 1983.

[12] D. S. Gardner, T. L. Michalka, T. W. Barbee, Jr., K. C. Saraswat, J. P. McVittie, and J. D. Meindl. In *Proc. First Intl. IEEE VLSI Multilevel Interconnection Conf.*, pp. 68–77, 1984. Awarded Best Paper.

[13] D. S. Gardner, T. L. Michalka, K. C. Saraswat, T. W. Barbee, Jr., J. P. McVittie, and J. D. Meindl. *Trans. on Electron Devices and J. Solid-State Circuits*, ED-32(2) and SC-20(1), 1985. Joint Special Issue on VLSI.

[14] S. C. P. Lim. *Semiconductor Intl.*, pp. 135–144, 1982.

[15] K. C. Cadien and D. L. Losee. *J. Vac. Sci. Technol.*, B 2(1), pp. 82–83, 1984. in Brief Reports and Comments.

[16] A. Rey, P. Noel, and P. Jeuch. In *1984 Proc. First Intl. IEEE VLSI Multilevel Interconnection Conf.*, p. 139, 1984.

[17] P. B. Ghate and J. C. Blair. *Thin Solid Films*, 55, p. 113, 1978.

[18] J. K. Howard, J. F. White, and P. S. Ho. *J. Appl. Phys.*, 49(7), p. 4083, 1978.

[19] F. M. d'Heurle and A. Gangulee. In *The Nature and Behavior of Grain Boundaries*, Metallurgical Society of AIME, Plenum Press, New York, pp. 339–370, 1972.

[20] K. Hinode, S. Iwata, and M. Ogirima. In *Extended Abstracts–Electrochemical Society*, 83-1, p. 678, 1983.

[21] T. J. Faith and C. P. Wu. *Appl. Phys. Lett.*, 45(4), pp. 470–472, 1984.

[22] D. S. Gardner, R. B. Beyers, T. L. Michalka, K. C. Saraswat, T. W. Barbee, Jr., and J. D. Meindl. In *Intl. Electron Devices Meeting Tech. Digest*, IEEE Electron Devices Society, pp. 114–117, 1984.

[23] D. S. Gardner. *Layered and Homogeneous Alloys of Aluminum for Integrated Circuits.* Ph.D. dissertation, Stanford University, 1987.

[24] D. S. Gardner, K. Saraswat, and T. W. Barbee, Jr. Layered and Homogeneous Films of Al and Al-Si with Ti and W for Multilevel Interconnects. U.S. Patent No. 4,673,623.

[25] W. V. Youdelis. *Metal Science*, 12(8), pp. 363–366, 1978.

[26] D. S. Gardner, T. L. Michalka, P. A. Flinn, T. W. Barbee, Jr., K. C. Saraswat, and J. D. Meindl. In *1985 Proc. Second Intl. IEEE VLSI Multilevel Interconnection Conf.*, pp. 102–113, 1985.

[27] R. E. Jones, Jr. and L. D. Smith. In *1985 Proc. Second Intl. IEEE VLSI Multilevel Interconnection Conf.*, pp. 194–200, 1985.

[28] B. W. Shen, T. Bonifield, and J. McPherson. In *1985 Proc. Second Intl. IEEE VLSI Multilevel Interconnection Conf.*, pp. 114–120, 1985.

[29] B. L. Draper, T. A. Hill, and H. B. Bell. In *1985 Proc. Second Intl. IEEE VLSI Multilevel Interconnection Conf.*, pp. 90–101, 1985.

[30] S. Toi and F. Choi. In *1985 Proc. Second Intl. IEEE VLSI Multilevel Interconnection Conf.*, pp. 138–144, 1985.

[31] M. Finetti, H. Ronkainen, M. Blomberg, and I. Suni. In *Thin Films-Interfaces and Phenomena*, Materials Res. Soc., 54, pp. 811–816, 1986.

[32] B. W. Shen, J. M. Anthony, P. Chang, J. Keenan, R. Matyi, and H. L. Tsai. In *Thin Films-Interfaces and Phenomena*, Materials Res. Soc., 54, pp. 103–108, 1986.

[33] S. Mak, T. Dahlstrom, and K. S. Ravindhran. In *1986 Proc. Third Intl. IEEE VLSI Multilevel Interconnection Conf.*, pp. 65–70, 1986.

[34] B. W. Shen, T. Bonifield, and R. Blumenthal. In *1986 Proc. Third Intl. IEEE VLSI Multilevel Interconnection Conf.*, pp. 191–197, 1986.

[35] Y. Wada. *J. Electrochem. Soc.*, 133(7), pp. 1432–1437, 1986.

[36] R. K. Nahar and N. M. Devashrayee. *Materials Letters*, 4(5,6,7), pp. 265–267, 1986.

[37] C. F. Dunn, F. R. Brotzen, and J. W. McPherson. *J. Electronic Materials*, 15(8), pp. 273–277, 1986.

[38] J. C. Sum, G. W. Ray, S. Hsu, P. J. Marcoux, J. Kruger, C. Lin, E. Liu, and S. Peng. In *Multilevel Metallization, Interconnection, and Contact Technologies*, The Electrochemical Society, 87-4, pp. 259–265, 1986.

[39] R. K. Nahar and N. M. Devashrayee. *Appl. Phys. Lett.*, 50(3), pp. 130–131, 1987.

[40] F. K. LeGoues, M. Wittmer, T. Kwok, H. C. W. Huang, and P. S. Ho. *J. Electrochem. Soc.*, 134(4), pp. 940–944, 1987.

[41] A. A. Brown, K. B. Affolter, S. R. Jennings, and P. J. Rosser. In *1987 Proc. Fourth Intl. IEEE VLSI Multilevel Interconnection Conf.*, pp. 426–433, 1987.

[42] R. K. Ball and A. G. Todd. *Thin Solid Films*, 149, pp. 269–282, 1987.

[43] H. Yamamoto, S. Fujii, T. Kakiuchi, K. Yano, and T. Fujita. In *Intl. Electron Devices Meeting Tech. Digest*, IEEE, pp. 205–208, 1987.

[44] H. Eggers and K. Hieber. In *Intl. Electron Devices Meeting Tech. Digest*, IEEE, pp. 200–204, 1987.

[45] S. Chambers. In *Extended Abstracts of the 172nd Meeting of the Electrochemical Society*, pp. 671–672, 1987.

[46] K. Hinode, N. Owada, and T. Terada. In *Proc. Third Intl. IEEE VLSI Multilevel Interconnection Conf.*, pp. 139–145, 1986.

[47] K. Hinode, N. Owada, and T. Terada. *IEEE Trans. on Electron Devices*, ED-34(3), pp. 700–705, 1987.

[48] S. Shima, T. Moriya, and M. Kashiwagi. In *1984 Proc. First Intl. IEEE VLSI Multilevel Interconnection Conf.*, p. 61, 1984.

[49] Y. Kamei, M. Kameda, and H. Nakayama. In *1984 Intl. Electron Devices Meeting Tech. Digest*, IEEE, pp. 138–141, 1984.

[50] K. Nakamura, S. S. Lau, M. Nicolet, and J. W. Mayer. *Appl. Phys. Lett.*, 28(5), pp. 277–280, 1976.

[51] M. Eizenberg, R. D. Thompson, and K. N. Tu. *J. Appl. Phys.*, 53(10), pp. 6891–6897, 1982.

[52] R. W. Bower. *Appl. Phys. Lett.*, 23(2), p. 99, 1973.

CHARACTERIZATION OF INTERFACIAL ROUGHNESS IN SEMICONDUCTOR
HETEROSTRUCTURES BY X-RAY REFLECTIVITY

A. KROL, C.J. SHER, H. RESAT, S.C. WORONICK, W. NG, Y.H. KAO
Department of Physics, State University of New York at Stony Brook, Stony
Brook, NY 11794
L.L. CHANG, J.M. HONG
IBM Thomas J. Watson Research Center, P.O. Box 218, Yorktown Heights, NY
10598

ABSTRACT

The reflection of monochromatic x-rays by a layered heterostructure can
be utilized as a nondestructive probe to obtain information on the interfacial roughness in the material. Interference between x-rays reflected
from the top surface and the interfaces can give rise to pronounced oscillations in the reflectivity as a function of the grazing incidence angle.
We have made use of this technique to investigate the interfacial roughness
in semiconductor heterostructures grown by molecular beam epitaxy.

INTRODUCTION

Many important physical properties of semiconductor heterostructures
and superlattices are affected by the interfacial roughness in the
material. In order to characterize and to control the material structures,
it is useful to determine the interfacial roughness in as-made layer
structures by nondestructive means. For this purpose, the angular dependence of grazing-angle x-ray reflection can be utilized to probe the interfacial roughness and to determine the thickness of film layers in
heterostructures [1].

For smooth planar interfaces, the electric field inside the layers can
be calculated from the usual Fresnel formulae. However, with the presence
of interfacial roughness, corrections to these equations are necessary. In
a first approximation, one may assume a scalar parameter σ, the rms
deviation from the mean plane boundary, to account for scattering at each
interface. In this approximation, the roughness effect appears in a way
somewhat similar to the Debye-Waller factor arising from static disorder.
By a comparison between experimental curves and theoretical calculations,
the roughness parameters and the thickness pertaining to each thin film
layer can be determined.

THEORY

We assume that a heterostructure consists of N layers of thin films and
a substrate with parallel mean planes at the interfaces; bulk inhomogenie-
ties, plasmon, shadowing and multiple scattering effects at the interfaces
are all neglected. Following the formulation by Vidal and Vincent [2], the
connection between the incident (E_i), specularly reflected (E_r), and
transmitted (E_t) electric fields can be written as:

$$\begin{pmatrix} E_i \\ E_r \end{pmatrix} = P \begin{pmatrix} E_t \\ 0 \end{pmatrix} \qquad (1)$$

Mat. Res. Soc. Symp. Proc. Vol. 103. ©1988 Materials Research Society

where P is a product of matrices:

$$P = I_o S_o T_1 I_1 S_1 \cdots\cdots T_N I_N S_N \tag{2}$$

The matrix I_j connects the electric fields on the two sides of the jth smooth mean planar interface, S_j accounts for the perturbation due to roughness at the jth interface, and T_j describes the phase change in the electric field as the wave traverses through the jth layer. The subscript o refers to the interface between vacuum and the top surface of the heterostructure. Details of these matrices will be described in a later publication [3]. The intensity of scattered radiation can be written as

$$\langle \rho\rho^* \rangle = \langle |E_{rs}/E_r|^2 \rangle \tag{3}$$

where ρ is the complex scattering coefficient [4], $\langle \rho\rho^* \rangle$ denotes the mean value, E_{rs} is the scattered electric field, and the bar over E means time average of the electric field.

In an actual experiment, one measures the total intensity due to both specularly and diffusely scattered radiation intercepted in a solid angle Ω at the detector. Also, the measured intensity can be complicated by the divergence and partial polarization of the incident beam. The predicted total intensity of scattered radiation becomes:

$$T(\theta) = \int_{\alpha_1}^{\alpha_2} \iint_{\Omega} d\alpha d\Omega \ D(\alpha) \langle |\bar{E}_r(\theta+\alpha)|^2 \rangle \langle \rho\rho^*(\theta+\alpha,\theta_s,\phi_s) \rangle \tag{4}$$

Here we have neglected the effect of polarization which was estimated to be less than 1% of the total intensity for the present case. $D(\alpha)$ in (4) accounts for the vertical divergence of the incoming beam; θ is the grazing incidence angle; θ_s and ϕ_s are the polar and azimuthal angles of the scattered beam; α_1 and α_2 define the limits of incident beam divergence, respectively. Finally, the total intensity $T(\theta)$ is subject to a correction due to a nonlinear response function of the detector $R[T(\theta)]$.

EXPERIMENTAL RESULTS AND DISCUSSION

Our experiment was performed at the National Synchrotron Light Source (NSLS) using the U-15 beamline, which is equipped with a toroidal grating monochromator with an energy resolution about 2eV (at 600 eV). The grazing incidence angle was varied by means of a high precision tilt stage with an accuracy of 0.1 mrad. The scattered radiation was detected by a fluorescence screen and a photomultiplier tube. The experimental setup is described in more detail in Ref. 1.

Two samples studied in this experiment were GaAs/AlAs (nominal thickness of GaAs = 250Å on a thick layer of AlAs) and GaAs/AlAs/GaAs heterostructures (nominal GaAs top layer thickness = 25Å, AlAs thickness = 200Å) made by an overgrowth on GaAs(100) substrate surface using molecular beam epitaxy.

We have measured the intensity profile of the collimated x-ray beam and estimated the vertical divergence of the incident beam to be smaller than 0.2 mrad. The effect of beam divergence is therefore negligible. The nonlinear response function of the fluorescence screen detector was found to be exponential below a critical intensity T', and linear above T'. The exact behavior around T' is not known. The experimental curves are compared with theoretical calculations from (4) with σ's arising from the interfacial roughness treated as adjustable parameters. Since the complex refractive index $n = 1 - \delta - i\beta$ for GaAs and AlAs in the soft x-ray wavelength region has not been accurately measured, we also attempted to determine these refractive indices from our experimental curves obtained at three different x-ray energies (500, 600 and 700eV). To assure the correct dependence on the x-ray energy E far away from the absorption edges, we have assumed that both δE^2 and βE^3 be constant in the curve fitting process. Results are depicted in Figs. 1 and 2 for the two samples studied in the present experiment.

In Figs. 1 and 2 the log of experimental normalized reflectivity data are shown as points, and theoretical calculations from (4) are shown as curves. The dashed section of the theoretical curves is caused by an over-simplified model of the nonlinear response function $R[T(\theta)]$ which has no significant effect on the physical consequence of the results. For sim-plicity, we have assumed that a sharp transition from linear to exponential response in R occurs at T' given by $\ln T = -3.6$. Other than the transition region around T', the overall fits to the experimental curves are quite satisfactory. The parameters used to obtain these fits are summarized in Table 1.

The thickness of the epilayers can be determined accurately from the reflectivity measurements, either by curve fitting or from the position of the minima in the oscillations. This is an x-ray analogue of the Newton's ring effect in optics. The absolute accuracy in the thickness determination is better than 7Å.

The overall uncertainty in the determination of the rms roughness parameter σ is around 5Å. In the present experiment, the rms roughness of the top surface of GaAs in both samples was found to be 10 ± 5 Å. The rms interfacial roughness between GaAs and AlAs was around 10 ± 5 Å when GaAs was grown on AlAs (as in GaAs/AlAs) and 17 ± 5 Å when AlAs was grown on GaAs in an inverted structure (as in GaAs/AlAs/GaAs). Errors in the determination of these roughness parameters may partly result from uncertainties in the refractive indices of GaAs and AlAs at the wave-lengths used in this experiment. It should be noted that the roughness parameters determined for 700eV contain larger errors because this energy is closer to the L absorption edges of the element Ga.

In summary, we have demonstrated that the angular dependence of grazing angle x-ray reflection can be employed to investigate the interfacial roughness in layered semiconductor heterostructures. The interfacial roughness between GaAs and AlAs in a GaAs/AlAs heterojunction is found to be different from that in an inverted structure.

This research is supported by ONR under grant No. N0001483K0675.

REFERENCES

1. S.C. Woronick, B.X. Yang, A. Krol, Y.H. Kao, H. Munekata, L.L. Chang and J.C. Phillips, Proc. III International Conf. Modulated Semicon-ductor Structures, Montpellier, France (1987).

2. B. Vidal and P. Vincent, Appl. Opt. 23, 1794 (1984).

3. A. Krol, C.J. Sher, and Y.H. Kao, to be published.

4. P. Beckmann and A. Spizzichino, The Scattering of Electromagnetic Waves from Rough Surfaces (Pergamon Press, New York, 1963).

Table 1

Parameters used to fit the experimental curves shown in Figs. 1 and 2

Sample	Energy (eV)	rms Roughness ($\AA$)			Layer Thickness ($\AA$)	
		σ_o	σ_1	σ_2	d_1	d_2
GaAs/AℓAs	500	14	15	–	235	–
	600	11	15	–	242	–
	700	5	10	–	241	–
GaAs/AℓAs/GaAs	500	14	10.5	20	28	202
	600	11	7.0	15	28	195
	700	8	3.5	17	20	209

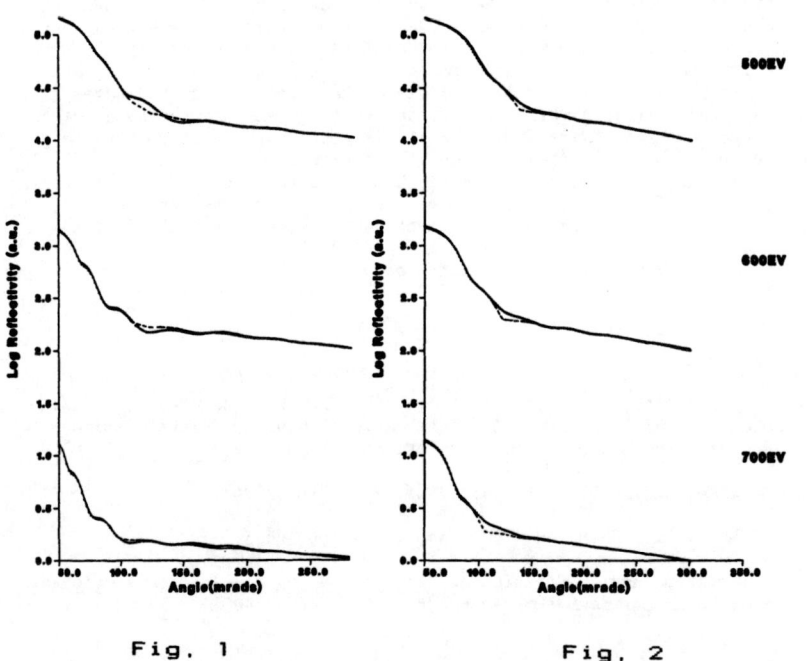

Fig. 1   Fig. 2

Figures 1 and 2 -- Log of normalized reflectivity (in arbitrary units) vs.
grazing angle for GaAs/AℓAs and GaAs/AℓAs/GaAs
heterostructures, respectively.

Author Index

Subject Index

CPSIA information can be obtained at www.ICGtesting.com
Printed in the USA
LVOW12s0728210514

386631LV00008B/263/P